COGNITIVE RELIABILITY AND
ERROR ANALYSIS METHOD

CREAM

Elsevier Science Internet Homepage

http://www.elsevier.nl (Europe)

http://www.elsevier.com (America)

http://www.elsevier.co.jp (Asia)

Full catalogue information on all books, journals and electronic products.

Elsevier Titles of Related Interest

FRANGOPOL, COROTIS & RACKWITZ
Reliability and Optimization of Structural Systems

GUEDES SOARES,
Advances in Safety & Reliability,
Proceedings of the ESREL'97 International Conference on Safety and Reliability.

Related Journals
Free specimen copy gladly sent on request: Elsevier Science Ltd, The Boulevard,
Langford Lane, Kidlington, Oxford, OX5 1GB, UK

Advances in Engineering Software
Computer Methods in Applied Mechanics and Engineering
Computers and Fluids
Engineering Analysis with Boundary Elements
Engineering Failure Analysis
Finite Elements in Analysis and Design
Probabilistic Engineering Mechanics
Reliability Engineering & System Safety
Structural Safety

COGNITIVE RELIABILITY AND ERROR ANALYSIS METHOD

CREAM

Erik Hollnagel
Institutt for Energiteknikk
Halden, Norway

ELSEVIER

UK   Elsevier Science Ltd, The Boulevard, Langford Lane, Kidlington, Oxford. OX5 1GB, England

USA   Elsevier Science Inc., 655 Avenue of the America, New York, 10100, USA

JAPAN   Elsevier Science, 9-15 Higashi-Azabu 1-chome, Minato-ku, Tokyo 106, Japan

First edition 1998

Library of Congress Cataloging in Publication Data
A catalogue record for this title is available from the Library of Congress.

British Library Cataloguing in Publication Data
A catalogue record for this title is available from the British Library.

ISBN 0-08-0428487

Transferred to digital printing 2005
Printed and bound by Antony Rowe Ltd, Eastbourne

TABLE OF CONTENTS

Foreword xiii

  A Fable.. xiii

  Acknowledgements.. xiv

CHAPTER 1: THE STATE OF HUMAN RELIABILITY ANALYSIS

1. INTRODUCTION 1

  1.1 The Pervasiveness Of Human Erroneous Actions... 2

  1.2 Human Actions As Causes 4

  1.3 Deterministic And Probabilistic Analyses.. 5

  1.4 Point-To-Point Analyses... 7

  1.5 Analysis And Prediction 8

2. SHORTCOMINGS OF FIRST-GENERATION HRA 9

  2.1 A Pragmatic Criticism 9

  2.2 A Principled Criticism 10

  2.3 THERP.. 10

  2.4 Time-Reliability Correlation And Human Cognitive Reliability 11

  2.5 The Reality Of The Human Error Probability 13

  2.6 Consequences Of The Criticism 14

  2.7 PSA-*cum*-HRA 14

3. COGNITIVE RELIABILITY AND ERROR ANALYSIS METHOD 15

  3.1 Cognitive 15

  3.2 Reliability 16

  3.3 "Error".. 16

  3.4 Analysis 17

  3.5 Method.. 17

  3.6 The Scientist And The Engineer 18

4. BACKGROUND OF THE BOOK... 18

  4.1 Structure Of The Book... 19

CHAPTER 2: THE NEED OF HRA

1. THE UBIQUITY OF ERRONEOUS ACTIONS.. 22

  1.1 Definitions Of "Human Error".. 23

  1.2 The Criterion Problem 25

  1.3 Performance Shortfall.. 26

  1.4 Volition.. 29

  1.5 Conclusions 31

2. THE ROLE OF HRA IN PSA 31

2.1 The PSA Sequence Model 33
2.2 The Consequences Of Human Actions 34
2.3 Data And Quantification 36
    2.3.1 The Decomposition Of Cognition 37
2.4 The Scope Of HRA 39

3. THE MODELLING OF ERRONEOUS ACTIONS 42

3.1 Omission And Commission 44
3.2 The Hunting Of The SNARK 45
3.3 Omission, Commission, And "Cognitive Error" 46
3.4 Overt And Covert Events 47
3.5 Phenotypes And Genotypes 48
3.6 "Cognitive Error" Defined 49

4. CHAPTER SUMMARY 50

CHAPTER 3: THE CONCEPTUAL IMPUISSANCE

1. THE CLASSIFICATION OF ERRONEOUS ACTIONS 52

1.1 Cause And Manifestation 53

2. TRADITIONAL HUMAN FACTORS APPROACHES 54

2.1 Descriptions Of Specific Psychological Causes 56
2.2 Descriptions Of General Psychological Causes 58

3. INFORMATION PROCESSING APPROACHES 59

3.1 Human Information Processing Models 59
    3.1.1 Quantitative Models Of Erroneous Actions 61
    3.1.2 Qualitative Models Of Erroneous Actions 61
3.2 Pedersen's Classification Of Error In Accident Causation63
3.3 Generic Error Modelling System 63
3.4 Rouse's Operator Error Classification Scheme 65
3.5 HEAT - Human Error Action Taxonomy 67
3.6 POET 68
3.7 NUPEC Classification System 68
3.8 Summary 69

4. THE COGNITIVE SYSTEMS ENGINEERING PERSPECTIVE 71

4.1 The Joint Cognitive Systems Paradigm 71
4.2 Contextual Determination 72
4.3 Socio-Technical Approaches 73

5. EVALUATION 73

5.1 Traditional Human Factors and Ergonomic Approaches 74
5.2 Information Processing Models 75
5.3 Cognitive Systems Engineering 75

6. THE SCHISM BETWEEN HRA AND PSYCHOLOGY 75

  6.1 Performance Analysis - Explaining The Past 77
  6.2 Performance Prediction - Divining The Future 79

7. CHAPTER SUMMARY 81

CHAPTER 4: A CONCEPTUAL FRAMEWORK

1. INTRODUCTION 83

2. THE NEED TO PREDICT.. 84

  2.1 Initiating Events And Response Potential 84
  2.2 Prediction For Interactive Systems 85

3. METHOD, CLASSIFICATION, MODEL... 86

  3.1 Method.. 86
  3.2 Classification Scheme.. 87
  3.3 Model.. 88
  3.4 The MCM Framework.. 88
  3.5 The Role Of Data... 89
  3.6 Data Analysis... 89

4. MODELLING OF COGNITION 92

  4.1 Modelling Traditions 92
  4.2 Micro-And Macro Cognition 93
  4.3 Cognitive Functions... 94
  4.4 Structural Models 95
      4.4.1 The Sequentiality Of Cognition 96
      4.4.2 Context Free Processes 98
  4.5 A Simple Model of Cognition (SMoC) 99

5. STANDARD CLASSIFICATION SCHEMES.. 100

  5.1 Factors Influencing Vulnerability To Error 101
  5.2 Classification In First-Generation HRA 103
  5.3 Classification In Human Information Processing... 105
  5.4 Classification In Cognitive Systems Engineering.. 106

6. PERFORMANCE SHAPING FACTORS AND COMMON PERFORMANCE CONDITIONS 107

  6.1 Performance Shaping Factors In THERP 107
  6.2 Classical Performance Shaping Factors... 108
  6.3 Error Modes And Error Models.. 110
  6.4 Specific Effects Of Performance Conditions.. 113
  6.5 Dependency Of Performance Conditions 115

7. CHAPTER SUMMARY 117

CHAPTER 5: HRA - THE FIRST GENERATION

1. RELIABILITY AND SAFETY ANALYSIS OF DYNAMIC PROCESS SYSTEMS 120

2. FIRST-GENERATION HRA APPROACHES 122
    2.1 Accident Investigation And Progression Analysis (AIPA) 124
        2.1.1 Method 124
        2.1.2 Classification Scheme 125
        2.1.3 Model 125
        2.1.4 Conclusion 125
    2.2 Confusion Matrix 125
        2.2.1 Method 125
        2.2.2 Classification Scheme 126
        2.2.3 Model 126
        2.2.4 Conclusion 126
    2.3 Operator Action Tree (OAT) 126
        2.3.1 Method 126
        2.3.2 Classification Scheme 127
        2.3.3 Model 127
        2.3.4 Conclusion 127
    2.4 Socio-Technical Assessment Of Human Reliability (STAHR) 127
        2.4.1 Method 128
        2.4.2 Classification Scheme 129
        2.4.3 Model 129
        2.4.4 Conclusion 129
    2.5 Technique For Human Error Rate Prediction (THERP) 129
        2.5.1 Method 130
        2.5.2 Classification Scheme 130
        2.5.3 Model 130
        2.5.4 Conclusion 130
    2.6 Expert Estimation 131
        2.6.1 Method 131
        2.6.2 Classification Scheme 131
        2.6.3 Model 132
        2.6.4 Conclusion 132
    2.7 Success Likelihood Index Method / Multi-Attribute Utility Decomposition (SLIM/MAUD) 132
        2.7.1 Method 132
        2.7.2 Classification Scheme 133
        2.7.3 Model 133
        2.7.4 Conclusion 133
    2.8 Human Cognitive Reliability (HCR) 133
        2.8.1 Method 133
        2.8.2 Classification Scheme 134
        2.8.3 Model 134
        2.8.4 Conclusion 134
    2.9 Maintenance Personnel Performance Simulation (MAPPS) 135
        2.9.1 Method 135
        2.9.2 Classification Scheme 135
        2.9.3 Model 135

2.9.4 Conclusion 135

3. CONCLUSIONS 136

3.1 Method Description 137
3.2 Classification Schemes 137
3.3 Operator Models 138
3.4 Design And Performance Analysis 139

4. HRA AND COGNITION: EXTENSIONS 140

4.1 Cognitive Environment Simulator (CES) 140
    4.1.1 Method 141
    4.1.2 Classification Scheme 141
    4.1.3 Model 141
    4.1.4 Conclusion 142
4.2 INTENT 142
    4.2.1 Method 142
    4.2.2 Classification Scheme 142
    4.2.3 Model 143
    4.2.4 Conclusion 143
4.3 Cognitive Event Tree System (COGENT) 143
    4.3.1 Method 143
    4.3.2 Classification Scheme 143
    4.3.3 Model 144
    4.3.4 Conclusion 144
4.4 EPRI Project On Methods For Addressing Human Error In Safety Analysis 144
    4.4.1 Method 144
    4.4.2 Classification Scheme 145
    4.4.3 Model 145
    4.4.4 Conclusion 145
4.5 Human Interaction Timeline (HITLINE) 145
    4.5.1 Method 145
    4.5.2 Classification Scheme 146
    4.5.3 Model 146
    4.5.4 Conclusion 146
4.6 A Technique For Human Error Analysis (ATHEANA) 146
    4.6.1 Method 146
    4.6.2 Classification Scheme 148
    4.6.3 Model 148
    4.6.4 Conclusion 148
4.7 Conclusions 148

5. CHAPTER SUMMARY 150

CHAPTER 6: CREAM - A SECOND GENERATION HRA METHOD

1. PRINCIPLES OF CREAM 151

1.1 Method Principles 152
1.2 Model Fundamentals 152

2. MODELS OF COGNITION.. 153

   2.1 A Simple Model Of Cognition 153
   2.2 Competence And Control 154
   2.3 Four Control Modes... 155

3. BASIC PRINCIPLES OF THE CLASSIFICATION SCHEME 157

   3.1 Causes And Effects... 157
   3.2 A Note On Terminology 160

4. CLASSIFICATION GROUPS 161

   4.1 Details Of Classification Groups 163
      4.1.1 Error Modes (Basic Phenotypes) 164
      4.1.2 Person Related Genotypes 166
      4.1.3 Technology Related Genotypes 170
      4.1.4 Organisation Related Genotypes... 173
      4.1.5 Summary 175

5. LINKS BETWEEN CLASSIFICATION GROUPS 176

   5.1 Consequent-Antecedent Relations In CREAM 177
      5.1.1 Error Modes (Phenotypes) 178
      5.1.2 Person Related Genotypes 179
      5.1.3 Technology Related Genotypes 181
      5.1.4 Organisation Related Genotypes... 182
   5.2 The Interdependency Of Consequents And Antecedents 184
   5.3 Direct And Indirect Consequent-Antecedent Links.................................... 186
   5.4 Context Dependence Of Classification Groups 187
   5.5 Possible Manifestations And Probable Causes.. 188

6. CHAPTER SUMMARY 189

CHAPTER 7: THE SEARCH FOR CAUSES: RETROSPECTIVE ANALYSIS

1. ANALYSIS AND STOP RULES 191

   1.1 Terminal And Non-Terminal Causes.. 193
   1.2 Analysis Of A Fictive Event.. 194
   1.3 Analysis Of A Real Event.. 195

2. OVERALL METHOD 198

   2.1 Context Description... 199
   2.2 Possible Error Modes... 200
   2.3 Probable Error Causes 202
   2.4 Detailed Analysis Of Main Task Steps... 203
   2.5 Going Beyond The Stop Rule... 206

3. EXAMPLE OF RETROSPECTIVE ANALYSIS... 207

   3.1 Tube Rupture 207
   3.2 Isolation Of Ruptured Steam Generator - How Soon? 209
   3.3 Event Analysis 210
      3.3.1 Describe Common Performance Conditions 210

3.3.2 Describe The Possible Error Modes..210
3.3.3 Describe The Probable Causes..212
3.3.4 Detailed Analysis Of Main Task Steps 212
3.3.5 Summary Of Analysis 214

4. CHAPTER SUMMARY 214

CHAPTER 8: QUALITATIVE PERFORMANCE PREDICTION

1. PRINCIPLES OF PERFORMANCE PREDICTION 216

1.1 Scenario Selection 216
1.2 The Role Of Context..217
1.3 Performance Prediction In First-Generation HRA 219
1.3 219
    1.3.1 Pre-Defined Sequence Of Events...220
1.4 Success And Failure 221
1.5 The Separation Between Analysis And Prediction..222

2. PREDICTIVE USE OF THE CLASSIFICATION SCHEME 223

2.1 Combinatorial Performance Prediction 223
2.2 Context Dependent Performance Prediction 225

3. PRINCIPLES OF QUALITATIVE PERFORMANCE PREDICTION 229

3.1 Forward Propagation From Antecedents To Consequents 229
3.2 Example: The Consequents Of Missing Information 229
3.3 Discussion..231

4. CHAPTER SUMMARY 232

CHAPTER 9: THE QUANTIFICATION OF PREDICTIONS

1. CREAM - BASIC METHOD..234

1.1 Construct The Event Sequence..235
1.2 Assess Common Performance Conditions 236
1.3 Determine The Probable Control Mode 239
1.4 The Control Mode For The Ginna Example...242

2. CREAM BASIC METHOD: AN EXAMPLE 242

2.1 Construct Event Sequence 242
2.2 Assess Common Performance Conditions 243
2.3 Determine The Probable Control Mode 245

3. CREAM - EXTENDED METHOD 245

3.1 Build A Cognitive Demands Profile...246
3.2 Identify Likely Cognitive Function Failures..249
3.3 Determine Failure Probability 251
3.4 Accounting For The Effects Of Common Performance Conditions On CFPs 252

4. EXTENDED CREAM METHOD: AN EXAMPLE..255

    4.1 Build A Cognitive Demands Profile..255
    4.2 Identify Likely Cognitive Function Failures..257
    4.3 Determine Failure Probability 258
    4.4 Incorporating Adjusted CFPs Into Event Trees 260

5. CHAPTER SUMMARY 260

References 262

Index 276

Foreword

"No man is an island entire of itself;"
John Donne (c. 1572–1631).

1. A FABLE

Once upon a time a bunch of well-meaning scientists started to get worried about the many errors that other people made. This was not just the slip of the tongue and the housewife everyday errors, or not even the many people that killed each other on the roads. Such things had become commonplace and nobody paid much attention to them. In particular, the newspapers rarely reported them on the front page. The scientists were rather concerned about errors in complex industrial systems where a single person could cause irreparable harm to many others. These were the kind of events that would grab the headlines and dominate the news - at least for a day or two.

They scientists were concerned about what they called the "man at the sharp end", by which they meant the people who were caught between the demands from complex technology and inadequate means they were given to achieve their tasks. Some scientists wanted to lighten their plight, others just wanted to calculate when they would fail.

The well-meaning scientists said to themselves, look we must do something about this. And since they knew that everything has a cause, they began to look for causes. This meant that they had to leave their laboratories and go out into the field.

When they came into the field, they looked and found two promontories. One was a hill called The MINd Field (although it had previously been called the MINE field) which looked as the results of the forces of nature and the other was a clearly man-made structure, though somewhat weather worn, called the HEAP.

In addition there was, of course, scientists they paid little attention therefore they did not notice it.

the whole environment, but being to that. They were in it, and

As many of the scientists had been trained in psychology, they decided to begin with the MINd-field. And they started digging. In the beginning work was easy, and they found lots of little relics in the upper layers. There was also some dirt, which they kept to themselves and only discussed during conference coffee breaks. But they did not find the chest with the treasure, i.e., the cause of all things, in search of which they had started their quest. As they began to dig deeper and deeper, they ran into rocks and boulders, and the work became very strenuous. So they held a conference. Some were of the opinion that they should continue digging and try to remove the boulders or get around them, in the certain conviction that the answer could be found further down. Others doubted that it was worth the effort and reasoned that they instead started to dig in the other place, the HEAP. So they split into two groups, and continued their work.

And here we are. The two groups are still going at it in different directions. Every now and then someone from either group will look up from their work and notice the world around them; and sometimes they will begin to wonder whether the cause of all things is to be found elsewhere. But soon they will turn their attention to the unfinished work and keep on digging and analysing.

This book is about what we found when we tried to dig deeper in the MINd field.

2. ACKNOWLEDGEMENTS

There are two types of acknowledgements, one that refers to the contents and one that refers to the context. In terms of contents, the ideas described in this book have been developed over a number of years and have benefited from the discussions and collaborations with and inspirations from a number of friends and colleagues. Chapter 1 gives a brief description of the background for the book in terms of the gradual development of the basic ideas. Here I will confine myself to thank a small number of people who, in one way or another, have been very important for the development and realisation of CREAM. It is unfortunately practically impossible to mention everyone, and it is decidedly unfair to the rest to single out a few. So with the risk of offending members from both camps, I will mention three groups that have been particularly important. One group includes the people, past and present, from the Institute for Systems, Informatics, and Safety (ISIS) at the Joint Research Centre in Ispra (Italy), in particular Carlo Cacciabue and Mauro Pedrali who were among the first to see the possibilities of the general approach described by CREAM and the power of the phenotype-genotype distinction. A second group includes my colleagues at Human Reliability Associates (UK), in particular Phil Marsden who was instrumental in the first development of the predictive method and who contributed significantly to the historical perspective. Many of the practical issues of CREAM were also refined in a constructive interaction with two customer organisations, the Institute for Nuclear Safety Systems (INSS) in Japan and the Korean Atomic Energy Research Institute (KAERI). A third group includes the PSA/HRA practitioners, in particular Ed Dougherty Jr., Mike Frank, Tony Spurgin, and John Wreathall. Their interest has been particularly encouraging for a poor psychologists who have ventured into the quagmire of the real world (and survived!). I am also indebted to David Woods who provided me with many succinct comments and asked more questions - in writing - than I could possibly answer. Finally, I would like to thank Dr. Singh from EPRI for reference material on the Method for Addressing Human Error in Safety Analysis.

In terms of the context, I am thinking of the tolerance, patience, and encouragement that every author needs from his family. Writing a book is a very egotistical enterprise where the author can disappear for long periods of time into a universe that often must seem impenetrable. My dear wife, Agnes, has not only been infinitely patient but has also provided me with both moral support and sustenance whenever needed. Without that I would not have been able to complete this book. Unfortunately, I dare not promise that this was the last time.

Chapter 1
The State Of Human Reliability Analysis

"Cheshire-Puss, would you tell me, please, which way I ought to go from here?"
"That depends a good deal on where you want to get to," said the Cat.
"I don't much care where----" said Alice.
"Then it doesn't matter which way you go," said the Cat.
"---so long as I get somewhere," Alice added as an explanation.
"Oh, you're sure to do that," said the Cat, "if you only walk long enough."
Lewis Carroll (1832-1898), Alice's Adventures in Wonderland (1865).

---

1. INTRODUCTION

Since the middle of this century the main perspective on how technological systems should be designed, built, operated, and maintained has changed dramatically. In the late 1940s the technological development had reached a state where the capabilities of the unaided human started to become a limiting factor for the performance of the overall system - although seen in terms of efficiency rather than of risk. In order to overcome that, the human factor was taken into account in the design of systems to ensure that the demands to human performance did not exceed the natural capabilities. The concern was initially focused on sensory-motor capabilities (Fitts, 1951) but was later extended to cover the so-called higher order human functions, in particular cognition as epitomised by the information processing descriptions of decision making and problem solving. The need to consider the human factor development was further motivated by a growing number of prominent accidents in technological systems (Casey, 1993) and by the changed role of technological systems in the Westernised society, both of which made the interaction between humans and machines a main issue.

Technological systems are found everywhere in modern society: in production, administration, transportation, healthcare, finance, etc., and always seem to be growing in complexity. In every case our way of living depends on the proper functioning of these systems, something that occasionally is used deftly in work disputes. It has for many years been clear that human actions constitute a major source of vulnerability to the integrity of interactive systems, in whatever field they are used, complex as well as simple. Inappropriate, incorrect or erroneous human actions are thus a cause of great concern. In the technical literature such actions are commonly referred to as "human errors", but as argued elsewhere (Hollnagel, 1993a; Woods et al., 1994) this term is quite misleading. A "human error" is best seen as a judgement in hindsight, as I will discuss more thoroughly in Chapter 2. The preferred term, and the one that will be used throughout this book, is human erroneous actions.

This view is valid even if the focus of interest shifts to the realm of organisations and the topic of organisational accidents (Reason, 1997). An organisation is basically made up of people and is the organisation of people and the people in the organisation, rather than the organisation as an abstract entity, which may constitute the cause of accidents. The perspective offered by this book is therefore applicable to organisational and technological accidents alike.

Most, if not all, of the technological systems depend on the interaction with people in order to function appropriately. Few of these systems are completely autonomous, although some - such as hydroelectric power plants - may function for extended periods of time without the need of human intervention. Throughout this book the terms "system" is used to refer to complex and dynamic technological systems where human interaction is essential for the proper functioning. The importance of human interaction was first recognised as part of the control of the system, but experience has shown that it is equally important in relation to design, implementation, and maintenance. Good examples are nuclear power plants, gas & oil production, aviation, ships, industrial production processes, and modern hospitals (especially surgery and intensive care), These systems have often been the focus of considerable concern because of large scale accidents. The principles of reliability and error analysis are however equally valid for seemingly simpler systems, such as automatic ticketing machines, private cars, food and drink machines, etc. In these cases the consequences of accidents may be smaller, but the potential for accidents may be the same. The examples that are used throughout the book are mostly related to the control of complex industrial systems, but other cases such as those described by Norman (1988), Reason (1997) or Woods et al. (1994) might equally well have been used.

1.1 The Pervasiveness Of Human Erroneous Actions

Although valid figures are difficult to obtain there seems to be a general agreement to attribute somewhere in the range 60-90% of all system failures to erroneous human actions (Hollnagel, 1993a), regardless of the domain. The concern for safe and efficient performance has, for natural reasons, been especially high in relation to the operation of nuclear power plants which also has seen the most widespread use of Probabilistic Safety Assessment (PSA) (or Probabilistic Risk Assessment, PRA) and Human Reliability Analysis (HRA). In a nuclear power setting estimates of the incidence of operator induced system failure tend to be towards the higher end of the scale. For instance, an analysis of 180 significant events in the nuclear power industry carried out by the Institute of Nuclear Power Operations (INPO) revealed that more than 51% of incidents were traceable to human performance problems (INPO, 1984 & 1985). The study indicated that design deficiencies (32%) and poorly manufactured equipment (7%) comprised the other major sources of system failure. It could reasonably be suggested that the two categories may also be considered instances of human erroneous actions, thus raising the contribution to 90%. Similarly, in March 1988 the U. S. nuclear industry reported on 2,180 mishaps in nuclear power-plant operation that had been reported to the Nuclear Regulatory Commission (NRC) during 1987. Again it was estimated that around half of these incidents were attributable in some way to human erroneous actions. (As an aside, it was also noted that of the 110 nuclear plants licensed to operate at full power, seven produced no electricity at all. The average power output from the remaining available plant was estimated to be less than 60% of expected capacity.)

The picture is the same in other domains, although data collection may be less comprehensive. A NASA analysis of 612 shuttle incidents in the period 1990-1993 showed that 66% of them were seen as caused by human erroneous action. Of the remaining causes 5% were attributed to faulty procedures, 8% to equipment failure, and 21% to the category "other" - including communication breakdown and poor training. The total number of cases where human actions were involved is therefore considerably higher, probably around 80-85%. (Included in the category of other causes was such things as lightning strikes;

although lightning is a natural phenomenon, the decision to launch under conditions when lightning can occur is made by humans. If this argument is taken to its extreme, the contribution of human actions becomes very close to, if not reaching, 100%.)

Analyses of car accidents show that technical defects account for 5-8% of the causes, while the rest are due to incorrect judgement, unanticipated behaviour, loss of consciousness, etc. In fully 47% of car accidents no other vehicles were present! In another field, an analysis of shipping accidents that occurred during the period 1987-1991 showed that 12% were due to structural error, 18% to equipment and mechanical error, and the rest to a variety of causes all involving human action (shore control, crew, pilot, officers) - which means that human erroneous action was seen as a cause in at least 70% of the cases. The same trend is found in aviation (e.g. Helmreich, 1997). Figure 1 summarises the trend in **attribution** of causes for accidents across a range of technical domains. The figure shows, firstly, that the number of data points has increased during the last 10-15 years and, secondly, that there is a clear trend towards higher estimates. The reader should keep in mind that this may well differ from the distribution of actual causes, to the extent that the actual causes can be found at all (Cojazzi & Pinola, 1994).

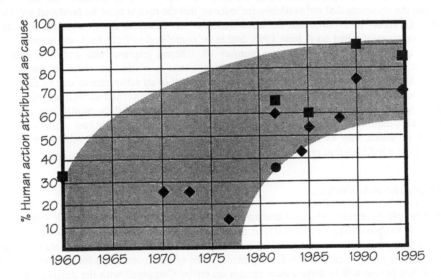

Figure 1: Number of accidents where human actions were seen as the main cause.

The trend shown in Figure 1 nevertheless does **not** mean that there is a growing number of cases where human actions actually are the cause of accidents. As argued elsewhere (Hollnagel, 1993a) it only means that there is an increase in the number of cases where the cause, rightly or wrongly, has been **attributed** to human actions. The percentage is derived from the fraction:

$$\frac{C_M}{C_M + C_T + C_O + C_{MT} + C_{MO} + C_{TO} + C_{MTO}}$$

where C_M are the causes attributed to human factors, C_T are the causes attributed to technological factors, C_O are the causes attributed to organisational factors, and where C_{MT}, etc., are causes attributed to a combination of the three main factors. The numerical value of this fraction will increase if either the numerator is increased or if the denominator is reduced. Both of these are possible explanations. Over the years the numerator has probably increased, not because humans have become more prone to make incorrect actions but because human actions have come more into focus. Since 1979 there has been a distinct change in the accepted "understanding" of the nature of accidents and their causes, which to a considerable extent determines how accidents are analysed and explained. As an example, something that previously was considered as a "pure" technical fault may now be seen as the result of a failure in maintenance or design (Reason, 1997). At the same time the denominator has decreased due to advances in engineering and technology. Technical components and subsystems have in general reached a very high level of reliability, and the technical systems are often designed to be partly or wholly fault tolerant. The same goes, to a more limited extent, for organisations.

When it comes to the interactions between the main factors, it is arguable whether they have increased or decreased. Something that may lead to an increase in the more complex causes is the changes in system constituents, leading to more complex systems with tighter couplings, and increased performance demands that reduce the operational safety margins. Another is the motivation that drives accident analysis, and the resources that are available, for instance that the results must be produced within a given time frame. (A recent example is that the investigation of the failure of the first Ariane-5 flight, which came to an unsuccessful end on 4 June 1996, had to be completed in little over one month.) The trend shown in Figure 1 can therefore have a number of different explanations and it is important to acknowledge that the determination of what is a main cause is neither an absolute nor an objective judgement. Any outcome, whether it is classified as an accident or an achievement, can be associated with a set of causes. In order to be able to explain accidents that **have** happened as well as predict accidents that **may** happen it is therefore necessary to improve the understanding of the multiplicity of factors that are at play. This requires a conceptual framework that adequately accounts for how actions are shaped by the context in which they take place, and which effectively enables the identification of the most important of the conditions that pragmatically can be said to be the causes.

1.2 Human Actions As Causes

As long as people have been designing artefacts to be used by others rather than just building to be used by themselves, the possibility that the artefacts may fail has been a concern. The design of an artefact or a system by its very nature involves a consideration of the possible situations that may arise, including the situations that may lead to unforeseen developments. If the system is designed for someone else then the reaction of that person will to some extent remain uncertain. Compared with the designer, the user may not know in detail how the system works - and in many cases there should not be a need to know this. The user may furthermore be uncertain about the significance of a specific event or about what the outcome of a specific development may be. The essence of design is therefore to think out the functions of the system in advance and to develop detailed plans for building it so that it will achieve its designated function under practically all conditions - including all likely (and some unlikely) actions by the user. Unless the designer of the system is excessively optimistic or self-assured, it is always necessary to consider what may happen if something fails or if the user handles the system incorrectly. The perfect design - or the perfect plan - is that which has taken every contingency into consideration. Yet as we know from the lore about the perfect crime, there is usually something that goes amiss.

In this respect real life is not different from fiction. Practice has shown that even the best designed systems fail every now and then - and that for complex systems it often happens with embarrassing regularity. Whenever a failure occurs for which the consequences are non-trivial, it becomes the start of a search for the causes. In a small number of cases the event is written off as due to an Act of God, i.e., it is seen as an unusual or unforeseeable manifestation of the forces of nature beyond the powers of human intervention, such as an earthquake or a bolt of lightning. (It may nevertheless rightly be argued that whereas an earthquake is unforeseeable, the decision to build e.g. a nuclear power plant in an earthquake prone area is the result of human activities. The adverse consequences could therefore have been avoided by building the plant in another place.) In the majority of cases, however, the search ends by identifying human action as the main - or sometimes even the only - cause.

System design naturally tries to make a system impervious the most likely causes. In particular, most hazardous systems are made resistant to the occurrence of single failures. As an example, the European Space Agency uses the following requirements (ESA PSS-01, p. 45):

- No single failure shall have a catastrophic or critical hazardous consequence.

- No single operator error shall have a catastrophic or critical hazardous consequence.

- No combination of a single failure and a human operational error shall have a catastrophic or critical hazardous consequence.

A system can in principle be made resilient to single failures by first making a list of everything that may have a significant effect on its functioning, using some acceptable definition of significance. (Note, however, that the third requirement of ESA refers to the combinations between single failure types.) Following that each item on the list is examined in greater detail in order to determine whether a possible failure will have unacceptable consequences for the system. If that is so, then the final step is to consider what can possibly be done to prevent this from happening by developing protection systems or barriers. (A similar approach is used for defence-in-depth.) In this process, human erroneous actions present a special problem, since human actions are inherently more difficult to regulate and predict than the functions of the technological components. Similarly, it is also more difficult to improve or change the human "component" or even to eliminate it, e.g. via increased automation.

1.3 Deterministic And Probabilistic Analyses

Due to the technological advances of the last 30 years, particularly with regard to engineered safety features, most hazardous systems are now resistant to single failures of either human actions or technical functions. In order to feel reasonable sure that a system will work it is, however, necessary to consider not only single failures but also what could happen if two or more failures occur at the same time, cf. the ESA requirements. It is easy to see that the number of possible combinations of failures that must be checked quickly becomes prohibitively large, at least if a complete and deterministic analysis is to be made. Many systems have therefore adopted the philosophy of defence-in-depth, which means that a number of barriers or safeguards must be breached before an accident occurs. Although this has had the overall effect of making it less likely that accidents occur - or rather, that specific initiating events have adverse consequences - it has not made it impossible. It is still quite possible for an accident to be caused by a combination of several conditions where each is necessary but none sufficient to bring about the accident by itself.

The growing complexity of systems necessary to meet the needs of end users provides ample opportunities for the harmful accumulation of such conditions. The increased automation and functional sophistication high-technology systems, such as nuclear power plants or fly-by-wire aircraft, also mean that they have become much more difficult to understand for the people who control and maintain them - as well as for the people who must design, build, and regulate them. These systems are therefore especially prone to rare, but potentially catastrophic, accidents that result from a combination of otherwise innocent events. This phenomenon has become known as **latent failures** - although it would be more appropriate to call it **latent failure conditions** - meaning that there are latent conditions in the system that may become contributing causes for an accident. Latent failure conditions are thus seen in contrast to **active failures**, i.e., failures of technological functions or human actions, which become the local triggering events that afterwards are identified as the immediate causes of an accident.

The defining feature of latent failure conditions is that they are present within the system as unnoticed conditions well before the onset of a recognisable accident sequence. The influence of latent failure conditions in systems that are low in risk but with possible high-hazard consequences, such as nuclear power plants, chemical process plants, modern aircraft, etc., is frequently seen as the cause of multiple-failure accidents (Reason, 1987 & 1990). The latent failure conditions can arise from oversights or failures in design, construction, procedures, maintenance, training, communication, human-machine interfaces and the like.

It is impossible in practice to make a **deterministic** analysis of even simple interactive systems, since this requires an examination of the links between every possible cause and every possible consequence. To illustrate that, consider Table 1 that shows the number of components that is characteristic of various types of processes. It is reasonable to assume that the number of links is at least as great as the number of components, and that it is probably considerably larger. Even if all the links could be enumerated, a deterministic analysis would further have to assume either that the links were not changed by the actual working conditions, or that the working conditions were known in sufficient detail. There are, however, several reasons why such knowledge must remain incomplete. For the technical parts of the system - the software and the hardware - the real performance conditions can only be predicted approximately (corresponding to **incomplete anticipation**); furthermore, the actual system will always be incomplete or deficient in relation to the design requirements and specifications (**incomplete implementation**). This is the case for all systems, except those which are trivially simple and which therefore usually are found only in laboratories or under experimental conditions.

Since a deterministic analysis is not possible, the common solution is to make a **probabilistic** analysis instead. In this the search is reduced to include only the links between significant causes and significant consequences - leaving completely aside the problem of finding a workable definition of what significance is.

Table 1: Characteristics of various process types (Alty et al., 1985)

Process type / domain	Number of information points (variables) in process	Main frequency of operator actions	Time allowed for operator action
Power distribution networks	~100 - 200.000	1/hour (usually clustered)	< 1 minute
Power generating stations	10.000 - 20.000	1/hour (usually clustered) more during start-up	10-30 minutes
Process industries	2.000 - 10.000	5-6/hour (sometimes clustered)	< 1 minute
Rolling mills	< 100	1/second	Direct
Aeroplanes.	100 - 300	1/minute (landing) 2-3/hour (cruising) 1/sec (manual flight)	Direct

The main objectives of probabilistic analyses are: (1) to reduce the probability that an untoward event occurs, and (2) to minimise the consequences of uncontrolled developments of accident conditions in terms of e.g. the potential for injuries and loss of human life, a negative impact on the environment, and damage to the facility itself. Probabilistic analyses of safety and risk are a natural way of compensating for the effects of incomplete anticipation and incomplete implementation. They do require a good understanding of the nature of how a failure may occur and how the effects may propagate through a system. The need to perform a probabilistic analysis, in particular for interactive systems, has brought two problems to the fore, cf. Cacciabue (1991). The first is how to account for the impact of human action on the process (planned and unplanned interventions), and the second is how to evaluate and/or validate the efficacy of the proposed design vis-à-vis the anticipated human performance. Human action failures can be due to a number of things such as the inherent variability of human performance (residual erroneous actions), inappropriate design of working environment and procedures (system induced erroneous actions), unanticipated events (external disturbances, component failures), and communication failures (between crew members or in the safety management system as a whole). Theories, models, and methods that can be applied to solve these problems are therefore in high demand.

1.4 Point-To-Point Analyses

Analyses of interactive systems, and in particular safety and risk analyses, traditionally use a number of different techniques that have been developed over the last fifty years. Common to all of them are the concepts of a cause (a source) and a consequence (a target). The **cause** denotes the origin or antecedent of the event under investigation, corresponding to the Aristotelian notion of an effective cause, while the **consequence** denotes the specific outcome or effect. The purpose of safety and risk analyses is to find the possible links between causes and consequences, in order to devise ways in which the activation of such links can be avoided. From an analytical point of view one can consider four different types:

- **One cause, one consequence**. Here the specific relations between a single cause and a single consequence are sought. The analysis can either start from the consequence or start from the cause. An example of the former is root cause analysis (Cojazzi, 1993) which tries to identify the single (root) cause of an event. An example of the latter is consequence propagation. However, one-to-one analyses are very specific and not really practical, since virtually all events involve either multiple causes or multiple consequences.

- **One cause, many consequences**. This approach is a cause-to-consequences investigation of the possible outcomes of a single cause. Typical techniques are Failure Mode and Effects Analysis (FMEA), or Sneak Analysis (SA). FMEA (e.g. Modarres, 1993) investigates how a single component failure may affect the performance of a technical system. An FMEA reviews all components, identifies all failure modes for each component, and tries to assess the effects for each of them in turn. A Sneak Analysis (Buratti & Godoy, 1982) identifies undesired functions or inhibitions of desired functions (without assuming component failures) and goes on to analyse the effects of the so-called sneaks. Commonly known variants are Sneak Circuit Analysis for electronic components, and Sneak Path Analysis for systems in general including software and - potentially - man-machine interaction. One-to-many approaches are mainly analytical, starting with a single event and trying to calculate or imagine what the possible outcomes may be.

- **Many causes, one consequence**. This type of analysis looks for the possible causes for a specific consequence and the approach moves from consequence or event to possible causes. Examples are Fault Tree Analysis (FTA), and Common Mode Failure analysis. This type of analysis is also characteristic of the general approach to accident analysis that looks for breaches of the defence-in-depth principle. The rationale is that once an (actual or hypothetical) event has occurred, the well-defined consequence can be used as a starting point of an analysis for tracing backwards to the causes. A more general type of analysis is Functional Analysis (FA) which is a goal-driven approach to identify the basic functional elements of the system.

- **Many causes, many consequences**. This is the investigation of the relations between multiple causes and multiple consequences. The approach can start either from the causes or the consequences, or indeed be a mixture of the two. Examples are Hazard and Operability Analysis (HAZOP), Cause-Consequence Analysis (CCA), and Action Error Mode Analysis (AEMA). All of these start by a single event (the seed) and try to identify the possible causes as well as the possible consequences; the analysis is thus bi-directional.

The common feature of point-to-point analyses is that they are based on a limited number of pre-defined causes or consequences - or seed events, in the case of the bi-directional approaches. This limitation is a strength for the methods because it enables the analyses to focus on relevant subsets of the domain, i.e., highly critical events (safety critical, mission critical). The focus is necessary because it is computationally impossible to consider all possible events, even for very small systems. (For instance, making an exhaustive test of a system with only ten inputs and ten outputs, each having two states, will require over 1.000.000 tests!) The limitation is, however, also a weakness because the quality of the outcome of the analysis is restricted by the completeness or scope of the selection. The set of initiating events and conditions must be both comprehensive and representative in order for the results to be practically applicable, and these conditions may be hard to establish in practice.

The alternative to point-to-point analyses is a dynamic analysis where the actual development of an event is developed as part of the analysis rather than prior to it. This obviously involves the use of simulators that are computational models of the system being analysed. The principles and practice of this method have for many years been studied as part of the DYLAM approach (Amendola, 1988) and has more recently been extended to include also a dynamic simulation model of the human operator (Hollnagel, 1995c; Hollnagel & Cacciabue, 1992; Kirwan, 1996).

1.5 Analysis And Prediction

In order to know or predict how a system may fail in the future, we must either know how it has failed in the past, or be able to identify the principal ways in which it can fail according to how it has been designed. On a fundamental level the two approaches are similar because design encapsulates experience. In order to know how a system has failed or how it can fail it is necessary to analyse its function in detail. In both cases the analysis serves to identify the dominant links between consequences and causes, and the outcome of the analysis constitutes the basis for making the predictions. These can either be made in a simple manner, by a parametric extrapolation of accumulated experience, or in a more complex manner by means of a modelling approach. In either case the past must be understood before the future can be predicted; prediction and analysis are thus inherently coupled. Predictions can nevertheless not be made simply by reversing the process of analysis. This has unintentionally been demonstrated by the difficulty in using the findings of "error psychology" to establish a forward link from causes to consequences.

HRA is concerned with a particular type of prediction, namely the prediction of the likelihood that a human action may fail, usually expressed in terms of the probability of a human erroneous action. In order to achieve this objective an elaborate edifice has been built, which in the 1990s has come under severe criticism. The unsatisfactory state of then dominating HRA approaches was presented by Ed Dougherty (1990) in a seminal paper that established the distinction between first-generation ("classical") and second-generation ("modern") HRA approaches. In the following years that led to a considerable number of other papers that substantiated and amplified Dougherty's arguments - although some people naturally disagreed. At present, however, there are few viable alternatives to the often criticised first-generation HRA approaches. This is perhaps not so surprising, since it is easier to describe a symptom than to prescribe a cure. Towards the end of the 1990s the general consensus of the HRA community is that something needs to be done, although there are only few very clear ideas about what it should be. The aim of this book is to offer a consistent and argued alternative that can serve as an example of a second-generation HRA approach - or at the very least as a transition between what we have and what we want. The book will therefore aim to describe a method rather than a theory, and it is intended as a practical guidance rather than an academic treatise. This does not mean that it will be devoid of theory and theorising, but rather that they will be kept to a necessary minimum.

2. SHORTCOMINGS OF FIRST-GENERATION HRA

2.1 A Pragmatic Criticism

The shortcomings of first-generation HRA approaches can be described in several ways. From the practitioner's point of view the shortcomings refer mainly to how the approaches are applied and in particular the insufficient theoretical basis. A comprehensive list of generally recognised shortcomings is reproduced below (Swain, 1990):

1. **Less-than-adequate data**: the scarcity of data on human performance that are useful for quantitative predictions of human behaviour in complex systems.

2. **Less-than-adequate agreement in use of expert-judgement methods**: there has not yet been a demonstration of satisfactory levels of between-expert consistency, much less of accuracy of predictions.

3. **Less-than-adequate calibration of simulator data**: because the simulator is not the real world, the problem remains of how raw data from training simulators can be modified to reflect real-world performance.

4. **Less-than-adequate proof of accuracy in HRAs**: demonstrations of the accuracy of HRAs for real-world predictions are almost non-existent. This goes particularly for non-routine tasks, e.g. time-dependent diagnosis and misdiagnosis probabilities in hypothetical accident sequences.

5. **Less-than-adequate psychological realism in some HRA approaches**: many HRA approaches are based on highly questionable assumptions about human behaviour and/or the assumptions inherent in the models are not traceable.

6. **Less-than-adequate treatment of some important performance shaping factors (PSFs)**: even in the better HRA approaches, PSFs such as managerial methods and attitudes, organisational factors, cultural differences, and irrational behaviour are not adequately treated.

(Swain went on to remark that "all of the above HRA inadequacies often lead to HRA analysts assessing deliberately higher estimates of HEPs and greater uncertainty bounds, to compensate, at least in part, for these problems" (Swain, 1990, p. 309). This is clearly not a desirable solution.)

It is necessary to consider the requirements to a potential second-generation HRA approach in the light of the current criticism. The above list of shortcomings, however, only looks at the problems from the practitioner's point of view and does not take the additional step of examining **why** these shortcomings exist. It does not question the assumptions behind first-generation HRA and therefore represents a shallow rather than a deep criticism. For instance, the first shortcoming - the less-than-adequate-data - assumes that there is a need for detailed data to support the calculation of HEPs. The nature and reality of this need must, however, be examined. This will take place in Chapters 2 to 5 that will define the principles of the analysis as well as describe how it is made. At the same time the basis for a second-generation HRA approach will gradually be developed.

2.2 A Principled Criticism

Although the above list of shortcomings of HRA is indisputably correct it reflects the bias of the practitioner. This means that the criticism focuses on points that have become issues of contention in practice. In addition to that it is possible to give a criticism of a more principled or theoretical nature, which differs from the above list and which possibly may be more serious. This criticism starts by questioning the assumptions behind first-generation HRA and can be illustrated by examining the basic notion of a reliability model. The purpose of PSA and HRA is to foresee, in a scientific manner, the events that are likely to happen in a system by making systematic predictions on the basis of specialised knowledge. First-generation HRA approaches are, however, less concerned with **what** people are likely to do than with **whether** they will succeed or fail. This is obviously to a very large extent driven by the needs of PSA, and in particular the conventional representation of actions as a binary event tree. The majority of first-generation HRA approaches make a distinction between success and failure of actions, and the objective is to calculate the probability of one or the other. In order to make such calculations consistently it is necessary to refer to a systematic and ordered description of the underlying causal functions. This description is referred to as a **reliability model** - or just a model - which represents the required knowledge about the target system, which in the case of HRA is the human operator.

While reliability models can be developed with relative ease for technological systems, it is extremely difficult to do so for humans. There are many reasons for that, first and foremost the fact that whereas technological systems are designed with a specific set of functions in mind and therefore have a know structure, humans have evolved naturally. The characteristics of human performance in a given context can be observed and used as a point of reference but despite more than a century of psychological research no-one really knows what basis human actions have in internal, mental or cognitive functions. Human reliability modelling cannot be deduced from fundamental psychological principles but must rather start by recognising the inherent variability of human performance, and come to terms with the situations where this variability leads to failures rather than successes.

One of the undisputed assumptions in all HRA approaches is that the quality of human performance depends on the conditions under which the tasks or activities are carried out. These conditions have generally been referred to as Performance Shaping Conditions or PSFs. If the reliability model is to serve as the link between the past and the future, i.e., as the basis for making predictions about likely performance, it must necessarily provide a set of principles for how the PSFs are to be taken into account. (A more extensive discussion of the notion of performance shaping factors is provided in Chapter 4.) The basic principle of the performance shaping factors is illustrated by Figure 2, which also identifies the role of the reliability model as an intermediate construct.

2.3 THERP

The role of a reliability model can be illustrated by considering two of the classical HRA approaches, the Technique for Human Error Rate Prediction or THERP (Swain & Guttman, 1985) and the Time-Reliability Correlation or TRC (Hall et al., 1982). As the name implies, a pivotal element of THERP is the legendary, and illusory, Human Error Probability (HEP). In the following I will use the term HEP as if this was meaningful, i.e., as if one could talk about the probability that a specific action went wrong. (I profoundly disagree with this, but will accept the assumption for the sake of the discussion.)

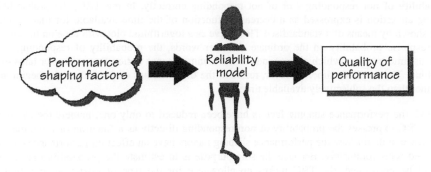

Figure 2: The role of the reliability model.

An often used solution for the reliability model is to apply a so-called linear model that describes the expected error rate as a function of the PSFs. In THERP the reliability model and the PSFs are represented as expressed by the following equation:

$$P_{EA} = HEP_{EA} * \sum_{k=1}^{N} PSF_k * W_k + C$$

where:

P_{EA} = the probability of a specific erroneous action,

HEP_{EA} = the corresponding human error probability,

C = numerical constant,

PSF_k = numerical value of PSF_k,

W_k = the weight of PSF_k, and

N = the number of PSFs.

The difference between the HEP_{EA} and the P_{EA} is that the former defines the probability of an erroneous action **before** considering the influence of the PSFs, while the latter defines the probability of an erroneous action **after** considering the influence of the PSFs. The actual probability that a specific type of action will go wrong is thus expressed as the basic HEP for that action type modified by the influence of the PSFs. The value of the P_{EA} is derived simply by multiplying the value of the HEP_{EA} with the weighted sum of the PSFs, i.e., by combining them in a linear fashion. The solution is psychologically vacuous in the sense that a model of the operator does not enter into it at all and the reliability model is nothing more than a rule of calculation. That is not by itself invalidating, provided that it is possible to establish empirically the parameters of the model. It does, however, make it semantically misleading to refer to performance shaping factors since there is no performance to speak of but only the single HEP. This solution is therefore also vulnerable to the criticisms listed above.

2.4 Time-Reliability Correlation And Human Cognitive Reliability

The Time-Reliability Correlation considers the quality of performance in terms of whether the person responds correctly in a given situation. This has for some reason traditionally been expressed in terms of the probability of not responding - or of not responding correctly. In the TRC, the probability of not performing an action is expressed as a decreasing function of the time available for that action, and is typically shown by means of a standardised TRC curve as a logarithmic plot with time on the abscissa and the non-response probability on the ordinate. In other words, the probability of responding (correctly) increases as time goes on, which corresponds to the common-sense notion that sooner or later something will be done. The available time, however, refers to the time that has elapsed since the beginning of the event, rather than the subjectively available time.

In the TRC, the performance shaping factors have been reduced to only one, namely the available time. Since the TRC expresses the probability of non-responding directly as a function of time, the reliability model that accounts for how the performance shaping factors have an effect on performance can therefore be dispensed with completely. Because the sole purpose is to estimate the probability that the required action will be performed, the TRC makes no allowance for the type of operation or action, for the readiness of the operator, for the working conditions, etc. Even though the value derived from the TRC can be modified by the effect of Performance Shaping Factors, there is no clear theoretical basis for doing that. The potential effect of a PSF must thereby be Platonic rather than Aristotelian.

In the mid-1980s the principle of the TRC was extended by combining it with a then popular information processing model, often referred to as the skill-based, rule-based, knowledge-based (or SRK) framework (Rasmussen, 1986). The resulting Human Cognitive Reliability (HCR) approach (Hannaman et al., 1984) advanced the TRC by providing a simple reliability model based on the SRK framework. (The underlying operator action event tree was also expanded to consider no-response, incorrect response, and success, hence using a trinary rather than a binary branching. However, in the end the HCR only provided the single non-response probability.)

Instead of using a single curve to show the relationship between elapsed time and non-response probability, the HCR used a set of three curves based on the Weibull distribution, corresponding to the hypothesis about three different types of performance or information processing. The objective of the HCR was to describe how two independent variables, the type of performance and the time available, could be used to find the non-response probability. The type of performance was determined by a set of decision rules that referred to salient task characteristics. These are shown in Figure 3 as an event (or decision) tree, where the downward branches correspond to a "no" answer. (This type of event tree is ubiquitous to PSA/HRA, and will be used throughout the book.)

The decision tree contains the basic features of a reliability model and also shows how some of the performance shaping factors - mainly the quality of the procedures and the level of training - may have an influence on performance. The HCR approach also makes it possible to include the effect of other PSFs. This, however, takes place **after** the type of information processing has been determined rather than before. This is intuitively wrong, since the degree of control the operator has over the situation should not be independent of the performance conditions.

Seen as HRA approaches, the TRC and HCR both have two important limitations that are characteristic shortcomings of first-generation HRA. Firstly, they look at the probability of a **single** event rather than at performance as a whole. Secondly, they treat the PSFs - in this case mainly time - as objective parameters without considering the means by which they affect performance. If either the TRC or the HCR represented true empirical relationships this would be acceptable. In the absence of that evidence the limitations mean that they should be used with great care.

The TRC and the HCR nicely illustrate the dilemma of HRA approaches. If there is no reliability model, it is difficult to account for how performance shaping factors exert their effect, except in a statistical sense.

If a reliability model is proposed, the assumptions that it makes about the nature of human performance must be carefully chosen, since they may easily invalidate the modelling attempts as the HCR demonstrated.

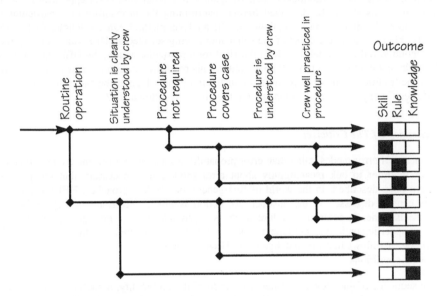

Figure 3: An event tree representation of the HCR decision rules.

2.5 The Reality Of The Human Error Probability

An alternative to the use of the HEP in THERP is to assume that the effect of the performance conditions alone determines the likelihood of a failure, i.e., that the PSFs are the signal and the HEPs merely the noise. This alternative is therefore based on the assumption that the variability between performance conditions is greater than the variability between the HEPs. This can be expressed as follows:

$$P_{EA} = f(\text{Performance Conditions})$$

Both the TRC and HCR can be seen as an example of this position, since in each the notion of a HEP for specific action types is missing. On the other hand, the TRC, and to a more limited extent the HCR, oversimplify the principle by focusing on time as the only or main independent variable. This extreme position nevertheless has something to recommend it, in particular the fact that it is simple to apply.

An intermediate position is to assume that some types of actions are more likely to go wrong than others, relative to the working conditions. According to this there may be differences in the probability of failure between one category of action and another, depending on the setting. (The probability of failure need, of course, not be expressed by an HEP.) This corresponds to the common-sense understanding of the differential vulnerability of actions, i.e., that under specified working conditions some actions are more likely to fail than others. This position clearly considers not just **whether** or not an action fails (in the PSA event tree terminology), but also **what** the type of failure will be: in other words the failure mode.

For example, in carrying out a task under time pressure, a person is more likely to forget a step than to make a planning error. It is therefore possible to propose the following expression:

$$P_{EA} = f(\text{Failure mode} \mid \text{Performance conditions})$$

This formulation points to two additional criticisms against first-generation HRA approaches. Firstly, the performance conditions should be described **before** determining the probability of a particular failure mode, and it should be done as an integral step of the analysis method. On the whole, first-generation HRA methods either calculate the effect of the PSFs after having selected an HEP value, or skip the HEP altogether. Secondly, rather than talking about the HEP for a particular action, the HEP should be related to a failure mode that corresponds to the type of action. This requires a notion of error modes that is more detailed than the traditional omission-commission distinction, and which at the same time gets rid of the multitude of specific or particular HEPs that are required by conventional HRA.

2.6 Consequences Of The Criticism

Using the concept of error modes rather than error probabilities ensures that actions are not considered in isolation. Yet in order to talk meaningfully about error modes, it is necessary that the performance condition have been described. In the world of first-generation HRA approaches, HEPs and omissions-commissions can be discussed in the abstract. An action is either performed or not performed - or something else is done. In contrast, a failure mode is a possible way of failing to perform an action correctly given specific conditions, but unless these conditions are specified it is futile to speculate about the failure mode. It follows that the first step of an HRA method must be an analysis and description of the performance conditions.

A further consequence of this view is that the calculation of the probability, if such is required, cannot be done in the usual simple manner. Just as the failure mode cannot be considered separately from the performance conditions, so the probability of an action failing cannot be calculated in isolation. In fact, it becomes meaningless even to entertain the notion of an action in the abstract as something that can either be performed or not be performed *per se*. The very notion of an action carries with it the concept of the performance conditions or working conditions.

Instead of referring to basic HEP values, an algorithm or a principle should be developed to calculate the probability of a specific failure mode given specific circumstances. Thus, rather than having a single value for each HEP, a table of values should be used for that failure mode given a range of conditions. In a more advanced form the table could be embedded in a calculating device or as a program. The entries of the table would not be point values but ranges, and the calculation would be a way of selecting the correct range, for a given set of conditions.

2.7 PSA-cum-HRA

The criticism of first-generation HRA (Dougherty, 1990) also had the consequence that the role of HRA as a part of PSA was put into doubt. The focus of HRA has traditionally been on human actions in control and operation of complex technical systems - in particular nuclear power plants. Here HRA has been carried out within the shell of PSA, which has focused on the plant as-built or as-operated. It is, however, not necessary that HRA must remain constrained by the PSA-*cum*-HRA construct. One reason for wanting to go beyond that is that the PSA-*cum*-HRA construct is not capable of modelling characteristics beyond the as-built or as-operated concept, and that it does not support the larger perspective where humans are involved in the design and construction of a system, in the operation and maintenance, and in the management (Hollnagel & Wreathall, 1996). Human reliability, or the lack thereof, can obviously play a role in every phase of a system's life-cycle, although the outcome of action failures in many cases may not be immediately visible. It is therefore necessary to develop a comprehensive understanding of human action in context, and this should be a natural part of a revised HRA approach.

PSAs are performed for a variety of reasons, from operational improvements to meeting safety-goal regulations, and the requirements imposed on the HRA methods are usually defined by the purpose of the PSA. Although it is reasonable to accept that HRA, even in an extended form, shall provide results that are useful to PSA it does not follow that HRA must model itself on PSA. The most conspicuous characteristic of the PSA is the description of the event sequence as a binary event tree. In the cases where an event involves a human interaction or intervention, HRA is supposed to provide the corresponding probability value - or probability interval. Yet the requirement to a specific type of outcome need not determine **how** this outcome is produced. HRA has all too willingly adopted the basic principles of the PSA, and almost exclusively described human actions in terms of event trees or operator action trees, i.e., in terms of products rather than processes.

This tendency has been particularly noticeable in the cases where there was a need to account for reliability in relation to cognition rather than manual action. It may to some extent be reasonable to describe the likelihood that a manual operation will succeed or fail, but the same does not apply to cognitive functions that are part of e.g. diagnosis or decision making. This has nevertheless often been done, possibly because of a lack of acceptable or possible alternatives. (To be fair, much of this development has been accomplished by people with an engineering rather than a psychological background, since few psychologists have responded to the problems of PSA-*cum*-HRA. The criticism is therefore as much directed against the sectarianism of psychologists, as the unfamiliarity of the engineers with cognitive psychology.) The consequences of the PSA-*cum*-HRA relation will be discussed more extensively in the following chapters.

3. COGNITIVE RELIABILITY AND ERROR ANALYSIS METHOD

It is common to include in the beginning of a book a short rationale for why it has been structured as it is and to add a summary of the chapters. (Sometimes this is extended to include also a reason - or an excuse - for writing it.) I shall follow the same time-honoured principle here, but deviate ever so slightly from the form. Instead of simply describing the structure of the book, I shall try to explain how the structure of the book is derived from the terms and concepts that make up CREAM.

CREAM is an acronym for Cognitive Reliability and Error Analysis Method. When I started to develop this approach some years ago, I found the acronym attractive because cream generally is something that is nice. On delving further I found, however, that the acronym was even more apt. Although the noun "cream" today means "the yellowish fatty component of unhomogenized milk that tends to accumulate at the surface", the word is derived from the Greek word *khrisma* that means unguent, a salve used for soothing or healing. In relation to the state of HRA this is certainly something that is needed, although it remains to be seen whether this particular CREAM will be the generally applicable unguent.

3.1 Cognitive

The first term "cognitive" is easy to explain. It is by now self-evident that any attempt of understanding human performance must include the role of human cognition. This is particularly so in the study of humans at work. The acceptance of the need to include cognition has generally spread from psychologists to engineers and practitioners of all types. If one goes back to the early to mid-1980s the situation was, however, completely different. In those days it was necessary to argue to technically minded people that cognition was important and to explain what cognition actually meant - in loose terms: that which went on in the head. It is, perhaps, also a sobering reminder that the notion of cognitive psychology itself did not catch on until the 1960s, although this is a different story.

Today we see the term cognitive applied in all possible ways, and there is a growing number of books, conferences, and journals dedicated to cognitive engineering, cognitive science, cognitive tools, cognitive ergonomics, cognitive systems, etc. (On a historical note, the term cognitive systems engineering was introduced as far back as in 1983, cf. Hollnagel & Woods, 1983. At the same time the European Association for Cognitive Ergonomics was also, but independently, established.)

In the context of CREAM, "cognitive" must be seen together with the second term "reliability". The term "cognitive" is not meant as a restriction in the sense that other factors are excluded but rather as a reminder to focus on the full complexity of the human mind. In particular, "cognitive" is not meant as the opposite of organisational or environmental. As a matter of fact I have argued at some length (Hollnagel, 1993a) that cognition, context, and control cannot be separated. Just as there can be no figure without a ground, there can be no cognition without context.

3.2 *Reliability*

The R stands for "reliability". The standard definition of reliability is the probability that a person will perform according to the requirements of the task for a specified period of time, To that is sometimes added that the person shall not perform any extraneous activity which can degrade the system (Swain & Guttmann, 1983), but this actually implied by the shorter definition. To wit, if a person performs according to the requirements of the task it implies the absence of extraneous and potentially disruptive actions.

As mentioned above, the two first terms should be seen together as "cognitive reliability" - or to be more precise, as the reliability of human cognition. (I shall, however, stay with the short term "cognitive reliability" although it strictly speaking means something else.) The importance of that for HRA contexts is that human performance is determined by human cognition (plus the technology, plus the organisation) and that it therefore is important to be able to account for cognitive reliability. In particular, it is important, in accordance with the above given definition, to provide a way of finding the probability of "cognitive reliability" - in other words to indicate (at least) the upper and lower bounds of the variability of human performance.

It is, of course, impossible in a fundamental sense to discuss human reliability without including cognition. Just as all performance in some sense is cognitive, any consideration of performance reliability must including cognition. The term "cognitive reliability" has, however, historically been used to emphasise that the focus is the complexity of human performance, rather than the simple components of behaviour. From an empiricist point of view one may refer to the occurrence of an omission as a simple action failure (similar to slips), but unless the focus of concern is strict reflexive action in a behaviourist sense, cognition is a necessary part of the study.

3.3 *"Error"*

The E stands for "error" - or rather for erroneous action (Hollnagel, 1993a; Woods et al., 1994). The possible meanings of the term "error" are discussed at length in Chapter 2. For now it suffices to say that the intention is that the study of "error" is a complement to the study of reliability. Whereas the concern for reliability mostly is directed towards predicting what is likely to happen in a future situation, the study of "error" is directed towards finding the causes or explanations for something that has happened, i.e., it is a retrospective rather than a predictive type of analysis. Academic psychologists, and like-minded engineers, have mostly been interested in developing systems to explain "human error". HRA practitioners have mostly been interested in finding ways of calculating the probabilities of action failures. The importance of this difference will be discussed in detail in Chapter 3.

Just as in the case for reliability, I shall not make a distinction between "error" and "cognitive error". Since an "error" is an action gone wrong - in the sense that the outcome was not the expected or desired one, and since all actions involve a modicum of cognition, all "errors" must also be cognitive. This means in particular that I will not make a distinction between "cognitive errors" and "non-cognitive errors" although this has often been done in traditional HRA, e.g. in juxtaposing cognitive and commission errors. This distinction was made because it was felt that cognition had not been properly treated by the traditional or first-generation HRA. While this is indisputably correct, the solution of introducing a new term called "cognitive error" is not the right answer. The right solution, as advocated in this book, is to realise that all erroneous actions are a result of cognition (and of technology and of organisation), and that the solution therefore is to revise the basis for HRA *per se*.

3.4 Analysis

The A stands for "analysis" - but it could conceivably also have stood for assessment, although the difference between the two terms is not insignificant. Analysis implies a separation or decomposition of a whole (the object of study) into smaller parts for the purpose of further study, better understanding, etc. Assessment, on the other hand, implies assigning a numerical value to something, indicating its worth - or in this case a numerical indication of the probability of an event. Analysis thus puts emphasis on the qualitative aspects of the study, while assessment puts emphasis on the quantitative.

HRA mostly takes place in the context of PSA as described above. This means that there is a strong emphasis on the quantitative side. HRA came into the world as a way of determining the probability estimates for human actions that were required by PSA/PRA. It is, however, important to emphasis the analysis rather than the assessment. Firstly, because there cannot be an assessment without a preceding analysis and, secondly, because the value and use of HRA mainly is in the improved understanding that results.

The emphasis on analysis rather than assessment is also important for the use of HRA in contexts different from PSA/PRA. It is somewhat disappointing to note that several of the fields where human action plays a central role, such as Human-Computer Interaction (HCI) or Computer Supported Co-operative Work (CSCW), pay little or no attention to the aspects of human reliability. The concern for the reliability of human performance should be a necessary part of any system design, and the proper methods for analysis and descriptions should be part of the general repertoire of system design tools.

3.5 Method

The final letter M stands for "method". The intention has been to develop a practically useful tool, i.e., one that is simple and cost-effective yet produces the required results. The emphasis is thus not on developing a new theory or model of human action, or a new set of concepts for man-machine interaction or cognition. The motivation for this book is primarily the practical concerns of HRA. To put it simply, since HRA is required - and in particular in a quantitative form - it is the responsibility of the scientific community to provide a method that is acceptable both to the practitioner and to the scientist. Otherwise practitioners will unavoidably develop their own methods, which scientists may not always see as adequate.

(An additional comment is that it is important to distinguish between method and methodology. The method refers to a specific tool, the methodology refers to the principles behind the tool. In this sense the M stands for Method rather than methodology. The whole of CREAM provides the principles and rationale, but the emphasis is on providing a straightforward method.)

3.6 The Scientist And The Engineer

The terms "scientists" and "engineers" can be used to refer to two clearly distinct groups of people - although it obviously is an oversimplification. There is nevertheless a very real difference between those who are concerned about getting a theory or model accepted by the scientific community, and those who are concerned about winning a bid for a contract or getting a safety case approved. Scientists are mainly interested in the scientific issues of (complex) technological systems, such as HCI, ergonomics, human information processing, cognitive science, etc. Practitioners are involved in building, operating, maintaining, and managing technological systems in the industry (process industry, energy production, transportation, manufacturing, etc.). Although there is a wide interest for the study of "human errors" among both groups, it is mostly practitioners who are seriously interested in HRA.

Practitioners have a recognised need for HRA approaches that are simple and efficient to use, and which produce results that are practically valid. Thus if a method works or has become accepted, there is usually little concern for theoretical niceties. Scientists, on the other hand, are more interested in the concepts and principles that are at the basis for the practical methods, and will be more concerned about whether an approach is theoretically correct and pure than whether it is cost-efficient or even practical to use. Many practitioners realise that they need input from scientists to improve their methods, but since they do not get much collaboration they often themselves delve into the field (and are consequently looked upon with scorn by "respectable" scientists). A case in point is the concern for "cognitive errors" that currently is the focus of much debate.

An important purpose of this book is to show that the distinction between scientists and practitioners is misguided and potentially harmful. Scientists have a serious need to learn from practitioners to get a proper understanding of what the problems are. If not, they are likely to develop models and theories that have little practical value. Practitioners, on the other hand, have a need to understand what is happening in the scientific community and how they can improve the basis for their practical methods. The analysis of human performance can nevertheless not be based on engineering principles alone, in isolation from psychological knowledge.

4. BACKGROUND OF THE BOOK

The background for the approach to human reliability analysis presented here is a line of work that started in the late 1980s as a survey of existing theories and models for human erroneous actions. It was generally felt that the then available approaches were insufficient both on the practical and the theoretical sides. To some extent this is still the case in 1997. These issues were extensively discussed in a NATO sponsored workshop in 1983 (Senders & Moray, 1991), and it is reasonable to see this workshop as an initiating event for much of the scientific interest in performance failures, at least in the sense of starting a more deliberate attempt to identify exactly what the problems were. The survey mentioned above was conducted as part of the work in developing an expert system for plan and error recognition (Hollnagel,

1988). The result of the survey was a clear realisation of the need to distinguish between manifestations and causes of erroneous actions.

This view was presented in 1988 as a proposal to distinguish between error phenotypes and error genotypes (cf. Hollnagel, 1990), and was later discussed more extensively in Hollnagel (1993b). In parallel to these theoretical clarifications, a suggestion to use the phenotype / genotype distinction to construct a workable classification scheme was developed by Hollnagel & Cacciabue (1991). The same paper also proposed the notion of the Simple Model of Cognition (SMoC) as a basis for the structure of the classification scheme. An important contribution to the contents of the classification scheme came from a project to develop a human reliability assessment method for the European Space Agency (Hollnagel et al., 1990). This project produced a highly useful summary of the many descriptive terms that had been developed through years of practical accident and error analysis, but which had rarely been systematically evaluated. A significant input to this project was the Action Error Analysis method developed by Robert Taylor of the Institute for Technical Systems Analysis (ITSA).

The situation in the early 1990s was therefore that an alternative to the traditional approaches to error analysis had begun to emerge, and that the ideas were viewed with interest by a growing number of people. The first serious attempt at practically applying the classification scheme and describing the associated method was made in 1992-93 by Mauro Pedrali as part of the work for a Ph.D. thesis (Pedrali, 1996). This work produced a detailed analysis of an aviation accident that demonstrated the value of the principles (Cacciabue et al., 1993). In parallel to that, the Institute for Systems Engineering and Informatics at the Joint Research Centre in Ispra carried out a project to survey and compare existing error taxonomies (Cojazzi et al., 1993). At the end of 1993 there was therefore a fairly well developed classification scheme and an associated method that had three characteristics. Firstly, it maintained the important distinction between phenotypes and genotypes as a way of structuring the description of an accident or event. Secondly, it made explicit use of an underlying cognitive model to structure the causal links, hence to provide consistent explanations. And thirdly, it included a relatively simple method of analysis that had proved valuable to unravel and understand even quite complex accidents.

The missing element was a way of applying the same basic components to performance prediction and HRA. It was strongly felt by many of those involved that the classification scheme, the model, and the method could be used as the main elements in an HRA approach that would be impervious to the criticisms raised against first-generation HRA (Dougherty, 1990). This book builds on and continues the previous work, and proposes an integrated method for analysis and prediction relating to human action failures, with special emphasis on the notion of cognitive reliability.

4.1 Structure Of The Book

Chapter 1 has provided a basic introduction to the issue of Human Reliability Analysis (HRA). The basic motivation for HRA is the pervasiveness of human erroneous actions, and the widespread tendency to see human action failures as the causes for accidents and incidents. The need better to understand the nature of the cause-effect dependencies that can explain the occurrence of undesirable events, hence also be the basis for preventing them, has generated a number of analysis techniques that will be described in Chapter 5. Here it is simply noted that the ability to make accurate predictions depends on the ability to analyse correctly past events, and that analysis and prediction therefore must go hand in hand.

The traditional approaches to HRA, which collectively have come to be know as first-generation HRA, became the target of widespread criticism in the beginning of the 1990s. The criticism came from some of the prominent practitioners of first-generation HRA and was generally based on pragmatic concerns. It is, however, also possible to evaluate first-generation HRA in a more principled way, based on the identification of the fundamental concepts. One issue is that few HRA approaches include an explicit reliability model that can "explain" how independent performance variables, such as available time, have an effect on performance reliability. It is also argued that the notion of a separate Human Error Probability (HEP) is psychologically misleading: actions do not occur in isolation, but only as part of a whole. It is therefore necessary to characterise the whole before analysing the individual actions. This has consequences for how the overall "error" probability can be computed.

Chapter 2 considers the ambiguity of the term "human error", which can refer to either causes, events, or consequences. It continues by discussing the relation between HRA and PSA. The ready acceptance by HRA of some of the important assumptions inherent in PSA has created problems that partly are artefacts. It is argued that since the need to understand the nature of human reliability clearly goes beyond the requirements of PSA, HRA should extend its scope and emerge as a field of practice in its own right. The chapter also discusses the need to make a clear distinction between what can be observed (the phenotypes) and what must be inferred (the genotypes). A consequence of this view is that the notion of a "cognitive error" in itself may be unnecessary.

Chapter 3 surveys a number of classification schemes that have been used to guide empirical investigations of erroneous actions. These are the traditional human factors approaches, the information processing approaches, and the cognitive systems engineering perspective. A key factor in each is the extent to which categories of erroneous action can be related to a viable model of human cognition. The information processing approach has been very successful in developing *post hoc* explanations of actions but it has been difficult to "translate" these to match the needs of HRA. Progress can only be made if models and methods are developed that overcome this schism by addressing the needs of "error" psychology and HRA at the same time.

Chapter 4 considers the issues involved in making predictions about performance in the future. These are difficult to make because the possible developments depend on complex interactions between technology, people, and organisations. In order for predictions to be reliable and valid they must be made using a well-defined method, an adequate classification scheme, and a sound model of the constituent phenomena. Together this constitutes a Method-Classification-Model (MCM) framework that is introduced in Chapter 4 and completed in Chapter 5.

In Chapter 5 a characterisation is given of nine frequently used first-generation approaches to HRA, using the MCM framework of Chapter 4 as a common basis. Each of the nine HRA approaches is thus characterised with respect to the classification scheme it employed, the explicitness of the method, and the characteristics of the underlying operator model. To complement this survey, six proposals for second-generation HRA approaches are also considered using the same principles. The candidate approaches differ considerably from each other due to the lack of a commonly agreed conceptual basis for HRA as well as to different purposes for their development. These efforts nevertheless represent a clear recognition of the basic problems and argue that the solution must try to address all four issues at the same time.

Chapter 6 begins the description of CREAM. The main principle of the method is that it is fully bi-directional, i.e., that the same principles can be applied for retrospective analysis and performance prediction. The model is based on a fundamental distinction between competence and control and the classification scheme clearly separates genotypes (causes) and phenotypes (manifestations), and furthermore proposes a non-hierarchical organisation of categories linked by means of the sub-categories called antecedents and consequents. The classification scheme and the model are described in detail in Chapter 6, while the method is explained in the remaining chapters.

Chapter 7 provides details of how CREAM can be applied for retrospective use, i.e., for the analysis of accidents and incidents, to find the likely causes for a given accident or event. One important feature is the definition of explicit stop rules for the analysis. Since the classification scheme necessarily is incomplete, the principles according to which the classification groups can be extended are also described. The retrospective analysis has four basic steps that are presented in detail and illustrated by two different examples.

Chapter 8 goes on to discuss the basic principles for performance prediction. A problem with most classification schemes is that performance prediction is reduced to a mechanical combination of the various categories. Since this eventually leads to a prediction of every possible occurrence, the practical value is limited. The prediction must therefore be constrained to provide only the outcomes that are likely given the circumstances or performance conditions. The general performance prediction in CREAM has six main steps that are presented on a general level as a preparation for the detailed method described in Chapter 9.

This last chapter presents CREAM as a full-blown second-generation HRA approach by describing how it can be used for performance prediction. This can be either as a basic type of performance prediction that serve as a screening to identify the events or situations that may require further analysis, or as an extended type of performance prediction that goes into more detail with specific events. The basic method uses a "mapping" of the performance conditions to the determine the characteristic control modes, which further can be described as probability intervals. The extended method uses a cognitive task analysis to identify the cognitive activities that are required of the operator. This can provide a profile of the cognitive demands of the task, which gives a first indication of where potential problem areas may be. The cognitive demands profile is then combined with a description of possible error modes to identify the likely cognitive function failures. These provide the qualitative results of the analysis, and may further be quantified using a table of nominal probabilities. The effects of the common performance conditions are described by using either the control mode found by the basic method, or by a more detailed calculation of specific adjustment values or weights. The various steps of the performance prediction are illustrated by an example.

Chapter 2

The Need Of HRA

1. THE UBIQUITY OF ERRONEOUS ACTIONS

This chapter will describe the domain of HRA and why there is a need of human reliability analysis. The general domain consists of the complex industrial systems that depend on human-machine interaction for their proper functioning - during operation, maintenance, and management. Since human actions clearly play a role everywhere one could, in principle, embrace the complete life-cycle and include aspects of design and implementation as well as of certification and decommissioning. The concern for human action and human factors, hence the need for HRA, has however been smaller in these phases, possibly because they have been less strongly linked with the occurrence of accidents.

Even with this narrowing of the domain there is a large and growing number of human-machine systems. Although such systems have traditionally been associated with process industries and energy production, they are now recognised also in transportation, finance, health care, communication, management, etc. Experience shows that in every case where a human action - including inaction or the lack of an action - may possibly be the initial cause of an event or influence how an event develops in either a positive or a negative way, there is a legitimate concern for a systematic way of addressing the issues of human reliability, hence a need of HRA. The concern for reliability thus goes beyond simply looking at human actions, but requires that human interaction with technology is seen in a specific context - the organisation or the socio-technical system.

In order to be able to describe and analyse human interaction with technology it is necessary to model or describe the functions of the human mind. This is specifically required if one is to give a credible explanation of how performance conditions may affect performance, hence to predict the effect of specific conditions. The model of the mind is an account of the substratum or mechanisms through which actions occur and how Performance Shaping Factors (PSF) can exert their influence. Such models may have various degrees of sophistication, going from black box input-output models, that are conceptually vacuous, to intricate and complicated information processing models. Any approach that intends to analyse and describe human performance, e.g., for the purpose of making predictions, must be able to account for the details of how human actions come about. Otherwise the application of various concepts, such as PSFs, approaches the practice of black magic. It follows from this, that an examination of the assumptions that are implied by the various approaches can be used to reveal their strengths and weaknesses. This continues the analysis of the list of acknowledged weaknesses that was begun in Chapter 1. That list was based on the accumulated grievances of practitioners (Swain, 1990). In this chapter the list of weaknesses will be based on a systematic examination of the principles behind the approaches, noting inconsistencies and contradictions. It follows, that a proper method must be able to answer these criticisms; the proposal for such a method is developed in the following chapters.

1.1 Definitions Of "Human Error"

Practically everyone who has published work on the topic of human fallibility in a work context has attempted, at one time or another, to provide a technical definition of the concept of "error". The net result of this effort has been a clear demonstration that it is extremely difficult to provide a precise technical definition, despite the fact that the term "error" has a relatively simplistic meaning in everyday life (Embrey, 1994; Kirwan, 1994; Reason, 1984; Senders & Moray, 1991; Singleton, 1973; Woods et al., 1994). One reason for this state of affairs is the lack of agreement regarding what constitutes the definitive qualities of the phenomena we commonly associate with the term "error". This lack of agreement is usually due to different premises or different points of departure. Some of the difficulties associated with providing an adequate definition of the concept of "error" have been discussed by Rasmussen, Duncan & Leplat (1978). They suggested that when the nature and origins of "error" are discussed from different professional points of view it soon becomes clear that the starting point for the analysis is often quite different. Thus, an engineer might prefer to view the human operator as a system component where successes and failures can be described in the same way as for equipment. Psychologists, on the other hand, often begin with the assumption that human behaviour is essentially purposive and that it can only be fully understood with reference to subjective goals and intentions. Finally, sociologists have traditionally ascribed the main error modes to features of the prevailing socio-technical system; in a sociological analysis items such as management style and organisational structure are therefore often seen as the mediating variables influencing error rates.

From another perspective the difficulties are due to the fact that "error" is not a meaningful term at all, or rather that it does not have one unique meaning. Despite the agreed usage of the word in daily language, "error" has been used in three quite different ways in the technical world. As I have argued previously (Hollnagel, 1993a; see also Senders & Moray, 1991, p. 19) the term "error" has historically been used to denote either the **cause** of something, the **event** itself (the action), or the **outcome** of the action.

- **Cause: The oil spill was caused by "human error".** Here the focus is on the cause (the "human error") of the outcome (the oil spill).

- **Event / action: "I forgot to check the water level".** Here the focus is on the action or process itself, whereas the outcome or the consequence is not considered. In some cases the outcome may not yet have occurred but the person may still feel that an "error" has been made, e.g. having forgotten to do something. Similarly, a forgotten item or action need not always lead directly to a manifest failure.

- **Consequence: "I made the error of putting salt in the coffee".** Here the focus is solely on the outcome, although the linguistic description is of the action. In this example, putting salt in the coffee is tantamount to making the coffee salt rather than sweet.

The differences in usage are illustrated in Figure 1. The purpose of that is not to make a linguistic or philosophical analysis, since clearly the notions of cause and consequence are involved to different levels with each usage. The intention is rather to point out that the term "error" commonly is used in several different ways. It is this ambiguity of common usage that makes the term unsuitable for more systematic analyses, and requires the introduction of a less ambiguous terminology.

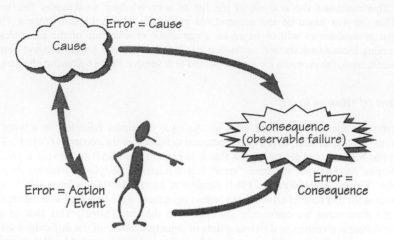

Figure 1: Three meanings of the term "error".

This multiplicity of usage has aptly been characterised by Woods et al. (1994), by noting that "error" is always a judgement in hindsight.

> "... the diversity of approaches to the topic of error is symptomatic that 'human error' is not a well defined category of human performance. Attributing error to the actions of some person, team, or organization is fundamentally a social and psychological process and not an objective, technical one."
> (Woods et al., 1994, p. xvii)

Or, as it is summarised towards the end of their book **"the label 'human error' is a judgment made in hindsight"** (Woods, et al., 1994, p. 200; emphasis added). An even better definition is that **"a 'human error' is the *post hoc* attribution of a cause to an observed outcome, where the cause refers to a human action or performance characteristic"**. In consequence of the above considerations I shall for the rest of this book endeavour not to use the term "error", but instead use the term erroneous action or performance failure (Hollnagel, 1993a). Should it become necessary to refer to "error", the term will be written with quotation marks.

Irrespective of the above differences it is useful to take a closer look at how one might conceivably define the term erroneous action. When viewed from the perspective of the acting individual there seem to be at least three intuitive parts to any definition of erroneous action:

♦ First, there needs to be a clearly specified **performance standard** or **criterion** against which a deviant response can be measured.

♦ Second, there must be an **event** or an **action** that results in a measurable **performance shortfall** such that the expected level of performance is not met by the acting agent.

♦ Third, there must be a degree of **volition** such that the person had the opportunity to act in a way that would not be considered erroneous. This implies that the considerations apply only to the person - or persons -directly involved in the handling of the system, i.e., the people at the sharp end (Woods et al., 1994).

Each of these characteristics of erroneous actions has proved controversial and the basic areas of disagreement are considered in more detail below. From the perspective of the acting person all of them are necessary conditions for an erroneous action to occur, although this may not be true for other points of view. Together they may also constitute a sufficient set of conditions.

1.2 The Criterion Problem

An erroneous action can only truly be said to occur when there is a clear performance standard that defines the criteria for an acceptable response. In an ideal world, performance criteria are specified prior to the erroneous action because otherwise the status of the act becomes nothing more than a value judgement or hindsight and therefore open to interpretation (Woods et al., 1994). The form of reasoning that lies behind statements of the type "the operation was a success but the patient died" represents a case in point. Here the "error" was the undesirable outcome (for the patient, at least) despite the fact that everything else went all right. In general, two broad classes of response criteria have been used by investigators: **externalised verifiable models** in which erroneous actions are defined relative to paradigms of objective correctness, hence relative to external criteria; and **internalised verifiable frameworks** in which erroneous actions are defined relative to the transient (impermanent) intentions, purposes and goal structures of the acting individual, hence relative to internal criteria.

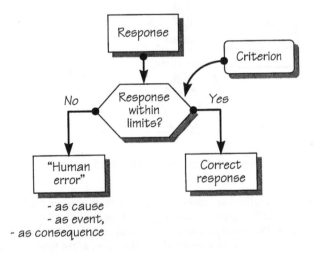

Figure 2: Error as an externalised category.

Reliability analysts have traditionally dealt with the criterion problem by using externalised measures such as system parameters as the standard for acceptable behaviour. Thus, following a line of argument proposed by Rigby (1970), Miller & Swain (1987) argued that human erroneous action should be defined as: "...any member of a set of responses that exceeds some limit of acceptability. It is an out-of-tolerance action where the limits of performance are defined by the system". This definition makes it clear that the erroneous action must refer to an observable and verifiable action, rather than to the person's own experience of having made an error (Figure 2). This definition thus corresponds to the categories of "error as cause" and "error as consequence", but not to the category of "error as action".

In contrast to the above, investigators working from the standpoint of cognitive psychology have naturally tended to prefer definitions of erroneous action that refer to internalised criteria such as the temporary intentions, purposes and goal structures of the acting individual. From this basis it is possible to define two basic ways that an action can go wrong. An erroneous action can occur either in the case where the intention to act was adequate but the actions failed to achieve the goal, or in the case where actions proceeded according to plan but the plan was inadequate (Reason & Mycielska, 1982), cf. Figure 3. In the former case the erroneous action is conventionally called as a **slip**, in the latter it is usually classed as a **mistake** (Norman, 1981). Note that in this case the erroneous action may be noted by the person before the actual consequence has occurred, although it commonly involves an anticipated consequence. It thus refers to the "error as action" category, rather than to any of the two others. (There must clearly be some kind of feedback that makes the person realise that an "erroneous action" has been made, but it will require a protracted discussion to go into that.)

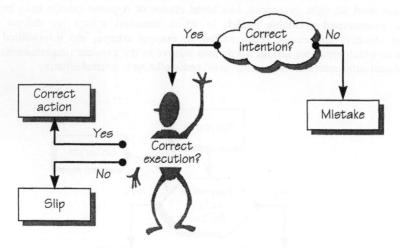

Figure 3: Error as a internalised category.

Although the use of an internalised or subjective criterion is *de rigueur* for the psychological study of erroneous actions, it is not really significant for the aims of human reliability analysis. The reason for that is simply that the focus of HRA is on actions that may lead to undesirable consequences, whether or not they are considered as erroneous by the perpetrator. Furthermore, at the time when an HRA is carried out it is impossible to assume anything about what the state of mind of the operators may be. The internalised criterion is therefore only of academic interest for HRA. This nevertheless does not mean that HRA has no interest in or need of modelling human cognition and human performance. It is, indeed, essential that predictions of human performance can refer to appropriate models. Yet the models should represent the essential (requisite) variety of human performance and of the joint cognitive system, rather than elegant but ephemeral theoretical constructions.

1.3 *Performance Shortfall*

Irrespective of whether one chooses to define erroneous actions relative to externalised or internalised criteria, most analysts agree that erroneous actions for the most part are negative events where there is some kind of failure to meet a pre-defined performance standard. Unintended or incorrectly performed actions that have positive consequences are usually called serendipitous and there is, strangely enough, less interest in why they happen. There is, however, considerably less of a consensus between investigators regarding how best to account for the psychological functions that can be used to explain erroneous actions. Some investigators have adopted a pessimistic interpretation of human performance capabilities, whilst others have attempted to account for the occurrence of erroneous action from within a framework of competent human performance. This presents two conflicting views on the basis for the undeniable performance shortfall.

It is in good agreement with the tenets of information processing psychology to have a pessimistic view of the human capabilities and conclude that erroneous actions provide strong evidence of "design defects" inherent within the information processing system. The classical expression of the notion of a limited capacity human information processor is epitomised by the so-called Fitts' List, which was originally used to allocate functions as part of the development of an air-traffic control system (Fitts, 1951). The principle is that a list or table is constructed of the strong and weak features of humans and machines, which is then used as a basis for assigning functions and responsibilities to the various system components. The underlying assumptions are: (1) that an operator can be described as a limited capacity information processing system, and (2) that if one subsystem cannot perform a task or a function, then the other can. This produces a perfunctory juxtaposition of operator and machine capabilities on a number of central criteria which inevitably leaves the operators in a rather passive role. More than twenty years ago this view was expressed by Vaughan & Maver (1972), when they outlined the following "performance defects" as part of a review of human decision making:

> "Man is typically slow to initiate action and conservative in his estimate of highly probable situations, and when he does act or accept a diagnosis, he is reluctant to change an established plan or a situational estimate when the available data indicates that he should. He is generally a poor diagnostician and does not learn by mere exposure to complex tasks but only when specific relationships that make up the complexity are explained. He is not particularly inventive and tends to adopt the first solution he finds. He finds it difficult to use more than one criterion at a time in evaluating actions and tends only to identify criteria that reflect favourably on the action he is developing. He tend to use only concrete, high confidence facts in his planning and prefers to ignore ambiguous or partial data rather than attempt to interpret them."

The above description corresponds well to that of the limited capacity decision maker. Decision making is usually assumed to be a formal process of reasoning that leads to the choice of the best alternative. According to this, the decision maker should first identify the alternatives, then compare them, and finally select the optimum one. Although few experts from the behavioural sciences seriously believe that this is an accurate account of how people make decisions, the normative description has nevertheless influenced most of the popular attempts to model human decision making. The common form of description thus describes decision making as going through a number of stages, as illustrated in Figure 4. Such models can be embellished with several feedback loops, and the number of steps or stages may differ from case to case. The basic principle is, however, always the same: that decision making can be adequately described in terms of a limited number of steps that are carried out in a prototypical sequence. The reasons for deficient or faulty decisions are consequently to be found in the information processing limitations for the various steps or functions in the process; an example is the impact of short-term memory limitations on the comparison / evaluation stage.

Workers following these traditions have been at the forefront of a research strand aimed as identifying the "design defects" of human cognition. Once identified it is assumed that guidelines can be developed to say where the human operator can, and cannot, be trusted with the care of complex plant and equipment (see for example Human Factors Reliability Group, 1985). Although the strict information processing view has been questioned during the late 1980s, the basic attitude has not changed in any significant way and it is still easy to find research and models from this school in the mid 1990s.

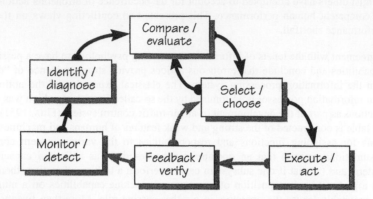

Figure 4: The prototypical stages of decision making.

A more optimistic viewpoint is found in a line of research that emphasises that most erroneous actions have their origins in processes that perform a useful and adaptive function. Such approaches take a much more beneficial view of tendencies for human cognition to go wrong and relate "error mechanisms" to processes that underpin intelligent cognitive functioning and especially the human ability to deal with complex ambiguous data forms that are characterised by a high degree of uncertainty. This view has an equally long history and several classical descriptions of decision making have challenged the orthodoxy of the rational decision maker - the *homo economicus*. Probably the best know is the description of satisficing by Simon (1957), but an equally interesting alternative is the "muddling through" description of decision making (Lindblom, 1959). Here decision making is seen as going through the following stages: (1) define the principal objective; (2) outline a few, obvious alternatives; (3) select an alternative that is a reasonable compromise between means and values; and finally (4) repeat the procedure if the result is unsatisfactory or if the situation changes too much.

It is not difficult to argue that one of the two descriptions summarised in Figure 5 is more realistic than the other in the sense that it corresponds better to what people actually do. In the "real" description the shortcomings are not due to specific information processing limitations but rather to the shape of the overall process. The muddling-through types of decision making or satisficing have evolved to be effective in real life situations. Although they fall short of the normative demands, they work well enough in a sufficiently large number of cases to guarantee the "survival" of the decision maker.

For human reliability analysis the choice is thus between an explanation based on capacity limits or design defects, and an explanation based on the adaptive nature of human behaviour. The latter accepts the notion of an inherent variability of human performance that has little to do with capacity limitations. The information processing or "design defect" approach can offer well-described models that can be used to explain the regular performance shortfalls. An example is the notion of the "fallible machine" described by Reason (1990). These models and theories have, however, had limited success in supporting the practice of HRA. The view of human performance as basically competent focuses more on the correspondence between capabilities and the situation or the demands. Human performance is competent because we are very good at identifying the relevant and regular features of a task and use that to optimise resource usage. Since the environment is constantly changing this "strategy" will inevitably lead to failures - on both a small and a large scale. Erroneous actions thus occur when the conditions and the skills do not meet - either on a grand scale (misdiagnosis) or on the small scale (performance slips, basic error modes). These differences must be reconciled in the basis for human reliability analysis, to the extent that this can be done without the need to solve a basic problem of psychology.

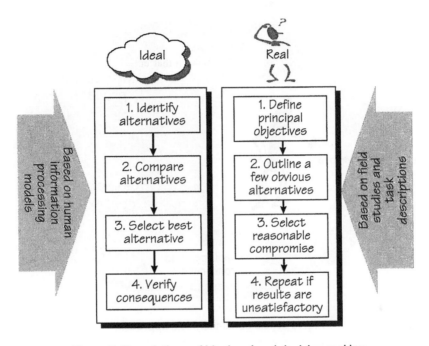

Figure 5: Descriptions of ideal and real decision making

1.4 Volition

The final definitive characteristic of erroneous actions concerns the requirement that the acting agent has the opportunity to perform in a way that would be considered appropriate. Thus as Zapf et al. (1990) have observed, if something was not avoidable by some action of the person, it is not reasonable to speak of erroneous actions. In such cases the cause should rather be attributed to events in the socio-technical system such as technological failures, management oversights, etc. Ultimately these may turn out to be due to an erroneous action or an incorrect judgement by a person, but that person will not be the one who was caught in the situation.

It is now generally recognised that a common factor in a large number of accidents is the organisational conditions. As the view on human erroneous actions gradually became more refined it was acknowledged that incidents usually evolved through a conjunction of several failures and factors, and that many of these were conditioned by the context. This has been developed into the notion of people at the "sharp end" of a system (Reason, 1990; Woods et al., 1994) as contrasted with people at the blunt end. The people at the sharp end are those who actually interact with the processes in their roles as operators, pilots, doctors, physicians, dealers, etc. They are also the people who are exposed to all the problems in the control of the system, and who often have to compensate for them during both normal and abnormal operating conditions. The people at the blunt end are those who are somehow responsible for the conditions met by people at the sharp end, but who are isolated from the actual operation. They can be managers, designers, regulators, analysts, system architects, instrument providers, etc.

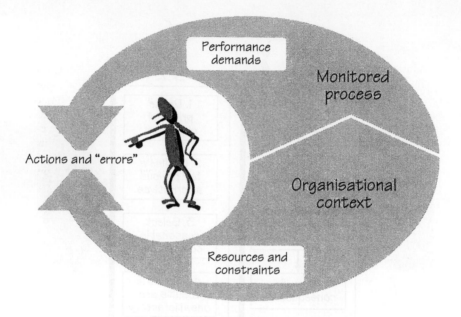

Figure 6: The socio-technical approach.

The socio-technical approach is illustrated in Figure 6. The basic idea is that human performance is a compromise between the demands from the monitored process and the resources and constraints that are part of the working environment. The two, demands and resources, meet at the sharp end, i.e., in the actions that people make (at the left-hand side of Figure 6). The demands come from the process, but by implication also from the organisational environment in which the process exists (the blunt end, at the right-hand side of Figure 6). The resources and the constraints are more explicitly given by the organisational context, e.g. in terms of procedures, rules, limits, tools, etc. The cognitive aspects of the actions are characterised using the notions of knowledge, attention dynamics, and strategic factors (Woods et al., 1994).

The socio-technical approach is in many ways the inverse of the classical information processing approach that concentrates on the internal human information processing mechanisms. The socio-technical approach therefore runs the same danger of focusing on one aspect of the situation, the context, and neglect the other. Even so the socio-technical approach is a valuable reminder that we need to consider both sides, i.e., to consider both cognition and context. At present the socio-technical approaches have not been developed to the stage where they can provide a potential explanation for human erroneous actions or be used in HRA. In fact, they tend to reduce the relative contribution of the operator's cognition to the benefit of the context. In the extreme case, the socio-technical approaches even play down the role of volition. When an undesirable event occurs it is because the cards have been stacked against the person at the sharp end, who almost is the innocent victim of organisational shortcomings and malfunctions. This view is obviously not very useful for human reliability analysis, and almost denies its importance. From the HRA perspective it does make sense to maintain that the person has an opportunity to perform the action correctly. Otherwise even the most ardent opponent of quantification must admit that the HEP in question must be 1.0. That can obviously not be the case. There must therefore be a compromise between the two views. It is to be hoped that in the future more mature socio-technical approaches will be developed which can effectively complement the current explanations of human erroneous actions.

1.5 Conclusions

For the purpose of HRA, as well as for the purpose of the discussion of human erroneous actions in general, it seems sensible that the concept of operator erroneous action is defined with reference to the observable characteristics of behaviour. Specifically, human erroneous actions are simply actions with undesirable system consequences. Such a definition is in keeping with the treatment of erroneous action in reliability analysis and avoids the potential confusion that can arise when discussing the causes and consequences of erroneous action (e.g., Hollnagel, 1993b). Moreover, a behavioural definition of the concept of erroneous action is neutral with regard to the issue of what sometimes causes human performance to fail. The term human erroneous action, or even "human error", does not imply that the action also is the cause. In relation to this debate I definitely favour the optimistic viewpoint of human performance capabilities and agree with many others that erroneous actions have their origins in processes that perform a useful and adaptive function in relation to most everyday activities. This suggests that the search for "error mechanisms" in the simple sense should be abandoned. The need is rather for an analysis of the complex interactions that occur between human cognition and the situations or context in which behaviour occurs (e.g., Hollnagel, 1993a, 1996; Woods et al. 1995).

2. THE ROLE OF HRA IN PSA

HRA has traditionally been closely coupled to Probabilistic Safety Assessment, (PSA) and there has been a strong emphasis on quantification. The PSA-*cum*-HRA construct has often discouraged behavioural scientists, and in particular psychologists and ergonomists, from even considering HRA a problem. That attitude is obviously not very constructive. The practical need for HRA has grown as part of the requirement to calculate more precisely the probability of an accident in order to guide resource investment, but there is a more fundamental need to improve the understanding of human action as a part of system design and in particular to develop models and methods for the analysis of the interaction between people and socio-technical systems. The two needs are not in conflict but can easily co-exit. Although the terms qualitative and quantitative sometimes are used as if they were antonyms, the difference is on the level of terminology rather than on the level of ontology. A quantitative analysis must always be preceded by a qualitative analysis and it is now widely recognised that a qualitative analysis in many cases may be quite sufficient for the purpose of HRA, although there still may be a residual need to provide numerical input to a PSA.

The reliability of a technical system can be analysed with different purposes in mind. Some characteristic analyses types are:

♦   Probabilistic Safety Analysis or Probabilistic Safety Assessment (PSA), where the purpose is to identify the sources of risk in a system.

- Hazard Operability Studies (HAZOPS), where the purpose is to identify sources of failure.

- Failure Modes, Effects and Criticality Analysis (FMECA), where the purpose is to identify the degree of loss of activity objectives - as well as the recovery potential.

In principle each analysis type has its own purpose. In practice it is, however, common to consider several purposes at the same time, for instance both risk analysis and recovery potential. The degree to which human action is included in the analysis also varies. FMECA and HAZOPS have often been carried out for the technological system alone without including the consequences of human actions. In cases where the target is an interactive system the outcome of such analyses is clearly incomplete. The degree to which human action is included in PSA depends on the level of the analysis, but since the target systems - such as nuclear power plants - usually involve human-machine interaction a consideration of human reliability cannot be avoided.

The purpose of a PSA is to identify and classify those elements where improper operation can result in a contingency. The possible contributors to the risk can come from many sources such as a technical system (a nuclear power plant), a natural phenomenon (an earthquake or a tornado), and human action. The outcome of a PSA can be one of the following (Dougherty & Fragola, 1988, p. 74):

- A **description** of equipment failures, human failures, and process events whose combination must occur before a specified hazard can occur. Due to the widespread adherence to the defence-in-depth principle it is rare that a single failure can be the cause of an incident.

- A **quantitative estimate** of the risk in one of the above terms, with some indication of the uncertainty of this estimate.

- Relative **quantitative measures** of the risk-importance of equipment, human factors, plant design parameters and policies, etc.

In the larger perspective humans are involved not only in the operation of a process, but also in the design, construction, maintenance, and management. Yet as long as HRA is carried out within the shell of PSA, the scope will be limited to human reliability during operation. The question is whether the scope of HRA must remain constrained by the PSA-*cum*-HRA construct. The answer should be a resounding NO, if only for the reason that this construct is too constrained by the PSA focus on operation. An action always takes place in a context, and the context is partly the outcome of preceding human activities in, for instance, design, maintenance, and management. It is therefore not enough for HRA to develop models of human actions during control and operation. It is necessary also to develop a comprehensive understanding of human action in context. This requires the integration of work from hitherto separate fields of investigation and effectively amounts to a redirection of HRA. For example, instead of making task analyses and concentrating on the limited context of man's direct functioning, HRA should consider a broader context including management goals and functions. In these domains the consequences of failures are, unfortunately more difficult to predict both because the mechanisms of propagation are more complex and because there may be considerable time delays involved. It is nevertheless still necessary to have an operational definition of what e.g. a "management error" is and to have a corresponding notion of consequences. This may be equivalent to making a PSA for organisations and management as well as for individual human actions (cf. Hollnagel & Wreathall, 1996).

If the purpose of PSA is to assess the effects of operational improvements, then HRA needs to focus on the traditional issues of layout and labelling, perhaps with an element of training and procedures thrown in. If, on the other hand, the purpose of PSA is to assess safety requirements, then it must consider much more complex "causes" of accidents, extending from flawed assumptions in the design and setting of technical specifications to improper or inadequate training for off-normal conditions. In such cases classical HRA methods such as THERP (Swain & Guttman, 1983), TRC (Hall et al., 1982), SLIM

(Embrey et al., 1984), etc., break down because they give context free **numbers** and not context dependent **"causes"**. This is an area where new HRA methods are seriously needed and where development should be encouraged and supported. Thus even if HRA is a tool within the PSA shell it needs to be tailored to the intended purpose of the end result, and though a set of methods may work at one level need it not work at another.

The influence of the PSA-*cum*-HRA construct on HRA can be seen by considering three fundamental elements of PSA namely: (1) models of events that can lead to a specified risk, particularly the propagation paths in the system; (2) models of the consequences of effects of the events; and (3) techniques and data that can be used to quantify the risk in probabilistic terms. Each of these has had a considerable effect on how HRA has developed, and can together be seen as the main causes of the present day problems.

2.1 The PSA Sequence Model

A particularly important type of model used in PSA is the sequence model. The purpose of the sequence model is to show the various possible sequences of events and their outcomes. Sequence models are usually represented as event trees; an example of a sequence model or an event tree for loss of off-site power is shown in Figure 7 (Orvis et al., 1991). The events in the sequences are described by the boxes in the top row: loss of off-site power, followed by reactor SCRAM, followed by manoeuvring the steam relief valves, etc. The horizontal tree diagram shows the possible developments of events; by convention, a successful event is represented by the upwards branch, whereas a failed event is represented by the downwards branch.

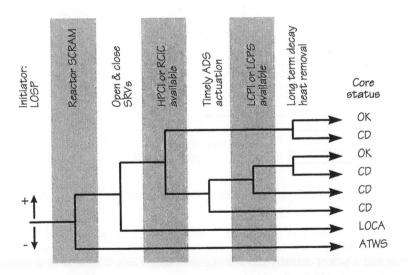

Figure 7: Sequence model for loss of off-site power.

Figure 7 shows that if the first event - reactor SCRAM - fails, then the outcome is an ATWS (Anticipated Transient Without SCRAM). If the SCRAM succeeds the next event is the manoeuvring of the SRVs. If this fails, the result is a LOCA (Loss of Cooling Accident). The remaining events in the sequence are analysed in the same way and the possible outcomes are used to draw the complete event tree that makes up the PSA sequence model. Many of the events represent an **interaction** between people and technology, i.e., points in time where operator actions are necessary if the event is to conclude successfully. HRA is traditionally only considered for events that represent such an interaction, and within the PSA-*cum*-HRA construct HRA is therefore restricted to such events. This disagrees with a socio-technical perspective that requires human performance to be considered for the task or sequence as a whole. The PSA-*cum*-HRA thus encourages a piecemeal analysis rather than an understanding of the overall context in which the events take place.

2.2 The Consequences Of Human Actions

Since most technical systems include and depend on human actions, it is necessary to include the effect of human actions in the PSA sequence model. This is traditionally done by constructing a fault tree for the event in question and using that as the basis for a specific analysis of human performance and its possible outcomes. Relative to a fault tree representation, the consequences of human actions can be introduced for almost every event. This can be illustrated by considering the generic fault tree, the so-called "anatomy of an accident" (Rasmussen & Jensen, 1973), which is shown in Figure 8.

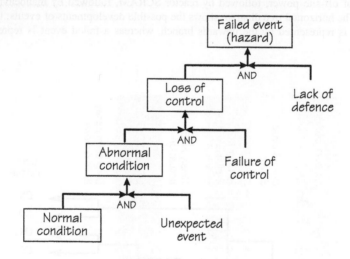

Figure 8: The anatomy of an accident.

The basic idea in the "anatomy of an accident" is that the final consequence - the failed event or the hazard - only occurs if a number of things go wrong. In Figure 8, the failed event occurs only if there is a loss of control **and** a lack of defence. The loss of control occurs only if there is an abnormal condition **and** a failure of control. And finally, the abnormal condition occurs only if there is a normal condition **and** an unexpected event. As an example, consider the spill at the Cadarache nuclear power plant illustrated in Figure 9. In this case the unexpected event was that someone forgot to turn off the tap for the eye rinsing basin. The failure of control was that in two cases the overflow alarms failed; this meant that the water was able to spill onto the floor and run into the sump without anyone noticing it. Finally, the lack of defence was that the sump was connected, by a mistake, to an outside rainwater tank rather than to a tank for industrial waste.

The "anatomy of an accident" representation is very convenient as a basis for considering how human action can have an impact on an accident, both in the sense of contributing to the accident or the propagation of it, and in the sense of being a mitigating factor. In relation to Figure 8 this means the human actions that can have an impact on the occurrence of an unexpected event, the failure of control, and/or the lack of defence. The event described by Figure 9 is a good illustration of that. The initiating event was clearly related to a human action. The failure of the overflow alarms could easily be due to inadequate maintenance, and the incorrect connection of the sump outlet is also due to in incorrectly performed human action.

In describing the consequences of human actions it is often important to make a distinction according to whether the effects of an action are recognised when the action occurs, or whether the action creates a latent condition. The analysis of numerous accidents has shown that the existence of latent conditions can often be the crucial factor that changes a simple mishap to a serious accident (Reason, 1990). Although both the failure of control and the lack of defence may initially be ascribed to malfunctions of the technology, such malfunctions rarely occur by themselves. For a very large number of cases they are due to inadequate maintenance, inappropriate planning, insufficient checking and inspection, etc. The PSA event tree is nominally concerned with the immediate effects of human action, i.e., the probability that an action fails during operation, but the scope should clearly be extended to consider the possible conditions that may exist in the system during the event, and the way in which they can have come about.

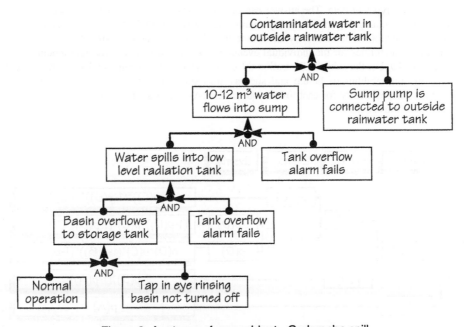

Figure 9: Anatomy of an accident - Cadarache spill.

The "anatomy of an accident" representation illustrates one way of describing the interaction between people and technology in a more detailed manner. Another approach is to perform a binary decomposition similar to the representation of the PSA event tree. (This will be illustrated below for the case of cognitive functions.) The main difference between the two is that the "anatomy of an accident" represents a general paradigm for **explaining** events that have occurred, i.e., for accident analysis (hence the name), whereas the binary event tree represents a paradigm for the **calculation** of probabilities. This difference between retrospective and predictive methods will be described further in Chapter 3. The difference has not been clearly acknowledge by HRA (or for that matter by cognitive psychology) and the first-generation HRA approaches have therefore been unable to reconcile it. The consequence of this has been that the PSA solution rather uncritically has been adopted as the model for HRA. As the following arguments will show this is probably the main reason for the stalemate that first-generation HRA approaches have reached.

2.3 Data And Quantification

Quantification has always been a stumbling block for HRA. In the list of shortcomings of first-generation HRA presented in Chapter 1, less-than-adequate data was a main problem. The same problem is found in the less-than-adequate calibration of simulator data and the less-than-adequate proof of accuracy. In this sense quantification is a stumbling block because it is difficult or impossible to obtain the "right" numbers required by the PSA. It is, however, also a stumbling block in the sense that many people with a behavioural science background find the idea of quantifying human actions unpalatable.

It is instructive to consider both why there is a need for quantifying human performance and why there is a need for specific types of data. The answer can in both cases be found by taking a second look at the PSA event tree. Without going into too many details, the practice of PSA strives to calculate a probability for each of the outcomes described by the event tree. This is done in a straightforward manner by combining the probabilities for the individual events. Consider, for instance, the fundamental event tree shown in Figure 10. For the sake of illustration the events have simple been labelled $Event_A$, $Event_B$, etc. In this tree the probabilities of failure and success are expressed for each event, and this is used to calculate the total probability for each of the outcomes. (As a check of the correctness the sum of the outcome probabilities equals one.)

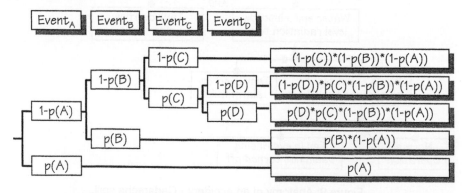

Figure 10: Event tree probability calculation.

The need to **quantify** human performance rises from the nature of the PSA event tree. The formalism of the event tree representation requires that for each event a probability can be assigned to both success and failure. In practice the concern is for the probability of failure, because the probability of success is calculated as one minus the probability of failure, i.e. $p(\text{success}) = 1 - p(\text{failure})$. When an event represents a human action, the probability of failing to perform that action - or rather the probability of not performing it correctly - must be found and entered into the event tree. In the case where the event represents an interaction between human and machine, or in cases where the event is analysed further by means of e.g. a fault tree it is, strictly speaking, not necessary that the more detailed analyses conform to the requirements of the PSA, except that they must produce a probability as the final outcome. The detailed analyses may thus be based on different principles, for instance with regard to how human performance is modelled. As the following section will show, it seems as if HRA practitioners on the whole have overlooked this point and have continued to apply the PSA principles of analysis for human actions as well.

The need for a specific **type** of data is also derived from the PSA event tree. Since the representation is expressed in terms of successes or failures, it has been necessary to describe human actions in the same way. This kind of binary classification is obviously much too simple to make any claims on psychological realism. First of all, there is an important difference between failing to perform an action and failing to perform it correctly. Furthermore, failing to accomplish an action may happen in many different ways and for many different reasons. Quantification attempts may therefore rightly be criticised on this account alone. The PSA event tree also considers each event or each action by itself. This has created a need for data on human "error" probabilities (HEP) for specific, characteristic types of action - such as the detection of a signal or the reading of a value. The need for such data is nevertheless quite artificial and should be recognised for what it is, namely an artefact of the PSA sequence model. It is therefore not surprising that the attempts to provide such data on the whole have been rather unsuccessful. Later in this chapter, as well as in following chapters, the nature of classification systems for human performance will be discussed and an alternative will be provided. It will be argued that the problem of less-than-adequate data to a considerable extent is an artefact of PSA. One consequence of choosing a different approach is therefore that the need for data changes, both in terms of the importance of data and their type.

2.3.1 The Decomposition Of Cognition

One area where the influence of PSA is particularly clear is in the decomposition of cognition. The custom of describing events in terms of the binary categories of success and failure can to some extent be defended in the case of manual actions such as manoeuvring a handle or pressing a button. Manual actions are clearly distinct, in the sense of being separable from each other, and can either be performed or not, depending on what criterion is used. It therefore makes some kind of sense to include them as nodes in an event tree description, of the kind shown in Figure 7. (The problems in classifying observable actions will be discussed in the last part of this chapter.)

In cases where the human interaction is a cognitive function or a mental act rather than a manual action, the use of the event tree description no longer makes sense. Human cognitive functions, such as diagnosis and decision making, and in particular the notion of a "cognitive error", play an increasingly important role in HRA. The main problem facing the HRA specialists working within the PSA-*cum*-HRA construct is therefore how to account for cognitive functions. The standard approach has been to decompose cognitive functions in their assumed components, and to describe their relations by means of a small event tree. Assume, for instance, that an operator action can be decomposed into the following four segments:

- **Problem identification,** where the operator must detect that something has happened, and that the situation deviates from what it should be. The operator must further identify or diagnose the situation.

- **Decision making,** where the operator must select an appropriate action to carry out, based on the preceding identification of the problem.

- **Execution,** where the operator must execute or perform the chosen action. The execution must be correct, i.e., according to the prescribed procedure or established practice.

- **Recovery,** which offers the operator the possibility of determining whether the action had the expected effect. If not, the action may have been incorrect, and the operator may have a chance to recover, i.e., correct it, provided the nature of the process and the characteristics of the system allow that.

This decomposition can be described graphically as shown in Figure 11, which furthermore consists of two parts. First, the decomposition into a response plus a possible recovery. Second, the decomposition of the response into smaller segments, according to the preferred conceptualisation of the nature of human action.

The first decomposition into RESPONSE + RECOVERY is dictated by the needs of a PSA. Since the PSA is interested only in the aggregate probability of a failure of the intervention, and since a *recovery* by the operator may affect that probability, it stands to reason that the possibility of a *recovery* must be included in the analysis of the manual intervention. The second decomposition is dictated by the need to account for some of the important cognitive constituents of a response. The approach illustrated by Figure 11 limits the decomposition to three segments, being *identification, decision,* and *execution.* This is probably the minimal number of segments that makes sense, and from a practical perspective there may be little reason to go into further detail, although several information processing models offer a far greater variety. Altogether, the two types of decomposition, taken separately, can therefore be justified by referring to accepted practice.

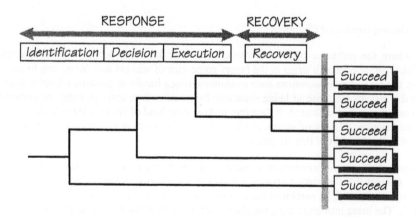

Figure 11: Decomposition of manual interventions

The way that the two types of decomposition are combined does, however, raise some questions. The reason for this is that the event tree in Figure 11 is reduced. A complete event tree would consist of all possible combinations of success and failure for each segment, yielding a total of $2^3 = 8$ outcomes; *recovery* should furthermore be associated with each of the possible outcomes thereby bringing the total to 16. From a psychological standpoint this would be reasonable, since an operator who made an incorrect *identification* still might continue to decide and act, unless the *identification* was recognised as being incorrect - in which case a *recovery* would take place. The same argument goes for an action following an incorrect *decision*.

There is an obvious need to reduce the event tree for practical reasons because the analysis of a complete event tree may require too much effort. In order to do make this reduction in a defensible way some consistent principles must be established. The reduction used in Figure 11 is mathematically correct because it produces the same total probabilities for success and failure as the complete tree. The reduction is, however, not psychologically plausible as it implies the following reasoning:

- An operator who has made an incorrect *identification*, does not bother to check it. Furthermore, the probability of making an incorrect *identification* is very much larger than the probability of detecting it later, e.g. after the consequent *execution*.

- An operator who has made an incorrect *decision*, does not bother to check it. Furthermore, the probability of making an incorrect *decision* is very much larger than the probability of detecting it later, e.g. after the consequent *execution*.

- The probability of making a correct *decision* after having made an incorrect *identification* is so small, that it can safely be neglected. The probability of making an incorrect *decision* after having made an incorrect *identification* is also so small, that it can safely be neglected.

On balance, the convenient use of the binary event tree to describe the details of responses and cognitive functions goes against a common-sense understanding of the nature of human action. It results in a description that is computationally simple but psychologically unrealistic. Quite apart from that it may be inherently difficult to find basic error probabilities for the results of the decomposition, i.e., identification, decision, and execution (and even recovery). The need for this kind of data is clearly determined by the way in which the decomposition is made, hence is an artefact of the composition. A revised consideration of how HRA should be performed, i.e., how human actions should be described and modelled, may therefore reduce the urgency of finding specific sets of data and perhaps even re-define the purpose of the quantification.

2.4 The Scope Of HRA

The study of human reliability can be seen as a specialised scientific sub-field, a kind of hybrid between psychology, ergonomics (human factors), engineering (hardware) reliability analysis, and system analysis. As the above discussion has indicated, the initial approach was to use the existing techniques for reliability analysis and extend them to include human actions. Thus Miller & Swain (1987) noted that:

> "The THERP approach uses conventional reliability technology modified to account for greater variability and interdependence of human performance as compared with that of equipment performance ... The procedures of THERP are similar to those employed in conventional reliability analysis, except that human task activities are substituted for equipment outputs."

Although the need to improve the description of human performance is quite clear, the influence from PSA and the engineering approach has been difficult to relinquish. The initial assumption in PSA was that the human operator could be described in the same manner as a machine. This turned out to be invalid except possibly for certain types of highly regular performance such as well-rehearsed skills. It was therefore necessary to find a more realistic approach and develop descriptions or models of human actions that could provide a better basis for e.g. system design, task analysis, etc. (Dougherty & Fragola, 1988). This meant that HRA had to expand the scope of description and to find inspiration in other fields such as psychology and cognitive science. At the same time it was necessary to maintain the specific relation to PSA. This created a dilemma, since the improvements to HRA were sought in sciences that could not easily accept the premises of PSA.

In relation to the PSA event tree, HRA serves the purpose of providing the necessary input in places where the event either is a human action or involves an interaction between humans and machines, cf. Figure 12. In this manner HRA easily comes to serve as a fault tree specialised with regard to human actions - and in particular human erroneous actions. Yet even though the impetus for the development of HRA to a large extent has come from the needs of PSA, it is necessary that HRA as a discipline or speciality transcends the needs of PSA. In order to answer the questions that come from a PSA, the study of human reliability must encompass the broader view of human actions. In other words, it is necessary to understand the nature of human action before one can understand the characteristics of erroneous actions.

As long as HRA remains within the shell of PSA the scope of HRA will be limited, **but so will also the scope of PSA**. This becomes obvious when the structure of PSA as a method is considered. For nuclear power stations, the purpose of a PSA is to analyse possible contributors to precisely defined types of consequences. A typical PSA will consist of the following steps (Dougherty & Fragola, 1988):

* Define the risk criterion or risk criteria. The risk criterion is used to determine which accident sequences should be included in the analysis.

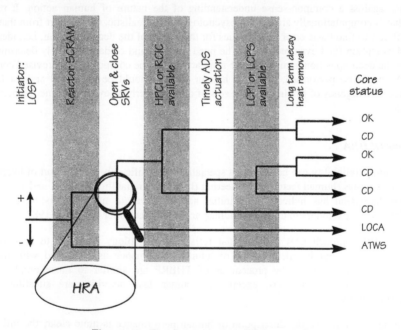

Figure 12: The link between PSA and HRA.

- Create a description or representation of systems that make up the plant. This representation should include a description of the interactions, in particular the interactions between the technical system and people.

- Define the associated hazards, i.e., the possible sources of damage or danger. The definition of the hazards may include information from the plant's operating history, if it is available, or information from similar plants. The outcome of this step is summarised in a list of the events that may initiate an accident, as well as any other event that must occur for the hazard to obtain.

- Define accident sequences that will lead to specific hazards. An accident sequence is usually described as an event tree. Each event may in turn be expanded and described in greater detail using e.g. fault trees or reliability block diagrams - or HRA if human actions are involved. The sequences may be developed using computer models to determine the success criteria, e.g. in terms of thermo-hydraulics.

- Evaluate the consequences of the accident sequence, i.e., of sequences that lead to failures.

- The accident sequence is used as a basis for determining the frequencies and uncertainties of the consequences. This range of values can be compiled into an aggregate estimate of risk or expressed in other ways that comply with the nature of the analysis. The essential outcome of a PSA is a quantitative expression of the overall risks in probabilistic terms.

For the current discussion the essential point is the description of the accident sequence as a PSA event tree. As mentioned above, the PSA event tree is a simplified representation of the accident sequence, and it is in particular simplified with regard to the human actions that may occur. The HRA is therefore limited to consider the human actions that are included in the event tree, and the quality of the analysis critically depends on the completeness and accuracy of the PSA event tree (Figure 13).

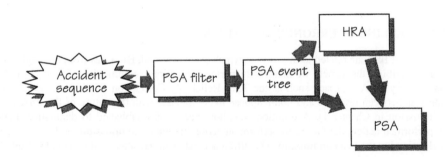

Figure 13: PSA-*cum*-HRA.

As indicated in Figure 13 both HRA and PSA are applied to the simplified description of the accident sequence by the PSA event tree. It is, however, feasible to analyse the accident sequence directly in terms of human performance and human cognition. This can be done by means of methods such as cognitive task analysis and cognitive profiling, which will be described in later chapters. When that is done it will lead not only to an improved basis for a cognitive reliability analysis, but possibly also to an improved basis for a PSA. In this view the PSA, as well as the HRA, is part of an Integrated Safety Analysis, as shown in Figure 14. The potential benefit of this approach has been realised by some of the newer HRA approaches described in Chapter 5.

This view of the scope of HRA removes the basic limitations that PSA has imposed on HRA. It effectively takes HRA out of the PSA shell, and puts both on more equal terms. By doing so it may also encourage a reconsideration of the basis for PSA, both when it comes to include the functioning of joint man-machine systems and to account for the effects of human action in a socio-technical system. It may turn out that HRA can contribute not only by providing the "human error" probabilities cherished by PSA, but also by enriching the descriptions of the accident sequence and - perhaps - by pointing to events that have been outside the reach of the conventional PSA event tree.

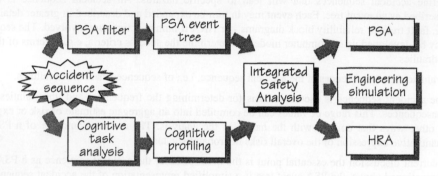

Figure 14: PSA and HRA as part of ISA.

3. THE MODELLING OF ERRONEOUS ACTIONS

Since the needs of PSA are related to the ways in which events can fail, HRA has from the beginning been directed at describing the variety of incorrectly performed actions, commonly referred to as "human error". In an early HRA approach such as the Time-Reliability Correlation (Hall et al., 1982), the only distinction was between response and non-response in compliance with the nature of the PSA event tree (cf. the discussion in Chapter 1). A tradition was, however, soon established to distinguish between a correctly performed action, the failure to perform an action known as an **omission,** and an unintended or unplanned action known as a **commission.** The difference between the latter is illustrated by Table 1.

Table 1: Errors of omission and commission (cf. Swain, 1982).

Errors of omission (intentional or unintentional)		Omits entire step
		Omits a step in task
Error of commission	Selection error	Selects wrong control
		Mispositions control (includes reversal errors, loose connection, etc.)
		Issues wrong command or information (via voice or writing)
	Error of Sequence	*No details given.*
	Time error:	Too early
		Too late
	Qualitative Error	Too little
		Too much

The dichotomy of omission-commission fits nicely with the representation of the event tree used by PSA, since in both cases a sequence of binary decisions is implied. This can be illustrated more clearly by showing how the error types traditionally used in HRA (cf. Table 1) can be expressed as resulting from a sequence of analytical decisions, cf. Figure 15. The figure also shows how the outcome for each path can be classified in terms of the simple response / non-response categories.

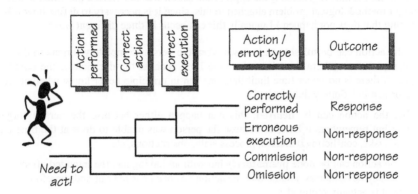

Figure 15: A pseudo event tree for omission-commission error classification.

Figure 15 represents a pseudo event tree rather than a proper event tree. The difference is that the ordering of the branches, corresponding to the boxes in the top row, shows a sequence of questions that could be asked to analyse an event, rather than a sequence of sub-events that together constitute an action. The order of the binary choices is arbitrary from the point of view of the nature of human action. This can easily be demonstrated by turning Figure 15 into a decision table; this representation removes the need to order the binary choices while producing the same result. A possible version of the decision table is shown in Table 2.

Table 2: A decision table for omission-commission error classification.

Action carried out	Y	Y	Y	Y	N	N	N	N
Correct action	Y	Y	N	N	Y	Y	N	N
Correct execution	Y	N	Y	N	Y	N	Y	N
Omission					♦	♦	♦	♦
Correctly performed action	♦							
Commission			♦	♦				
Erroneous execution		♦						

Errors of omission have usually been considered as a well-defined set, whereas errors of commission have been seen as less clearly described, hence including several different error modes. To put it simply, there is apparently only one way in which something can be omitted, while there are many ways in which something can be done incorrectly or inappropriately. In both cases there can obviously be many different reasons for what happens. In reality the issues are less easy to resolve, because the categories overlap. This can be illustrated by taking a closer look at the criteria for an omission.

3.1 Omission And Commission

An omission is defined as the failure to perform an action. It is assumed by this definition that the action is the required or appropriate action and that it takes place at the right time or within the right time interval. An omission is thus registered if the appropriate action was not carried out when it was required. (There is a slight methodological problem attached to this, since it is necessary to define how it is possible to note something that does not happen.) Logically this can occur in three different ways.

- Firstly, the action can be **missing** as in cases when the person completely forgets to do something. (Even here, however, there must be some time limit such as the duration of the complete event sequence. If there is no upper time limit then one can in principle never know whether an action is missing or just indefinitely delayed.)

- Secondly, the action can be **delayed**; this can happen either because the person forgets to do something but remembers it later, or because the person was unable to do it at the time due to e.g. competing tasks, conflicting demands, process state, obstructions, etc.

- Thirdly, the action can be done **prematurely** but without having had the intended effect at the time; the person may nevertheless assume that the action has been carried out and therefore omit it at the time when it is actually required.

A fourth possibility is that the action has been **replaced** by another action; this may happen for a number of reasons, e.g. that the person has lost control, that instructions or indicators are misleading, etc. If the person has done something that was unplanned or unintended then, for all practical purposes, a commission has occurred. Thus, making a commission by doing something that was unplanned or unintended effectively prevents doing that which was required (at the required time). A commission therefore logically implies an omission, in the sense it is not possible to make a commission without also making an omission! The four possibilities are illustrated in Figure 16.

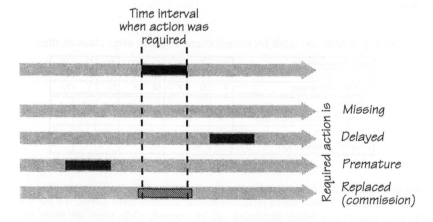

Figure 16: Four variations of omission.

This discussion shows that the seemingly simple terminology that traditionally has been used by HRA imposes some problems of definition. These problems are not only important for the purists but have some very practical implications. If HRA is to be capable of answering the needs of PSA, then it must be possible to account in detail for how incorrectly performed actions can come about. This requires that a clear separation can be made between the categories of erroneous actions that constitute the basis for qualitative analyses and performance prediction - and eventually for quantification. The above example shows that an omission is not a single or unique category but that an omission can come about in several different ways. In Chapter 6 the discussion of the classification of erroneous actions will be enlarged, as a basis for proposing a specific classification scheme for HRA.

3.2 The Hunting Of The Snark

In the traditional modelling of erroneous actions, omission and commission appear on the same level as basic ways in which an action can fail (cf. Table 1 and Figure 15; see also Barriere et al., 1994). First-generation HRA approaches have for a long time focused on omissions as the prototypical erroneous action, perhaps because this easily fits with the notion of a non-response as described e.g. by the TRC (Hall et al., 1982). Commissions have received less attention and have mostly been used as a conceptual garbage can for everything that could not be classified as an "error of omission". During the 1990s the need was realised to complete the picture by considering also the cases where people made incorrect or inappropriate actions, rather than just omitted an action. This has led to several efforts to develop descriptions and models of "errors of commission".

A recent example of that is Barriere et al., (1995) where an error of commission is defined as "an overt, unsafe act that, when taken, leads to a change in plant configuration with the consequence of a degraded plant state" (p. 7-3). This definition contains a reference to both the event / action and the consequence. It is also a rather sweeping definition that falls short of all the criteria discussed in the beginning of this chapter, and which basically includes in an error of commission everything that cannot be classified as an omission. This is, however, no coincidence as the authors acknowledge that the concept of an error of commission is characteristic of the way of thinking found in PRA/PSA rather than of human behaviour (Ibid., p. 7-3). The detailed classification they produce is therefore a mixture of categories referring to both events and consequences and does not propose a psychological or behavioural classification scheme.

As shown by the preceding analysis the concept of an omission is not as simple as it may seem. Not only can there be several types of omission but an "error of omission" is also logically implied by an "error of commission". It is therefore reasonable to suspect that the notion of an "error of commission" has more to it than meets the eye, in particular that it is not a simple or single category of action.

Much of the reason for the problems in unravelling the precise meaning of omission and commission is that these concepts refer to a theoretical rather than a behavioural distinction. Even though the basis for first-generation HRA approaches basically is phenomenological, in the sense that they refer to prominent surface features of performance, the categories used are theoretical rather than behavioural. Omission and commission are presented as opposites, literally meaning not doing something and doing something respectively, and this juxtaposition reflects the binary nature of the PSA event tree (this is discussed further in Chapter 3). The attempt of clarifying the category of "errors of commission" must have a behavioural basis because it must bring together a number of distinct error modes (manifestations) under the single category of commissions. For example, choice of a non-required action, repetition of an action, inversion of two actions, and timing errors have all been proposed as examples of a commission. Yet these have very little in common as behavioural categories, except the fact that they all "prevent" the required action from being carried out. In particular, they have no common aetiology. It is therefore neither very easy nor particularly meaningful to lump them together under a single label.

The search for the "commission" is not unlike the hunting of the Snark, which, in the end, turned out to be a Boojum instead. Since omission-commission with some justification can be seen as theoretical constructs, HRA may possibly be better off by leaving them alone. Instead, an effort should be made systematically to analyse and categorise the possible failure modes, for instance in relation to a specific model of behaviour, and thereby circumvent the omission-commission dichotomy altogether. This exercise will be carried out in Chapter 6.

3.3 Omission, Commission, And "Cognitive Error"

From the PSA perspective "all" that is needed is a clarification of how an action can fail supplemented with a reasonable estimate of the probability for the failure. Due to the nature of the event tree it seemed reasonable to make a simple distinction between correct actions and failures, or between response and non-response. As long as HRA could provide answers on this level of description, there was little motivation to delve deeper into how erroneous actions came about. HRA initially accepted the premises of PSA but extended the descriptions of the non-response category to include the categories of omission and commission. These categories were phenomenologically meaningful because they referred to something that could easily be observed: a person could either do something correctly, do it incorrectly, or not do it at all. The categories are, unfortunately, also limited because they refer to overt and observable events - without even considering the terminological ambiguity discussed above. This limitation became a problem when HRA practitioners, with the help of psychologists, tried to improve the understanding of the causes of erroneous actions. When erroneous actions were classified by internalised rather than externalised criteria the role of human cognition came to the fore, in particular the way in which cognition could fail. The distinction between a slip and a mistake, for instance, referred to the person's intention or the purpose of the actions (Norman, 1981). This made it necessary to speculate about how intentions could be formed, how problems could be understood, and how plans could be made. In other words, it was necessary to consider human cognition and in particular human decision making.

From the HRA practitioner's perspective the immediate solution was to introduce a new category which appropriately enough was called "cognitive error". This made good sense insofar as everyone knew that misunderstandings could occur, that wrong decisions could be made, that people could jump to conclusions, or that they simply could assume something without bothering to check it out. Surely, the reasoning went, if there is cognition and cognitive functions or processes, then there must also be "cognitive errors" in the cases where these functions go wrong. In the PSA analysis of disturbances, diagnosis is an important type of human action or event - in the meaning of deciding what to do as a precursor to actually doing it. The same is the case from the perspective of human performance in process control; here the decision process soon became the prototypical instance of information processing and erroneous actions were explained as the outcome of an incorrectly performed decision process (cf. Figure 4). Incorrect diagnosis or decision failure therefore quickly became synonymous with "cognitive error".

The addition of "cognitive error" to the basic categories of omission and commission seemed innocent enough. Attempts to include "cognitive error" in the existing methods and approaches, however, soon made it clear that some problems remained. The fundamental problem is that the usual categories describe the **manifestations** of actions, but that a "cognitive error" is the assumed **cause** of a manifestation rather than the manifestation itself. Going back to the discussion in the beginning of this chapter, the omission-commission distinction represents events and/or consequences but not causes. Yet even if we can refer to diagnosis as a PSA event, and consequently also to misdiagnosis as the event gone wrong, we cannot see or observe the misdiagnosis in the same way as we can e.g. an incorrect action. We may be able to infer from later events that the diagnosis has gone wrong - at least as a highly plausible hypothesis; but that means precisely that the misdiagnosis or the "cognitive error" is a cause and not a manifestation. The problem becomes quite clear when we consider the classification shown in Table 1, as well as the pseudo-event tree representation in Figure 15. Try as we may there is no simple way in which "cognitive error" can be added to the other categories. If the logic of the pseudo-event tree or the decision table (Table 2) is strictly observed, a "cognitive error" is the same as a commission. This follows from the fact that, according to the classification, the action is carried out but it is not the correct action. However, a "cognitive error" is obviously not the same as a commission, since if that was the case there would have be no reason to introduce the new term. The difference is that a commission is a manifestation, while a "cognitive error" is a cause. As such, the "cognitive error" can be the cause of an omission as well as of a commission.

It is always convenient to characterise major trends or traditions by a few simple terms. The advantage is that comparisons are easier to make and that differences can be highlighted. The disadvantage is that such characterisations invariably leave out important details and that they in some cases may even distort the description. Having said that I will nevertheless run the risk of applying this principle, not only in the rest of this chapter but throughout the book. I do fully recognise the dangers in doing so, but hope that the reader will keep the purpose of the simplification in mind and therefore not take the descriptions at face value.

In this spirit the current state of HRA can be characterised as a transition from "omission to cognition". By this I mean that whereas first-generation HRA refers to a fundamental distinction between omission and commission in the description of erroneous actions, potential second-generation HRA must go beyond that and address the thorny issue of human cognition. This has on several occasions been posed as the problem of "cognitive error", implying that there is a special category of erroneous actions that relate to failures of human cognition. I do not agree with this view. Rather, human actions - erroneous or otherwise - are all to some extent cognitive, which means that they cannot be properly described or understood without referring to the characteristics of human cognition.

3.4 Overt And Covert Events

Regardless of whether the "omission-commission" classification is represented as an event tree, a pseudo-event tree, or a decision table, it is easy to see that the addition of new categories of "error" requires great care. The "omission-commission" categories have their basis in the manifest aspects of performance, and were introduced many years before the consideration of human cognition became generally accepted. If, for the sake of argument, we assume the existence of a specific type of erroneous action called "cognitive error", then the question becomes where the "cognitive error" should be put in relation to the other categories, i.e., what is the criterion for identifying a "cognitive error".

The notion of a "cognitive error" probably stems from the intuitive distinction between **overt** and **covert** actions. Overt actions are those that can be seen or observed, such as information seeking and manipulation, and for these the categories of omission and commission can be applied, cf. above. In contrast, covert actions cannot be observed in others, although we can "observe" them in ourselves (through introspection). We know that the covert actions take place as steps in reasoning, evaluation of hypotheses in diagnosis, weighing of alternatives in decision making, etc., but from an analytical point of view we have to infer their existence. In process control and PSA considerable emphasis is put on the diagnoses that operators often have to make. In many cases, a sequence of operations or a procedure is preceded by a diagnosis, which in essence means that the operator must decide or determine what the current situation is and what should be done about it. A diagnosis can be correct or incorrect, and it is therefore tempting to consider an incorrect diagnosis a "cognitive error", since it is beyond dispute that making the diagnosis involves cognition. The problem, however, is that whereas e.g. shutting a valve is a unitary action, making a diagnosis is not. The outcome of a diagnosis may be right or wrong, but that does not mean that the way it happened was in any way uncomplicated. Making a diagnosis may span considerable time and involve a number of steps or actions, including consultation with other operators or other sources of information. It is therefore an oversimplification to treat the failed diagnosis as a "cognitive error", since there are so many ways in which it could have happened. In particular, there is no unitary or unique element of cognition that is pivotal for the diagnosis.

3.5 Phenotypes And Genotypes

The distinction between cause and effect or manifestation is very important, both to the psychology of action and to human reliability analysis. This distinction will be developed in greater detail in Chapter 6, using the terms of phenotype and genotype to represent two fundamentally different ways to consider erroneous actions (cf. also Hollnagel, 1993b). The **phenotype** is concerned with the manifestation of an erroneous action, i.e., how it appears in overt action and how it can be observed, hence with the empirical basis for a classification. The **genotype** is concerned with the possible causes such as the functional characteristics of the human cognitive system that are assumed to contribute to an erroneous action - or in some cases even to be the complete cause! In general the genotype refers to the set of causes that are deemed sufficient to explain a failed action or an accident, regardless of whether these causes are of a specific type - e.g. cognitive or organisational. The phenotypes and genotypes correspond to two different sets of categories, one for what can be observed and one for what must be inferred. In cases where a strong or well-established theory exists there is little risk in using the latter to define the categories for the former, hence to combine observations and inferences to some extent. In cases where a strong theory is absent, this combination should be avoided as far as possible. The most glaring example is the behavioural study of erroneous actions - whether it is called psychology, human factors, or cognitive engineering. Here the failure to maintain a distinction between the two categories has led to much confusion and misguided problem solving (cf. Hollnagel, 1993b). The fundamental problem is that there are no strong theories of human action in general - although there has been no lack of candidates. A similar caution should be shown in the analysis of organisational accidents, where the very term implies that a specific set of causes are at play. In general, it is highly advisable to keep observation and inference separate in empirical investigations.

In the most basic form, the classification scheme should express the relations between the phenotypes (manifestations) and the genotypes (causes). It is, however, important from the start to make clear that the phenotypes are not caused by distinct genotypes, but that they rather are the result of an **interaction** between multiple genotypes - corresponding to the fact that there rarely is a single cause for an event. Since the study of human performance very often has focused on people in a socio-technical system, there has been tendency to make a distinction between genotypes associated with people, such as misunderstanding, and genotypes associated with the socio-technical system, such as double-bind performance criteria. This is commonly expressed as a distinction between the causes and the context, as shown in Figure 17. The analysis of an event, the search for possible causes, naturally goes in the direction from phenotypes to genotypes, while performance prediction goes in the direction from genotypes to phenotypes.

It is, however, misleading to maintain a distinction between causes and context. On the contrary, the context is itself a set of causes and the *sine qua non* for human performance. The specific purpose of an analysis may warrant that the focus is directed at the genotypes that are associated with a specific part of the human-*cum*-socio-technical system, since it is practically impossible to consider everything and to give everything equal weight and importance. The pragmatic trade-off should, however, not mislead us to assume that the constraints voluntarily imposed by the analysis correspond to absolute features of the world we live in.

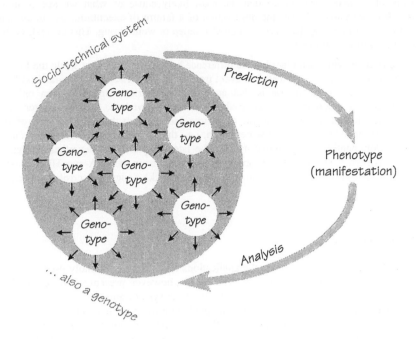

Figure 17: Phenotypes result from the interaction between multiple genotypes.

3.6 "Cognitive Error" Defined

Following the preceding considerations, a "cognitive error" can be defined in two different ways, as follows:

- **The failure of an activity that is predominantly of a cognitive nature.** This corresponds to the use of other common terms for erroneous actions, such as omission or commission, which in practice denote an observable or noticeable event. In all these cases the erroneous action in question manifests itself as the failure to perform a specific activity according to norms. In contrast to the categories of omissions and commissions, however, cognitive activities constitute a **set of activities** rather than a single activity. This set of cognitive activities is usually taken to include diagnosis, decision making, reasoning, planning, remembering, recognition, etc. - or in other words activities that involve thinking rather than doing. An incorrect diagnosis is an example of a "cognitive error", but although it may be characteristic it is by no means the only one. The membership of the set of cognitive activities can often be defined by referring to a specific theory or model of cognition. Furthermore, the definition acknowledges that cognition may be a part of all actions, but that in some cases - such as simple observations or movements - it plays a lesser role.

- **The inferred cause of an activity that fails.** This definition recognises the fact that cognition is a mental function, hence not observable to others than the person. According to this point of view we can only observe a limited number of error modes, which describe the possible manifestations of incorrectly performed actions. For instance, while we cannot **see** an incorrect diagnosis, we can

infer from observable performance that the most likely cause of what we see is an incorrect diagnosis. It is debatable whether the observation of a failure of execution, i.e., an error mode, is sufficient basis for inferring that the cognition also failed or went wrong. This depends on how rigid a model of causation is applied.

The two definitions agree with each other but the difference between them is that while the former can be applied to cognitive activities as represented in event trees or as identified steps (operations) from a task analysis, the latter cannot. (The reason is that an event tree, strictly speaking, can only represent observable events. Since e.g. a diagnosis is inferred rather than observed, the "cognitive error" must relate to an underlying fault model rather than to the event tree itself.) From the point of view of HRA it is therefore more convenient to adopt the first definition, i.e., that a "cognitive error" is the failure of an activity that is predominantly of a cognitive nature. The later chapters will nevertheless argue that the notion of a "cognitive error" is yet another example of an artefact from the PSA-*cum*-HRA construct. The development of an alternative HRA approach may therefore make "cognitive errors" superfluous as a separate concept.

4. CHAPTER SUMMARY

The need for HRA is little realised outside the applications that have traditionally used HRA, and is in particular neglected in the scientific community,. The need is, however, present whenever or wherever an interactive system is being designed, since the reliability of the system **must** be a concern. Furthermore, prediction is a necessary ingredient of design and the prediction must include considerations of how the user(s) of the system is (are) likely to behave. In the scientific community most of the efforts have, however, been spent on event analysis to develop explanations - sometimes elegant and often rather elaborate - for the causes of erroneous actions.

Chapter 2 began by considering the terminological ambiguity of the term "human error", which can refer to either causes, events, or consequences. It was argued that there were three parts to the definition of an erroneous action from the perspective of the acting individual: (1) a performance criterion, (2) an identifiable action, and (3) a degree of volition or freedom, so that the erroneous action was not inevitable. The notion of a performance shortfall is critical to the modelling of erroneous actions, such as whether the outcome is due to capacity limits / design defects or whether it is due to the adaptive nature of human performance. It was concluded that erroneous actions must be defined with reference to the observable characteristics of human performance, rather than to a theory or a model.

HRA has traditionally been pursued as an "appendix" to PSA, and has therefore obediently accepted some of the important assumptions inherent in PSA, primarily the notion of the PSA logic model or event tree and the need for particularised data for isolated events - the Human Error Probabilities. Together with the PSA-like decomposition of cognition that is found in most first-generation HRA approaches this has created a data problem that essentially is an artefact of the underlying assumptions. Rather than remain a part of PSA and accept the limitations that this implies, HRA should extend its scope and emerge as a field of practice in its own right. The need to understand the nature of human reliability clearly goes beyond the requirements of PSA.

The chapter concluded by an initial discussion of the modelling of erroneous actions. It was argued that the traditional distinction between omission-commission is too limited and that it is partly inconsistent. It is therefore insufficient as a basis for considering the more complex issues related to the concept of cognitive functions and "cognitive errors". Erroneous actions are basically overt events whereas the causes usually are covert. It is therefore useful to make a clear distinction between what can be observed and what must be inferred - or between phenotypes (manifestations) and genotypes (causes). One consequence of this view, to be elaborated in later chapters, is that the notion of a "cognitive error" in itself may be unnecessary.

Chapter 3
The Conceptual Impuissance

Science is the knowledge of consequences, and dependence of one fact upon another.
Thomas Hobbes (1588-1679)

1. THE CLASSIFICATION OF ERRONEOUS ACTIONS

In relation to the study of erroneous actions, whether as a part of academic psychological research or reliability analysis, there are two main practical concerns. One is the **retrospective analysis** of events that have happened, the search for explanations and causes. The retrospective analysis is usually qualitative, trying to determine the combination of causes and conditions that led to the observed event. The other is the **prediction** of future events, in particular how a specific incident is likely to develop and how the development may be affected by human actions. Predictions can be either qualitative or quantitative, in the latter case focused on the probability that a specific outcome will result from a specific set of causes or a specific initiating event. It appears that academic scientists and psychologists have mostly concerned themselves with the retrospective analysis, building sometimes elaborate theoretical models and edifices to account for the "mechanisms of error" (e.g. Rasmussen, 1986). The practitioners have, obviously, been more interested in the predictions and particularly in the probabilities that can be assigned to an outcome (e.g. Dougherty & Fragola, 1988). Each direction has consequently developed according to its own objectives and concerns and the exchange of ideas and methods has been rather limited. This is quite disappointing considering that the object of study is the same, i.e., the variability of human action. The only real difference is whether it is looked at in retrospect or in anticipation. This lack of a substantial interaction has led to what appropriately can be called a conceptual impuissance or abstruseness.

This chapter begins by presenting distinctive examples of existing approaches to the classification of human erroneous actions. The presentation looks at approaches from three different perspectives: traditional human factors, information processing psychology, and cognitive systems engineering. The presentation aims to show what is characteristic for each perspective, but does not claim to be comprehensive. The purpose is to show how the specific historic development has lead to the current stalemate rather than to provide a comprehensive survey of "error psychology". This is followed by a discussion of the principal differences between retrospective performance analysis and performance prediction, and how these are supported by the existing approaches.

In the general view the Law of Causality reigns supreme. The Law of Causality states that there must be a cause - or possibly several causes - for the events we observe, and much ingenuity has been spent on devising ways in which to find the causes. Part of that ingenuity has been used to develop classification schemes that provide an orderly description of the possible causes, although this objective usually is defeated by the multifariousness of reality. The present chapter will simply go through a number of these in order to illustrate what the current thinking is. The following chapter will build on that by critically analysing the principles of these classification schemes.

(It is common to refer to such classification schemes as taxonomies - specifically as error taxonomies. I shall, however, refrain from doing that here, for the very good reason that they are not taxonomies. A taxonomy is the classification of living things in an ordered system that indicates natural relationships. The classification systems that describe types of events or possible causes of events are neither concerned with living things, nor does it seem likely that there is an underlying natural relationship since the objects are linguistic constructs rather than things. It is therefore both less pretentious and more appropriate to talk about classification schemes - or simply classifications. As we shall see, these classification schemes are often based on very different principles, which makes it even less appropriate to refer to them as taxonomies.)

1.1 Cause And Manifestation

It is essential that the analysis of erroneous actions makes a distinction between **causes** and **manifestations**, regardless of whether it is for a retrospective or a predictive purpose. In the retrospective analysis, manifestations are the consequences that have been observed and causes are the explanations that are found for what has been observed. The retrospective analysis thus starts from the manifestations and tries to identify the causes. (Once again it is important to remember that the causes are ascribed after the fact.) In prediction, causes are the initiating events and manifestations are the possible outcomes. In this case the starting point is, in principle, the causes and the purpose is to identify the probable or possible outcomes. Disregarding for the moment the differences between retrospective analysis and prediction, we can illustrate the relation between cause and manifestation as shown in Figure 1.

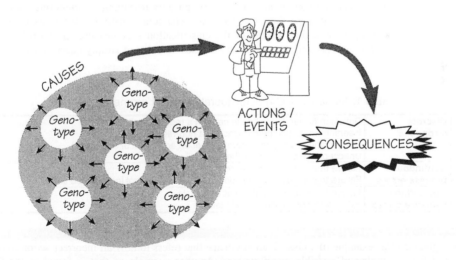

Figure 1: The relation between cause(s) and manifestation.

The distinction between causes and manifestations is clearly important for a classification scheme. The whole purpose of analysing an incident or an accident is to find an acceptable set of causes that can explain the consequences or effects. It follows, that if the terminology and the classification scheme do not embody or support a clear separation between causes and manifestations, then the outcome of the analysis is bound to be inadequate. This was illustrated in the Chapter 2 by the discussion about the possible meanings of the term "error". As I will show in this chapter, few of the existing classification schemes consistently make this important distinction. Using a relatively simple differentiation between stages of development, the discussion will focus on classification schemes from traditional human factors, from information processing psychology, and from cognitive systems engineering.

2. TRADITIONAL HUMAN FACTORS APPROACHES

Several attempts have been made to develop descriptions of operator behaviour that characterise "error" events in terms of their basic behavioural components, for instance Embrey (1992); Gagné (1965); McCormick & Tiffin (1974); Payne & Altman (1962); Rook (1962); and Swain (1963). Although these have been spread over many years, they were all based on a traditional human factors view. This means that they were more concerned with the perceptual-motor than the cognitive aspects of human action. It was furthermore assumed that humans and machines could be described by a set of common terms, which were mainly derived from the engineering disciplines. It was quite natural for traditional human factors to look to the engineering disciplines for concepts and models. On the one hand the need for ergonomics had come from the engineering disciplines rather than from psychology. On the other hand academic psychology had not paid much attention to the problems of working life, and was therefore unable to suggest any useful concepts.

In one early scheme proposed by Altman (1964), for example, the observable characteristics of erroneous behaviour were differentiated according to three general types of work activity: (a) activities involving discrete acts, (b) tasks involving continuous process, (c) and jobs involving a monitoring function, cf. Table 1. Altman suggested that these three categories of behaviour constituted the basis of an "error model" of observable operator action. (Many of the classification schemes use the term "model" to describe themselves. Strictly speaking they are, however, not models but simply quasi-organised lists of categories.)

Table 1: Altman's (1964) behavioural model of erroneous actions

Discrete erroneous actions	Omission errors (the failure to perform a required act)
	Insertion errors (the performance of a non-required action)
	Sequencing errors (the performance of an action out of sequence)
	Unacceptable performance (below quality action)
Continuous process errors	Failure to achieve desired state in available time
	Failure to maintain desirable degree of system control in permitted time
Monitoring / vigilance errors	Failure to detect relevant stimuli or signals
	False detection of stimuli or signals

Table 1 gives a clear example of a classification scheme that refers to how an incorrect action may show itself and lists a number of possible manifestations. Another example of that is found in the Critical Inflight Event Model (Rockwell & Giffen, 1987) which describes the "errors" that can occur during flight (Table 2).

Table 2: Critical inflight event model

1	Inadequate pre-flight checks.	2	Fails to recognise early warnings of problems
3	Fails to do sequence check	4	Decides to fly despite system discrepancies
5	Fails to recognise early warnings	6	Fails to monitor instrument readings
7	Fails to notice small discrepancies in flight sensations	8	Fails to notice lack of agreement of related instruments
9	Diagnostic error	10	Error in estimation of urgency
11	Improper corrective action	12	Poor emergency flying skills

Although the basis for any analysis must be the observed manifestations of the erroneous actions, it is common to have both manifestations and causes as part of the classification schemes. Table 1 and Table 2 both provide good examples of that. In Table 1 the first group (discrete erroneous actions) contains manifestations whereas the second and third groups (continuous process error + monitoring/vigilance errors) contain a mixture of manifestations and causes. Of the twelve events listed in Table 2 only numbers 2, 3, 5, 6, 7, and 8 are proper error modes; all of these are actually omissions in the classical meaning of the term. The remaining events are more in the nature of causes, in the sense that they are difficult to observe and in most cases must be inferred. Consider, for instance, number 10 "error in estimation of urgency". This is clearly not something that can be easily observed, although a highly experienced observer might be able to make the inference on the spot. In most cases the category represents the conclusion from an inference, hence an attributed cause rather than a manifestation.

Table 3: The PHEA classification of erroneous actions

Planning errors	P1	Incorrect plan executed	P3	Correct plan executed, but too soon or too late
	P2	Correct but inappropriate plan executed	P4	Correct plan but in wrong order
Operation errors	O1	Operation too long/too short	O6	Right operation, wrong object
	O2	Operation incorrectly timed	O7	Wrong operation, right object
	O3	Operation in wrong direction	O8	Operation omitted
	O4	Operation too little/too much	O9	Operation incomplete
	O5	Misalignment		
Checking errors	C1	Check omitted	C4	Wrong check on right object
	C2	Check incomplete	C5	Check incorrectly timed
	C3	Right check on wrong object		
Retrieval errors	R1	Information not obtained	R3	Information retrieval incomplete
	R2	Wrong information obtained		
Communi- cation errors	T1	Information not communicated	T3	Information communication incomplete
	T2	Wrong information communicated		
Selection errors	S1	Selection omitted	S2	Wrong selection made

An example of a more recent and much more detailed classification scheme is the one that is used in the Predictive Human Error Analysis (PHEA) technique (Embrey, 1992). PHEA has been developed to aid reliability analysts perform risk assessments in high risk process control environments. PHEA has been used on several occasions to quantify the risk posed to the integrity of complex man-machine systems by the actions of the human component. Described in overview, PHEA breaks down erroneous actions into six major categories that are further subdivided to identify the basic error types. The major dimensions of the PHEA classification scheme are shown in Table 3.

Although this classification scheme clearly identifies many of the ways in which an operator action can fail, the categories do not refer to observable manifestations - for example, P2, C2, R3, T3, and S2. Only the category of operation errors refers directly to manifestations, and even here some of the categories, such as O6 and O7, require a fair amount of inference to be made. On the whole, the erroneous actions described by this table refer to the mental process of planning with the associated cognitive functions, hence to attributed causes rather than manifestations.

To a large degree these descriptions are all variants of a scheme first proposed by Alan Swain in the early 1960s (Swain, 1963) which later has been used in a number of guises (see for example Swain, 1982). In essence, Swain proposed that a distinction could be made between: (a) errors of omission, as the failure to perform a required operation, (b) errors of commission, as an action wrongly performed, and (c) extraneous errors, as when the wrong act is performed. Of these three error modes, errors of omission and commission were viewed as being the most complex. The categories of erroneous actions originating from these sources were summarised in Chapter 2, Table 1.

Each of the three classification schemes outlined above serve to illustrate two important points that can be made in relation to the traditional human factors or ergonomic approaches to the topic of erroneous actions.

*   Despite minor variations in detail there remains considerable agreement between the various schemes concerning the issue of what constitutes the basic categories of erroneous actions when discussed in terms of observable behaviour. In many ways, the omission-commission distinction originally proposed by Swain and his colleagues have remained the "market standard" in human reliability analysis. The strength of this classification is its apparent simplicity. The practical weakness is that the basic framework often needs to be modified to take account of special constraints imposed by an actual operating environment. The theoretical or conceptual weaknesses were discussed in Chapter 2.

*   There is typically a large measure of agreement between judges when they are asked to assign "errors" to these relatively limited behavioural categories. This finding suggests that behavioural models of erroneous actions, such as the ones identified above, possess a high degree of face validity.

The main weakness of the traditional human factors approach is the limited ability to characterise events in terms of their psychological features, partly because the underlying models were poorly developed. Contrary to common belief there appears to be no one-to-one mapping of the external manifestations of erroneous behaviour onto the categories of "failure" that are presumed to occur in human information processing (e.g. Norman, 1983; Reason, 1990). Rather, all the available evidence suggests that erroneous actions that appear in the same behavioural categories frequently arise from quite different psychological causes, while erroneous actions with different external manifestations may share the same or similar psychological origins, e.g. lack-of-knowledge, failure of attention, etc. For this reason a number of influential psychologists and engineers began a line of research in the mid-1970s to mid-1980s aimed at specifying the psychological bases of predictable erroneous actions, although still well within the realm of traditional human factors. A precursor to that was the general attempt to associate causes and manifestations that was evident in many fields, and which particularly concerned practitioners.

2.1 Descriptions Of Specific Psychological Causes

One solution to enhance the descriptions was to include deliberately in the classification schemes possible psychological causes of manifest erroneous actions. The proposals for the description of causes fall into two major types. One type contains causes that are specific to a particular domain. In such cases it is of relatively little importance if the classification scheme refers to an underlying psychological model, or indeed if the causes are theoretically comprehensive. Here specificity is more important than generality, as the main objective is to produce a classification scheme that is highly efficient within a given domain, both in terms of identifying frequent causes and in terms of being easy to use for domain experts. The other type contains classification schemes that are associated with a commonly accepted psychological theory, although the theory itself may not always be explicitly identified. The case where this involves assumptions about the nature of human action - as an information processing model or as a cognitive model - will be treated in the following section.

Table 4 shows five examples of descriptions of psychological causes that have been used to account for observed "errors". Four of these are from aviation, while the fifth is from the nuclear domain. Table 4 only provides the main categories; each example is described in more detailed in the source documents.

The proposed causes listed in Table 4 vary both in their nature and in the level of detail. Some aim to address a broad range of events (Feggetter, 1982; Stoklosa, 1983), while others are clearly focused on a narrow set of occurrences (Billings & Cheaney, 1981; Caeser, 1987). Table 5 shows another example that also is taken from the aviation domain. This describes not only the psychological causes (in the sense of individual causes), but also other factors that may contribute to the occurrence of the event. The causes given in Table 5 aim to be comprehensive, and do in fact represent a mixture between a specific and generic classification scheme.

Table 4: Examples of specific psychological causes.

Human performance factors (Stoklosa, 1983).	Behaviour Medical	Operational Task	Equipment design Environmental
Information transfer problems (Billings & Cheaney, 1981)	Instructions Errors involved in briefing or relief controllers Human errors associated with co-ordination failures		
Human failure (aviation) (Caeser, 1987)	Active failure (aware) Passive failure (unaware)	Proficiency of failures Crew incapacitation	
Accident investigation checklist development (Feggetter, 1982)	Cognitive system Social system		
	Situational system	Physical Environmental stress Ergonomic aspects	
Managing human performance (SAE, 1987)	Behavioural aspects of sensing and mental processing Error evoked by sensing and mental process problems Verbal and written communications Defects in training contents and methods Work place environment		

On the whole, specific classification schemes can achieve high efficiency and specificity in the identification of possible causes. It is no coincidence that these classification schemes are from the domain of aviation rather than nuclear power production. In the former the focus is on understanding fully the causes of reported incidents and accidents. In the latter the focus is on determining the probabilities for scenarios of possible accident. The main limitation with the classification schemes is the relatively loose association with established psychological theory; the classification scheme in Table 4 is probably the most explicit in that respect with its emphasis on decision making. This limitation means that there may be problems in using the classification scheme for performance prediction, and also that the relation between analysis and design may be weak. The lack of a general conceptual basis makes it difficult to recommend specific countermeasures based on the outcome of the analysis whether to the interface, the task definition, the training, etc. The lack of generality in the concepts used by the classification schemes also means that transfer of results from one analysis to another may be difficult.

Table 5: Causes of pilot related aircraft accidents (Kowalsky et al., 1974)

Critical condition categories	Experience Crew co-ordination Air Traffic Control Work/rest (fatigue) Airport	Distraction Neglect "Decisions" Machine (plane) Weather
Critical decision categories	Decision resulting from out of tolerance conditions Decisions based on erroneous sensory input Decisions delayed Decision process biased by necessity to meet schedule Incorrect weighting of sensory inputs or responses to a contingency Incorrect choice of two alternatives based on available information Correct decisions Overloaded or rushed situation for making decisions Desperation or self-preservation decisions	

2.2 Descriptions Of General Psychological Causes

The other proposal contains classification schemes that are of a general nature. This means that they use terms and concepts that are associated with a commonly accepted psychological theory, although the theory itself may not always be explicitly identified.

There are two main groups of classification schemes that have been generally accepted. One group puts emphasis on the **manifestation** of faulty actions, although it does not amount to a full description of error modes or phenotypes. Examples of that are shown in Table 6.

Table 6: General classification of manifestations.

Source	Proposed classification	
Slips and mistakes Reason (1985)	Skill-based slips Rule-based mistakes Knowledge-based mistakes	
Categorisation of action slips (Norman, 1981).	Action slips	
	Errors of omission	
	Errors of commission	
	Errors of substitution	Error in formation of intention Faulty activation of schemas Faulty triggering of active schemas

The other group puts emphasis on a description of the **prototypical information processing** that is assumed to be the substratum for human action. Here the ubiquitous model is the step-by-step description of decision making which has been popularised by Rasmussen (1986). An example of that is shown in Table 7.

These two groups are only mentioned briefly here, since they will be dealt with in more detail in the following sections. One common characteristic is that they are relatively independent of a specific application or domain. This is clearly demonstrated by the classification schemes shown in Table 7;

although they have been proposed for three different applications, they are basically variations of a single underlying principle, which in this case is the traditional sequential information processing model.

Table 7: General classification of causes.

Human error classification scheme		
General classification scheme (Rouse & Rouse, 1983)	Troubleshooting live aircraft power plants (Johnson & Rouse, 1982)	Supertanker engine control room (Van Eckhout & Rouse, 1981)
Observation of system state	Observation of system state	Observation of system state
Choice of hypothesis	Choice of hypothesis	Identification of fault
Testing of hypothesis		
Choice of goal		Choice of goal
Choice of procedure	Choice of procedure	Choice of procedure
Execution of procedure	Execution of procedure	Execution of procedure
	Consequence of previous error	Contributing factors

3. INFORMATION PROCESSING APPROACHES

As described in Chapter 2 the need for HRA came from the practical applications, and HRA was therefore seen as an engineering discipline that made use of mainstream psychological concepts. The first-generation HRA approaches were deeply rooted in the stimulus-(organism)-response (S-O-R) paradigm that dominated psychology at the time, particularly in the US. Machines and processes had since the early 1940s been described in terms of flows of information and control, and the introduction of information processing psychology seemed to make it possible to describe the operator in the same way. In the mid-1960s the information processing approach began to change mainstream psychology but it took many years before it was generally adopted by psychology and even longer before it became common knowledge in related disciplines such as HRA. Although the information processing metaphor served to enrich the understanding of the human mind, the basic dependence on a stimulus or event as the starting point for processing was retained. The information processing approach can thus be seen as a way of extending and enlarging the "O" in the S-O-R paradigm. Yet although most of the details were developed for the information processes that were assumed to take place between stimulus and response, the fundamental principle of a sequence with a clearly defined beginning and end was not abandoned. A few people such as Miller et al., (1960) and Neisser (1976) realised the limitation of the sequential information processing paradigm at an early stage, but the mainstream of academic psychology has failed to do so until fairly recently. The situation was probably less extreme in Europe where the established national traditions in many cases moderated the influence from mainstream US psychology. In relation to the classification and description of erroneous actions and the analysis of human reliability, the transition from behavioural to information processing approaches nevertheless caused something of a revolution.

3.1 Human Information Processing Models

Models of human behaviour aiming at providing explanations of erroneous actions in terms of the malfunction of psychological mechanisms have been influenced by two things. One is the adoption of the digital computer as the primary model for human cognition (e.g. Neisser, 1967; Newell & Simon, 1972; Reason, 1979; Reitman, 1965; Simon, 1979). The other is the belief that the methods of information theory can be meaningfully applied to the analysis of mental activities of all types (e.g. Anderson, 1980; Attneave, 1959; Legge, 1975; Lindsay & Norman, 1976; Thompson, 1968). In essence, an analysis of human performance from an information processing standpoint aims to trace the flow of information through a series of cognitive stages that are presumed to mediate between a stimulus and response. Resultant theories are conventionally expressed in terms of information flow diagrams analogous to those which are prepared when developing a specification for a computer program during software development.

Information processing investigations can be conducted using either quantitative or qualitative methods. Described in general terms, quantitative methods involve measuring the performance of a person under the controlled conditions of the psychological laboratory. Qualitative models, on the other hand, are usually developed on the basis of observations of human performance under real or simulated conditions, that is, behaviour in the "natural problem solving habitat" (Woods & Roth, 1988). The relative advantages of these two approaches are of central importance in the behavioural sciences and have been discussed at length elsewhere (e.g. Bruce, 1985; Neisser, 1982). In relation to the present discussion, however, the important point is that the two approaches have produced models of human performance and explanations of "error" that are quite distinct.

The quantitative line of research has tended to produce models of human performance that focus upon the limited capacity of the communication channels that, according to the models, connect the various information processes. Psychological explanations of erroneous actions have typically been given in terms of concepts borrowed from communication theory, such as information overload, together with their psychological correlates, such as cognitive strain. According to this view erroneous actions are typically due to capacity limitations of the hypothetical human information processor. In contrast to this, qualitative lines of research have tended to produce models that emphasise that the process of reacting to environmental signals can involve several stages or layers of information processing. The amount of processing a person will actually undertake is presumed to vary depending upon a person's familiarity with the task being performed. According to this view erroneous actions are typically characterised as "over adaptations" where the individual places too much reliance on his or her past experience.

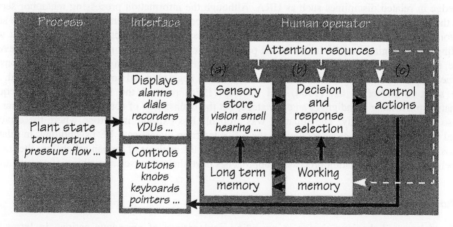

Figure 2: Wickens' (1984) models of human information processing applied to the human-machine interface.

3.1.1 Quantitative Models Of Erroneous Actions

A highly characteristic example of a quantitative analysis model has been developed by Wickens (1984; 1987). In this model, shown in Figure 2, information processing is conceptualised as passing through three basic stages: (a) a **perceptual** stage involving the detection of an input signal and a recognition of the stimulus material; (b) a **judgmental** stage in which a decision must be made on the basis of that information, relying where necessary on the availability of working memory; and (c) a **response stage** in which a response must be selected and executed. Each of these stages has optimal performance limits (as determined by limited attention resources) and when the limits are exceeded the processing in that stage may fail, leading to an overall "error". The determination of optimal performance limits and the various error forms that emerge when the limits are exceeded were estimated using data obtained from laboratory-based psychological experiments.

While frameworks such as Wickens' (1984) model have not been universally accepted they nevertheless represent of a large set of theoretical models that have been developed on the basis of "error" data elicited from experiments conducted in the psychological laboratory. A typical "error experiment" is exemplified by Parasuraman's (1979) attempt to explain vigilance decrement in terms of working memory defects, while Signal Detection Theory (SDT), response latency and the speed-accuracy trade-off paradigms have provided appropriate methodological tools.

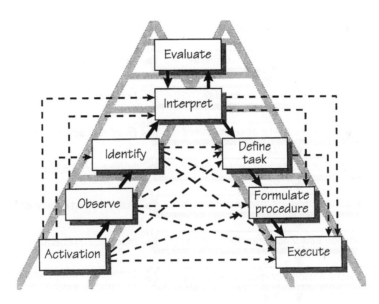

Figure 3: The step-ladder model for decision making.

3.1.2 Qualitative Models Of Erroneous Actions

The step-ladder model (SLM) of decision-making (Rasmussen & Jensen, 1974; Rasmussen, 1986) is probably the best known example of a qualitative information processing model, and has been widely used as the basis for error classification schemes. The step-ladder model proposes that there is a normal and expected sequence of information processing stages which people engage in when performing a problem solving or decision-making task, but that there also are many situations where people do not perform according to the ideal case. To exemplify this an eight-stage model of information processing was developed, as shown in Figure 3.

A central theme of the step-ladder model is the idea that short-cuts between cognitive stages represent an efficient form of information processing because they can reduce the amount of cognitive effort that needs to be invested in the performance of a task. Borrowing a phrase first used by Gagné (1965), Rasmussen talked of people "shunting" certain mental operations whereby varying amounts of information processing could be avoided depending on the operators' familiarity with the task. This is shown by the dotted arrows in Figure 3. For operators working in highly familiar circumstances several types of behavioural short-cuts have been identified (Rasmussen & Jensen, 1974). However, these strategies can also increase a person's vulnerability to making "errors" because they depend on both the appropriateness of past experience and the ability to match it correctly to the current situation.

It is not hard to see that the step-ladder model provides an attractive way of generating an explanation for erroneous actions. This can be accomplished by relating a failure of a specific information processing step with a set of manifestations. As the following sections demonstrate, this opportunity has not gone unnoticed.

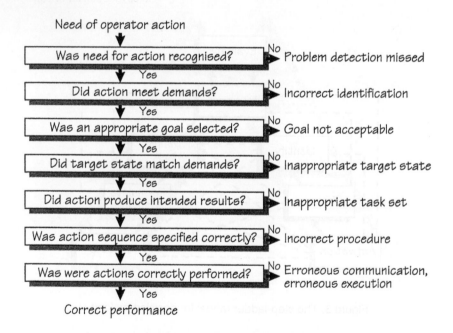

Figure 4: Pedersen's (1985) guide for error identification.

3.2 Pedersen's Classification Of Error In Accident Causation

Pedersen (1985) made an interesting attempt to develop the error classification component of the step-ladder model by using it as the main constituent of a taxonomy designed for incident investigation. The resultant classification of erroneous actions can best be illustrated as a question-answer list such as the one shown in Figure 4. This figure has been slightly modified from the original to be consistent with the representation of the step-ladder model shown in Figure 3.

In this classification scheme, the set of manifestations is exhaustive in the sense that there is one for each of the information processing steps. The labels for the manifestations also bear a strong resemblance to the contents of the associated decision making step. Pedersen, however, made no attempt of relating this set of manifestations to the pragmatic lists that had been developed by people in the field. The manifestations therefore reflected the structure of the model rather anything else. This proposal therefore did not lead to any significant change in practice.

3.3 Generic Error Modelling System

Reason (1990) used a variation of the step-ladder model as the technical basis for the Generic Error Modelling System (GEMS). The objective was to develop a context-free model of erroneous actions. Reason suggested that the emphasis on cognitive factors, as opposed to environmental or context related factors, would permit the error classification embodied within GEMS to be applied to the analysis of "error" in a variety of industrial situations (e.g. nuclear power, process-control and aviation, etc. In a later book Reason (1997) has developed a corresponding description of the organisational factors). In essence, GEMS extended a second important feature related to the step-ladder model, namely the assumption that it is possible to distinguish among three types of performance that correspond to separate levels of cognitive functioning. This is commonly known as the skill-based, rule-based, knowledge-based (SRK) framework. A representation of the SRK framework is shown in Figure 5.

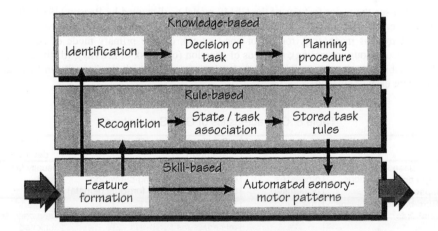

Figure 5: The SRK framework.

In the SRK framework, skill-based behaviour was assumed to be characteristic of highly familiar situations where the control of individual actions was delegated to stored patterns of "pre-programmed" motor sequences operating with little or no attention resources. The performance of routine tasks in familiar situations, on the other hand, was seen as rule-based. In this case the attainment of goals was presumed to require the development and maintenance of an action whereby conscious control of action is required at critical choice points plan (Reason, 1976; 1979). Finally, the knowledge-based level covered situations where the individual had to rely upon resource-limited forms of information processing, such as reasoning. In the context of HRA, Hannaman & Spurgin (1984) made two important points in relation to the quality of performance at the knowledge-based level. Firstly, that in general, knowledge-based reasoning is expected to be more prone to misjudgements and mistakes due to inherent limitations in the capacity of the human information processing system. Secondly, that the identification of problem solutions will take longer because of the need to rely on resource intensive cognitive activities such as deductive reasoning, rather than recognition.

The error component of the SRK framework has been discussed at length by Reason & Embrey (1985). The classical GEMS model used of the distinction between slips and mistakes superimposed onto the SRK framework. The resultant model gave rise to four basic error types conceptualised in terms of three information processing failures: skilled-based slips and lapses, rule-based mistakes and knowledge-based mistakes.

At the lowest level of the system were skill-based slips and lapses. Reason (1990) defined these as "errors" where the action deviates from current intention due either to incorrect execution (e.g. a slip) or memory storage failure (e.g. a lapse). In contrast to this, mistakes were defined as "deficiencies or failures in the judgmental and/or inferential processes involved in the selection of an objective". Like slips and lapses, mistakes were viewed as falling into two broad types. Rule-based mistakes occurred from the inappropriate application of diagnostic rules of the form: "IF *<condition>*, THEN *<action>*". Knowledge-based mistakes occurred whenever people had no ready knowledge to apply to the situation that they faced. Thus in contrast to failures at the skill-based and rule-based levels, knowledge-base failures were supposed to reflect inherent qualities in novice performance.

Table 8 summarises the major features of the GEMS model as discussed by Reason (1987). It shows that each error type can be distinguished according to five factors: (a) the type of activity being performed at the time the "error" is made, (e.g. routine or non-routine); (b) the primary mode of cognitive control (attention or unconscious processing); (c) focus of attention (on a task or activity); (d) the dominant error form (strong habit intrusions or variable); and, (e) the ease with which the "error" can be detected and corrected (either easy or difficult).

Table 8: Characteristics of GEMS error types

	Activity	Mode of control	Focus of attention	Error forms	Error detection
Skill-based slips and lapses	Routine actions	Mainly automatic processes (schemata)	On something other than the task at hand	Largely predictable "strong-but-wrong" error forms (schemata)	Usually fairly rapid
Rule-based mistakes	Problem-solving	(rules)	Directed at problem related issues	(rules)	Hard, and often only achieved
Knowledge-based mistakes		Resource-limited conscious processes		Variable	with help from others

While Table 8 shows the conditions under which a specific type of erroneous action is likely to occur, it does not relate that to a list of manifestations. In GEMS this is achieved by Table 9. Here the various types of slips identified by Reason are shown opposite the skill-based and rule-based columns, while categories of cognitive malfunction implicated in operator mistakes are shown in the column opposite knowledge-based mistakes.

In terms of the cause-manifestation criterion, the GEMS classification shown in Table 9 clearly puts the emphasis on causes. In fact, none of the categories describe manifestations. This is not very surprising, since the GEMS model was based on the step-ladder model, hence on a quite detailed description of hypothetical information processing mechanisms rather than on observable actions. Despite the considerable number of cases where these classification schemes have been tried in practice they have not been completely vindicated. Both the step-ladder model and the GEMS model remain the result of laboratory research, as opposed to field research. Some examples of that are presented in the following.

Table 9: Major error types proposed within GEMS (Reason, 1990)

Cognitive Control Mode	Error Type	
Skill-based	Recency of prior use Frequency of prior use	Environmental signals Shared "schema" properties Concurrent plans
Rule-based	Mind-set Knowledge availability	Matching bias Oversimplification Over-confidence
Knowledge-based	Selectivity errors Short term memory limitations Bounded rationality Thematic vagabonding	Encystment Reasoning by analogy Errors of deductive logic Incomplete mental model Inaccurate mental model

3.4 Rouse's Operator Error Classification Scheme

Another classification scheme based on the step-ladder model was proposed by Rouse & Rouse (1983; see also Senders & Moray, 1991). This scheme explicitly tried to combine the approach to error classification supported by the step-ladder model with the more traditional methods of error classification used by reliability analysts. The result was a classification scheme based on information processing, designed to be used to identify the probable causes of erroneous actions.

The cognitive modelling component of Rouses' scheme is shown in outline form in Figure 6. The basic model is similar in both form and content to the step-ladder model, although some of the information processing stages have been modified. Described in overview, the model proposed that during normal operations the operator of a human-machine system cycled through a sequence of activities that involved at least three stages: (a) observation of a system state, (b) the selection of a procedure, and (c) the subsequent execution of that procedure. In this interpretation it should be noted that the term "procedure" was used by Rouse in a generic sense to include forms of script-based reasoning where the operators followed a pre-established pattern of actions from memory. Conversely, Rouse proposed that when one or more state variables had values outside the normal range, the situation might be considered abnormal and under these conditions the operator would usually engage in problem solving, involving the formulation and testing of hypotheses. In this mode of operation the stages of information processing were presumed to be particularly vulnerable to failure due to the high task demands placed on human cognition by problem solving and decision-making activities.

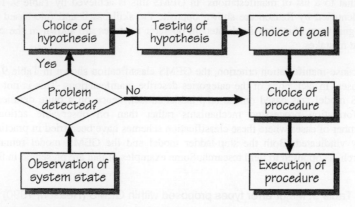

Figure 6: Rouses' conceptual model of the human operator.

Rouse used the model of operator behaviour shown in Figure 6, to guide the definition of possible failure modes in human performance. In an interesting departure from the work of both Pedersen (1985) and Reason (1990), Rouse introduced a two-fold classification of error causation that identified a general category (error causation relative to the operator model) as well as a specific category (relative to the behavioural episode). Table 10 reproduces the general and specific categories of error causation proposed by Rouse, that are related to the cognitive model.

Table 10: Rouse's proposed classification scheme

General Category	Specific Category	
Observation of system states	a. excessive b. misinterpreted c. incorrect	d. incomplete e. inappropriate f. lack
Choice of hypotheses	a. inconsistent with observations b. consistent, but unlikely	c. consistent, but costly d. functionally irrelevant
Testing of hypotheses	a. incomplete b. false acceptance of wrong hypothesis	c. false rejection of correct hypothesis d. lack
Choice of goal	a. incomplete b. incorrect	c. unnecessary d. lack
Choice of procedure	a. incomplete b. incorrect	c. unnecessary d. lack
Execution of procedure	a. Step omitted b. Step repeated c. Step added d. Steps out of sequence	e. inappropriate timing f. incorrect discrete position g. incorrect continuous range h. incomplete i. unrelated inappropriate action

3.5 HEAT - Human Error Action Taxonomy

Bagnara et al., (1989) attempted to strengthen the methodological component of taxonomies based upon an information processing standpoint. Their work applied the principles of a knowledge-based system (KBS) approach to make inferences about the nature of causes of erroneous actions in an industrial situation. As in the schemes considered above, the model of human malfunction utilised in the HEAT classification scheme was a variant of the step-ladder model to which had been added three further classes of error causation which Bagnara et al. labelled: (a) decision-making, (b) socio-organisational condition and (c) external situation. These additional classes correspond approximately to the range of factors identified as being implemented in error causation, as described by the CSNI classification system (Rasmussen et al., 1981).

A general flavour of the type of error classification that was provided by the HEAT approach is shown in Table 11. This table reproduces the types of "error" identified as arising from a consideration of the general category "problems with human performance".

Table 11: Example of the HEAT taxonomy

Human Performance Category of Failures			
Phenomeno-logical appearance	*Time*	Faulty activity Proper Activity	
	Task not performed due to	Task omission Act omission Inaccurate performance Inappropriate timing	Actions in wrong sequence Other Not applicable
	Erroneous act on system	Wrong act, right equipment Wrong equipment	Wrong time Other Not applicable
Cognitive Function	*Detection*		
	Identification of system state	Faulty or incomplete monitoring of system state Faulty or incomplete assessment of system state	
	Decision	Selection of goals Selection of system target state Selection of task	
	Action	Specifying the procedure Carrying out the action	
	Evaluation	Outcome inappropriate to goal Outcome inappropriately related to action	
Cognitive Control Mechanism		Knowledge-based Skill-based Rule-based	

HEAT clearly attempted to make a distinction between manifestations, called "phenomenological appearance", and causes, called "cognitive function", although the latter was mixed with the quite vaguely defined notion of a cognitive control mechanism. This distinction was maintained in the major sub-categories (shown with *italics* in Table 11), and also on the next level of detail although some of the terms were inconsistently applied. Thus, a specific form of "erroneous act on system" was called "wrong time", although it would have been more natural to include this in the sub-category of "time". HEAT nevertheless represented a classification scheme that emphasised the ways in which erroneous actions manifest themselves. The main shortcoming was an incomplete description of how causes and manifestations were related.

3.6 POET

As mentioned several times in the preceding, a distinction is traditionally made between slips and mistakes or lapses (Norman, 1981; Reason, 1979). The difference is that "slips result from automatic behavior, when subconscious actions that are intended to satisfy our goals get waylaid en route. Mistakes result from conscious deliberations." (Norman, 1988, p. 105). In his book on the Psychology Of Everyday Things (POET), Norman (1988) extended the classification as shown in Table 12.

This classification scheme is incomplete because it does not cover all possible manifestations. For instance, failed actions due to incorrect timing or wrong use of force are not included. (It is also quite uncommon to consider mode error as a slip.) Furthermore, mistakes are described as a single category without further specification - in contrast to e.g. the GEMS classification in Table 8. (A possible distinction is between the "gulf of evaluation" and "gulf of execution" although these refer to issues of comprehension rather than mistakes *per se*.) Table 12 is also a mixture of phenotypes and genotypes, as are indeed the categories of slips and mistakes themselves (Hollnagel, 1993a). The POET classification was proposed in the context of system design, rather than in relation to human reliability analysis. It would, however, have been reasonable to expect a higher degree of overlap between the two.

Table 12: POET classification of slips and mistakes

Type	Sub-type	Description
Slips	Capture error	Habit/stereotype take-over
	Description error	Incomplete specification
	Data driven error	Dominant stimulus driven
	Associative activation error	Freudian slip
	Loss-of-action error	Forget intention
	Mode error	Mistaken system state
Mistakes		

3.7 NUPEC Classification System

The last example in this section is a classification scheme developed with the explicit purpose of analysing human erroneous actions in the field of nuclear power plant operation and maintenance (Furuta, 1995). Although the classification scheme nominally is based on the step-ladder model, it goes far beyond that and is closer to the socio-technical approaches described later. As seen from Table 13, the NUPEC scheme has five categories of main causes, none of which refer explicitly to information processing functions. Instead the five main causes encompass the essential features of the total work environment, i.e., man, technology, and the organisation - often referred to as the MTO space. The NUPEC scheme represents a very pragmatic approach to classification of erroneous actions; it is not strongly linked to a specific psychological model, but rather reflects the growing concern for understanding all the factors that may lead to an erroneous action. Unfortunately it also resembles many of the other classification schemes by not maintaining a strong separation between manifestations and causes. On close inspection, Table 13 actually contains only causes or performance influencing factors. The NUPEC classification system is nevertheless unique in trying to provide a global characterisation of the context of the work, and is significantly different from e.g. the CSNI classification scheme (Rasmussen et al., 1981).

The NUPEC scheme has been used to analyse a large number of incidents from Japanese nuclear power plants, covering the period of 1966 - 1989. The outcome of these studies has been used to develop a set of guidelines for identifying causes of human erroneous actions.

3.8 Summary

Models of human information processing have been used on numerous occasions to account for the regularities in human performance - whether correct or erroneous. Some information processing models have been based on empirical data from experiments carried out in the psychological laboratory, while others have been based on observations of naturally occurring activities in a real-world processing environment. Irrespective of these methodological differences it is still possible to recognise three common assumptions that information processing models use to answer questions about the nature of causal characteristics in human cognition that (e.g. Kruglanski & Azjen, 1983).

First, information processing models of human behaviour assume that there are **reliable criteria of validity** against which it is possible to measure a deviant response (cf. the discussion of the criterion problem in Chapter 2). On some occasions the performance standards employed are externalised, as in cases where actual performance is measured against the prescriptions of normative models of rational decision-making or system performance. On other occasions the standard of acceptable performance is determined subjectively in accordance with the person's intentions at the time the failure occurred.

Table 13: The NUPEC classification scheme (Furuta, 1995).

Main causes		Detailed causes
Individual characteristics	Psychological stress	Excessive demands / Fear / Boredom
	Physiological stress	Temperature / Humidity / Fatigue
	Subjective factors	Habits / Subjective judgement / Personal matters & conditions
	Work performance incapability	Insufficient knowledge / Insufficient expertise / Insufficient skill training / Insufficient experience
	Physical configuration	Anthropometric mismatch
Task characteristics	Task difficulties	Difficulties of judgement / Difficulties of prediction
	Workload inadequacies	Time constraints / Excessive task load / Resource inadequacies
	Irregular working time	Shift work / Work schedule
	Parallel work	Unexpected tasks / Simultaneous tasks
Work (site) environment	MMI inadequacies	Configuration of control panel & indicators / Inadequate equipment
	Work place inadequacies	Elevated work space / Narrow work space / Dangerous places
	Work condition inadequacies	Humidity / High radiation / Noise / Temperature / Inadequate lighting
	Special equipment	Safety gear & equipment / Protective clothing
Work (team) environment	Team organisation inadequacies	Inadequate team formation / Unclear job description / Inadequate communication / Inadequate organisation
	Inadequacies in instruction or supervision	Wrong / negligent instruction of supervisors / Wrong directives / Inconsistent instructions / Inadequate supervisory chain of command
	work Inappropriate team	Lack of awareness of potential for human error / Violation of team norms / Lack of team cohesiveness
Management characteristics	Inadequate education and training	Deficiencies related to general knowledge / Deficiencies related to specific knowledge & skills
	Inadequate work planning	Inadequate managerial rule / Inadequate procedure / work drawing / Inadequate initial planning / Inappropriate change of planning
	Lack of incentive	Inadequate evaluation / appreciation of efforts

Second, all information processing models assume - to a greater or lesser extent - that psychological factors affect information processing and act to bias responses away from the standards considered appropriate. As previously suggested the most commonly cited factors implicated in erroneous actions are the various information processing limitations that are presumed to make human performance intrinsically sub-optimal. These limitations are brought into play whenever there is too much information available from the environment to be processed within a particular period of time. (Other factors play a role when there is too little information available, in a situation that has been characterised as "information underflow", cf. Reason, 1987.) Other psychological factors that are believed to degrade the quality of human information processing are "emotional charge" (e.g. Reason, 1990) and performance influencing factors (PIFs) such as fatigue and stress (Embrey, 1980; Swain & Guttman, 1983).

Finally, and some would argue most importantly, the human information processing system is assumed to comprise a diverse range of limitations that are invoked under particular information processing conditions. For an example of this see Figure 7 that shows the Error Model Based on Empirical Data (EMBED) proposed by Senders, Moray & Smiley (1985). Figure 7 illustrates the range of information processing limitations that are presumed to underpin "error" in information processing models. Thus, according to the model, failures of attention are likely to occur whenever there is too much attention-grabbing information present in the environment, decision-making is likely to fail whenever judgements of a certain type are to be made (e.g. the estimation of quantitative data), the information processing system is predisposed to "error" whenever a rapid diagnosis of the situation needs to be made, and so on.

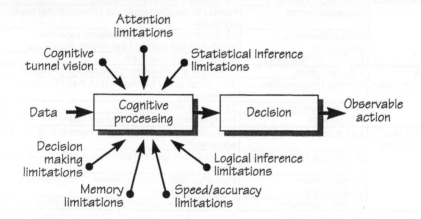

Figure 7: The range of information processing limitations in information processing models.

The first of the three assumptions is probably not tenable but neither is it necessary. Although performance standards can be useful as a comparison for actual performance, there is no need to give an absolute judgement of whether a single response or action is correct or incorrect. The attribution of "error" is a complex process of psychological and social judgement (Woods, et al., 1994), rather than a measurement of the quality of information processing.

The second assumption is true, but focuses on the negative aspects. There are many factors, psychological and otherwise, that have an impact on human action but they need not lead to a performance detriment. Which of these are important depend on the point of view. If information processing is seen as essential, then information processing limitations will obviously be of great concern. The important point is whether a distinction is made between external performance influencing factors and internal information processes. In the perspective adopted by this book, this distinction is in itself misleading because it propagates a simple mechanical metaphor. The focus should be rather on the interplay and dependency between cognition, context, and control, and means should be found to describe this adequately.

It follows from this line of reasoning that the third assumption is no longer necessary. The limited capacity of humans - in thought and in action - certainly plays a role. But the assumption of specific information processing limits is only relevant if an information processing mechanism is postulated. In the cognitive perspective, a more global approach is advocated.

4. THE COGNITIVE SYSTEMS ENGINEERING PERSPECTIVE

A third class of model which more recently has been applied to describe and explain human erroneous actions is based on the cognitive systems engineering (CSE) perspective (Hollnagel & Woods, 1983; Woods, 1986; Hollnagel, 1993a; Woods et al., 1994). Cognitive systems engineering harbours two important assumptions regarding the analysis of human performance in a work setting. First, it assumes that interactions between the human agent and automated control systems are best viewed in terms of a **joint cognitive system**. Second, the approach advocates a position in which the behaviour of the human operator (and therefore also possible erroneous actions) are seen as being shaped primarily by the socio-technical **context** in which behaviour occurs, rather than by the peculiarities of an internal information processing system. Each of these assumptions is discussed in more detail below, while a description of the approach based on cognitive systems engineering is given in the following chapters.

4.1 The Joint Cognitive Systems Paradigm

The assumption that complex man-machine combinations are best conceptualised in terms of a joint cognitive system is a theme which has been prominent in the work of David Woods and his colleagues (e.g. Woods, 1986; Woods & Roth, 1988; Woods, Roth & Pople, 1990; Woods et al., 1994). The idea was particularly influential in the development of the Cognitive Environment Simulator, or CES, and has more recently found favour in a line of work in an aeronautics domain aimed at making intelligent decision-aids more effective team players (e.g. Malin et al, 1991). In essence, the joint cognitive systems paradigm acknowledges the consequences of the proliferation of computing technology in control systems coupled with a rapid increase in the power of the computer. This has brought about a situation in which it is no longer reasonable to view the relative roles of the human operator and supporting control systems in terms of Fitts' list type tables that purport to show the relative merits of men and machines (e.g. Fitts, 1951; Hollnagel, 1995a). In the view of cognitive systems engineering it is more appropriate to consider the machine element a cognitive agent in its own right with the ability to make decisions about the current state of the process. Such systems can, for instance, respond to certain classes of process events without the need for active intervention by the operator.

Some of the ideas implicit in the notion of a joint cognitive system can be exemplified with reference to the supervisory control model of MMI proposed by Sheridan & Hennessy (1984; see Figure 8). In supervisory control the machine is thought of as a discrete autonomous cognitive agent with the ability to monitor and change process parameters. The human operator, however, remains in ultimate control of the system and is able to override the automatic control processes whenever a situation requires intervention. Thus, according to Boy (1987), in a system structured according to the supervisory control model the machine acts essentially as an "intelligent assistant" whose main role is to aid operators in the control of a physical process.

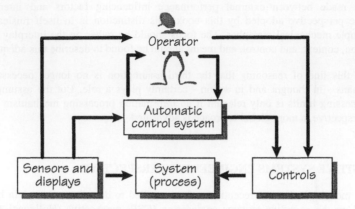

Figure 8: Example of a joint cognitive system - the "supervisory" control mode.

The notion of the joint cognitive system implies, firstly, that the process (or machine) and the operator should be modelled on equal terms and, secondly, that the coupling of the two models is necessary to appreciate and analyse the details of the interaction. In other words, modelling the human operator as a system in itself is not sufficient. This is the basic reason why classical information processing models are inadequate for analyses of human erroneous action. Although the context or environment is present as input (signals, messages, disturbances) this representation is not rich enough to capture the dynamics and complexity of the interaction. That can only be achieved by providing a coupled model of the human-machine system, and by making the models of either part equally rich. Several projects have, and are still, using this approach, e.g. Corker et al., (1986), Cacciabue et al., (1992), Hollnagel et al., (1992), Roth et al., (1992).

4.2 Contextual Determination

The second element of the cognitive systems approach is more directly relevant to the present discussion and concerns the assumption made within CSE regarding the nature of human cognition. In contrast to the information processing view, which assumes that all information processing activities are essentially reactive, the CSE perspective is based on the premise that cognition is an **active** process that is shaped by the operator's goals and the prevailing situation or context. When this latter interpretation is used as the background for describing the behaviour of the human operator, it is more appropriate to focus upon the global characteristics of human performance (both correct and incorrect responses according to specific environmental circumstances), than to confine the focus of attention to an analysis of malfunctions of presumed cognitive mechanisms (Hollnagel, 1993a; Woods et al., 1994). The implications of the cognitive systems engineering perspective have been used to guide the definition of contextual models of operator behaviour. An example of a model of this type is exemplified by the contextual control model (COCOM; Hollnagel, 1993a; 1997), which will be discussed in greater detail in a later chapter.

The need to replace sequential information processing models with contextual models has also been argued by Bainbridge (1993), who has made the point that sequential models assume that behaviour occurs as a simple perception-decision-action sequence. Since such a description leaves out essential features of human cognition, such as the use of knowledge / competence and the adaptability to changing circumstances it is neither useful as a basis for analysing performance nor for explaining erroneous actions.

4.3 Socio-Technical Approaches

Since the late 1980s a further approach to modelling of erroneous actions has been on the rise. In contrast to the approaches described so far, this alternative places the focus on the organisational and operational context in which the work takes place. It can therefore appropriately be called a socio-technical approach.

The main background for this approach was the recognition that organisational conditions were a common factor in a large number of accidents. As the view on erroneous actions gradually became more refined it was acknowledged that incidents evolved through a conjunction of several failures and factors, and that many of these were conditioned by the context. This evolved into the notion of people at the "sharp end" of a system, as described in Chapter 2.

The socio-technical approach has recently been presented in a comprehensive and competent way by Reason (1997). Rather than try to condense the numerous issues that characterise this approach, the reader is strongly recommended to read the above mentioned work. The present book puts the focus squarely on the actions of people at the sharp end, but from a cognitive systems engineering perspective rather than human action as information processing. There is therefore no inherent conflict with the socio-technical perspective, but only a difference in emphasis.

5. EVALUATION

This section will evaluate each of the three main approaches described in the preceding in relation to the following six criteria:

- Analytic capability, which refers to the ability of each approach to support a retrospective analysis of events involving human erroneous actions. The specific outcome of a retrospective analysis should be a description of the characteristics of **human cognition** that are included in the set of assumed causes.

- Predictive capability, which refers to the capability of each approach to predict the **probable type** of erroneous actions (phenotype) in specific situations. If possible, predictions should also be made of the likely magnitude or severity of the erroneous actions. None of the models are actually very good for making predictions, because predictions require both a valid model and a reliable data base. While better models are gradually emerging a reliable data base still awaits a concerted effort.

- Technical content, as the extent to which models generated from within each approach are grounded in a **clearly identifiable model** of human action.

- The relation to and/or dependence on existing classification schemes, as the extent to which each of the three approaches is linked to viable systems for **classifying** the **erroneous actions** that occur in a real-world processing environment.

- Practicality of each approach, which refers to the ease with which each approach can be turned into a **practical method** or made operational.

- Finally, the relative **costs and benefits** that are associated with each approach.

Summary results for the evaluation are shown in Table 14. The assignments are based on the subjective evaluation of a small group of subject matter experts. Some readers may obviously disagree with the assignments; it is important, however, that the reasons for the assignments are made clear. A more detailed discussion of the aspects referring to the technical content and the relation to existing classification schemes is deferred to chapters later in the book.

The lack of a clear distinction between error modes and causes is characteristic of most classification schemes. This may lead to practical problems in defining unequivocal event reporting schemes. It is clearly important for the quality of event recording that there is a separation between observation and analysis, even though both steps may be performed by the same person and in a single step. This separation, however, requires as an absolute minimum that the categories used for event description have a minimal but well-defined overlap with the categories used to describe the causes. If that is not the case, it becomes difficult to ensure that the chain of inferences that lead from manifestations to causes is distinct, and therefore also difficult to verify the conclusions. The imprecision that is a result of mixing error modes and causes may go some way towards explaining why event reporting schemes usually only have limited success - and why it is very difficult to combine different types of event reporting.

Table 14: Summary of the evaluation of the main approaches to error classification.

	Traditional HF & ergonomic models	Information processing models	Cognitive systems engineering models
Analytic capability	*Low:* Models are very sparse or even non-existing.	*Medium:* Models can be quite detailed, but have a very narrow scope.	*Medium-high:* Models are detailed and reflect the context-cognition interaction.
Predictive capability	*Medium:* Good, mostly because the categories are very simple.	*Low:* Not very good, because the models are biased towards internal processing.	*Medium:* Good, because the influence of the context is accounted for.
Technical basis	*Low:* Models are very sparse or even non-existing.	*Medium:* Models often have a narrow scope and reflect an experimental bias.	*High:* Models are detailed and address the context-cognition interaction.
Relation to existing taxonomies	*Medium-high:* The models are often identical to the taxonomies	*Medium-high:* The taxonomies are often derived from the models.	*Medium-high:* Models explicitly try to address dominant manifestations.
Practicality	*Medium:* Models and method have been developed in parallel.	*Medium:* Practicality may be limited by lack of explicitness in model details.	*High:* The models have been made with applications in mind, i.e., are often pragmatic.
Cost / effectiveness	*Medium-high:* The simple structure of the models is easy to operationalise, but also limits the efficiency.	*Low-medium:* It may take a considerable effort to transfer the models to real-life contexts.	*Medium-high:* Same as above

5.1 Traditional Human Factors and Ergonomic Approaches

Models of erroneous actions developed within a human factors and ergonomic tradition provide a useful framework for a reasonably effective classification of erroneous actions in terms of their external manifestations, and have also served as a basis for the development of many practical tools. This is especially the case where the objective of analysis is to **predict** the occurrence of **simple** erroneous actions that may occur in process control and man-machine interaction. An example of simple erroneous actions in human performance is the traditional distinction between omissions and commissions. Although more complex models of erroneous actions have been specified at the level of observable behaviour, the basic limitation of the approach is that it provides little information regarding the possible psychological causes. Thus, the analytic capability of traditional human factors approaches is typically quite low and the models contain few general principles that can be used to construct a description of failures in terms of underlying cognitive functions.

5.2 Information Processing Models

Information processing models tend to score quite high on criteria relating to technical basis and analytic capability, but score low in relation to the issue of how easily they can be converted to useful and practical tools for performance prediction. The analytic capability of information processing models derives mainly from the large number of general statements relating to "error tendencies" in human cognition that are typically incorporated into such models. The validity of the results therefore depends on the validity of these statements. As an example, it is commonly accepted that short-term memory only is capable of processing a strictly limited number of chunks of information and that information processing demands in excess of this amount therefore will cause the short term memory system to fail. Although information processing models do permit general predictions to be made regarding the likelihood that certain types of erroneous actions will occur, the extent to which such predictions transfer to a real-world processing environment is unclear. A statement that is correct for one context, say restricted laboratory conditions, need not be correct for another, say a familiar work process. Analysis of accidents and near-miss reports, for example, frequently describe situations where information processing failures occurred in situations that were well within the performance capabilities of the human operator. Conversely, there are many other well-documented instances where operators have succeeded to control a situation where information processing models predicted failure.

5.3 Cognitive Systems Engineering

The evaluation of models of erroneous actions generated from the standpoint of cognitive systems engineering suggested that such models tend to perform quite well across the range of evaluation criteria. In none of the cases considered do CSE-based models appear to be weak, although they still fall short of the ideal with regard to their current ability to make predictions concerning likely error modes. It should be noted, however, that CSE-based models are not bettered in this respect by either ergonomic or existing information processing models. In addition to this, the results of the analysis suggested that these models are particularly strong in terms of their technical content because they tend to be based on viable and well-articulated models of human action. In this respect CSE-based models appear to offer the opportunity for the development of practical tools for reliability and error analysis and the subsequent reduction of erroneous actions.

6. THE SCHISM BETWEEN HRA AND PSYCHOLOGY

HRA is concerned with an important aspect of human performance, namely iability. It would therefore be reasonable to assume that HRA had been based on and maintained a close relation to those aspects of psychology that deal with human action at work and with the nature of human action in general. Among those disciplines are the so-called "error psychology", cognitive psychology, cognitive science, cognitive systems engineering, human factors engineering in general, industrial psychology, etc. Although the concern for HRA arose in the US at a time when cognitive psychology as we presently know it was not yet fully developed, the progress made during the 1980s, and in particular the interest for all things cognitive, should presumably have led to a close interaction between HRA and psychology. Experience, however, shows that the interaction has been rather minimal, for a number of reasons that will be discussed below.

One main reason for this lack of interaction has already been hinted at. It is basically that whereas HRA practitioners are interested in **predicting** events, cognitive psychologists (using the term rather broadly to include also systems engineers and others who have delved into these matters) are mostly interested in **explaining** events - in particular in developing elegant models that can account for prototypical behaviour. Due to the different backgrounds and starting points, the two fields differ considerably in their concepts and methods. The essence of this difference is shown in Figure 9:

The development of HRA was strongly influenced by the engineering foundation of PSA and the main concepts and methods were initially taken over from PSA and the engineering disciplines. In particular, the representation of events was cast as the standard event tree - or sometimes an operator action tree. The methods were essentially quantitative, with the general aim being how to provide probability estimates for specific events. The focus was on the probability of the end events or final consequences, but in order to provide that it was necessary to know also the probabilities of the individual events in the sequence. Finally, the underlying model of the system was probabilistic or stochastic in PSA, due to the practical obstacles in making a deterministic analysis. In a similar vein, the models used by HRA were generally probabilistic, because they had to fit into the PSA framework.

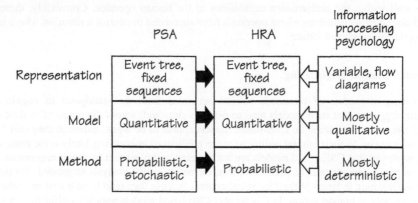

Figure 9: Differences between PSA, HRA, and information processing psychology.

If we look at the psychological basis for HRA - i.e., where it should or could have found its inspiration - the picture is quite different. Firstly, the representation of events is usually in the form of flow charts or flow diagrams as a short-hand for verbal descriptions, and the form is quite variable. Another important difference is that a flow chart need not be confined to a binary branching structure, nor need it be strictly hierarchical (in fact, it can be a network). Secondly, the methods of cognitive psychology (and of psychology in general) are mostly qualitative, i.e., they focus on descriptions and explanations rather than on measurements. Finally, the models are usually deterministic. A good example of that is the step-ladder model and the other information processing models described above. (This actually creates some serious problems because it is not straightforward to use deterministic models for probabilistic purposes.)

Altogether this means that first-generation HRA had some problems in relating to mainstream psychology - whether it was information processing psychology or cognitive psychology. Even if HRA had developed methods that used flow charts rather than event trees, that used deterministic rather than probabilistic models, and that would produce qualitative rather than quantitative results, they would be of little use for PSA. The demands of the PSA community are so strong that genuine alternatives to the established approaches have rarely been seriously attempted. Instead, some of the general features of information processing psychology such as the notion of a "cognitive error" and the ubiquitous skill-based, rule-based, knowledge-based framework have been adapted to fit with HRA methods, although often in a rather shallow manner.

There thus exists a real schism between HRA and psychology. As suggested above, this may be mainly due to the fact that HRA is interested in **predictions** while psychology on the whole is interested in **explanations**, cf. Figure 10. HRA starts from the event tree descriptions with the aim to find the probabilities corresponding to specific human "errors". Information processing psychology starts from a qualitative description of an event, and attempts to develop a cogent explanation, using a specific psychological theory or model. As long as this schism exists the theoretical foundation for HRA must remain incomplete. There does not seem to be any easy ways out of this, at least not in terms of making cosmetic modifications to either HRA or cognitive psychology. It therefore behooves the concerned practitioners and psychologists to identify the requirements to a HRA method, without taken too much for granted. In particular, the concern should be for the **purpose** of PSA rather than the **current practice** of PSA. Similarly, information processing psychology is at present in a state of growing turmoil, and it is quite likely that the dominating paradigm - of the mind as an information processing system - may be obsolete before too long.

Since this schism is important, the differences between explanation and prediction need to be considered in additional detail.

6.1 Performance Analysis - Explaining The Past

The analysis of erroneous actions and the search for plausible psychological mechanisms are by themselves interesting topics, since they clearly reveal the influence from occidental notions of determinism and causality. As described above, a main assumption is that there always is a clear cause-effect relation. This assumption is reinforced by other fields such as physics, engineering, and medicine as well as by the general literature and cultural stereotypes (as e.g. in psychoanalysis). Performance analysis thus amounts to finding the cause or causes of an observed event, in particular how a specific sequence of information processing could bring about the observed result.

The linking of a cause with an event is, however, not simply an exercise in logical deduction but a rather complex social and psychological judgement process, not least when human actions are involved (Woods et al., 1994). There is, in practice, no single root cause - nor a single pathway that leads to the consequence, if that pathway is to include the important facets of cognition and system interaction. In particular, a "human error" is invariably the attribution of a cause after the fact. A good event record or accident analysis may identify a number of events that clearly did happen, for instance from a flight recording, an incident report or another kind of registration. Many studies have nevertheless shown that highly similar conditions do not necessarily lead to the same results. The question is therefore why the differences in outcomes exist, and how they can be accounted for. The belief in the principle of causality, supported by deterministic psychological models of the mind, may tempt us to look for causal chains that weave together internal and external processes. But we should realise the illusory nature of such explanations.

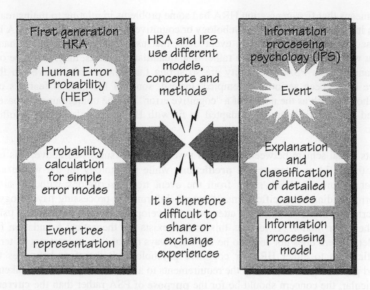

Figure 10: The schism between HRA and psychology

In the case of information processing models, the focus is on internal "mechanisms" and the methods aim at explaining the flow of control or of information through the models. Psychology has excelled in developing complex and highly plausible explanation for all kinds of phenomena - not only in cognitive psychology but also in psychoanalysis, developmental psychology, etc. The weakness of all such explanations is that they are *post hoc*, and that they as a rule have little predictive power. (From the point of view of first-generation HRA a further weakness is that they show little concern for quantification.) The descriptions of the stages or steps of the information processing models and of the links between them are usually given in rather general terms. This lack of precision makes it possible in practically every case to provide a plausible explanation by making a number of assumptions. Yet these assumptions limit the value of the explanations. Consider, for instance, the cases where actions fail although they should be well within the user's performance capability, such as many of the repetitions and omissions described in Reason & Mycielska (1982). Such action failures can be explained by referring to specialised "mechanisms" such as program counter failures, unresolved intentions, and anticipatory leaps. Although each explanation taken by itself is plausible - at least in a psychological sense - they are of limited general value because they depend on an unknown factor, i.e., the special "mechanism" or condition. In order to complete the explanation, we need to go one step further and demonstrate why this special "mechanism" played a role in the specific case, and so on.

As I have argued earlier in this chapter, information processing models are further weakened because the categories they use often mix manifestations and causes. Consider, for instance, the definitions of a slip and a mistake. A slip is defined as an action not in accord with the actor's intention, while a mistake is defined as an incorrect intention (Senders & Moray, 1991, p. 27). Yet for any given event all we have to go by as observers (rather than as actors) is the outcome or the manifestation. To describe that manifestation as a slip or a mistake implies a cause of which we - as observers - can know very little. The concepts of slips and mistakes are therefore of limited value as categories of observable manifestations. To describe the cause as a slip or a mistake may accurately characterise the person's state of mind when the action took place, but to decide whether or not the person had the wrong intention may not be practical as a starting point for an analysis. It would be more appropriate to determine whether the person had understood the situation correctly or misunderstood it. If the situation had been misunderstood, then further questions could be asked about why this was so.

In the case of models based on cognitive systems engineering, the focus is on the link between error modes (manifestations) and a set of likely causes rather than on possible internal "mechanisms". The causes do not refer exclusively to mental or cognitive functions, but include the domains of man, technology, and organisation taken together. Performance analysis is therefore seen as the complement of performance prediction, and both the classification schemes and the methods aim at supporting a bi-directional use. This will be described in detail in the following chapters.

6.2 Performance Prediction - Divining The Future

Explaining the past means constructing a sometimes rather intricate set of links that will lead from an observed manifestation to an acceptable set of causes. In this case knowledge of what actually did happen is a great help. Discovering what may happen in the future is more difficult since practically anything can happen - and very unlikely or even outright "impossible" events seem to occur with a depressing regularity. There is a very large - and possibly infinite - number of consequences that are the **possible** result of a set of causes, but there is a much smaller number of consequences that are **probable** - and an even smaller number that are highly probable. Performance prediction is therefore very often limited to finding the probabilities that a certain set of consequences will obtain.

One of the differences between HRA and psychology is the nature of the descriptions. In psychology the descriptions are generally qualitative, and the models are qualitative as well. In HRA the desired descriptions are quantitative, and models are needed to support that - even though one cannot really say that an event tree in itself is quantitative. (Qualitative means that something is expressed in terms of its characteristics or properties - using verbal or symbolic descriptions; quantitative means that something is expressed in terms of quantities or amounts - using numbers. Probabilities are, however, not quantities as such but rather arbitrary numerical descriptors that are interpreted according to a specific set of rules.) On the face of it, qualitative and quantitative predictions are different outcomes, and practitioners have often sought either one or the other. In the search for explanations, the general nature of the descriptions of the stages of the information processing models was an advantage because it provided a high degree of flexibility. In the prediction of consequences, the very same feature becomes a disadvantage because the negative side of flexibility is imprecision. Predictions are needed in order to prepare for future events, and the basis for the predictions must be precise.

For PSA-*cum*-HRA the aim is to calculate the probability that certain outcomes will occur. It follows that one must therefore begin by describing the various outcomes that can occur, i.e., the set of possible outcomes. As shown by Figure 11, the set of possible outcomes is determined by the initiating event and by how the system responds, typically described in terms of what the operators do, how the interface functions, and how the process behaves. All, of course, unfolds in the socio-technical context. (For the specific interests of HRA it is normal to focus on the primary influences, i.e., the operators and the technology, whereas the organisation is seen as a secondary and indirect but by no means unimportant influence.) On the basis of the qualitative description of the set of possible events it is then possible to go on to make a quantitative performance prediction. This is traditionally done by developing an event tree for a small set of sequences, and by assigning properly weighted event probabilities.

On reflection it becomes clear, however, that qualitative and quantitative predictions have a lot in common - and in fact that qualitative and quantitative descriptions simply are two aspects of the same thing (cf. below). Specifically, a quantitative description - such as a probability measure - must be based on a qualitative description. Quantification can only be done for something that has been clearly identified and described, and this description must necessarily be qualitative. The quantities must be quantities of **something**, and the something must be previously described.

The focus of first-generation HRA is on simple manifestations, i.e., the outcomes are described in very simple terms (omissions, commissions, extraneous actions). This apparently reduced the complexity of finding the associated probabilities, and in particular made it easier to define the necessary empirical data - although it has proven to be very difficult to collect such data. If there only are few behavioural categories and only few focal actions or events (typically detection, diagnosis, and execution), then the complexity of the prediction is considerably reduced. In this case causal models are not really required, since the event probabilities can be found by empirical studies alone. (Or more pessimistically, since we cannot develop valid models by which the probabilities can be generated, we have to rely on empirical data instead.)

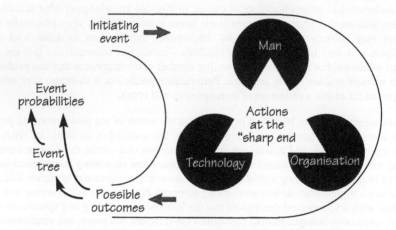

Figure 11: Qualitative and quantitative performance prediction.

The focus of information processing models is, as said, on detailed explanations of erroneous action types. The interest for predictions is very limited, and is mostly confined to general expressions about the types of erroneous actions that are possible (but with less concern for whether they are probable). Even in the case of such simple manifestations as slips and lapses, predictions are of limited value because slips and lapses occupy an uncertain middle ground between causes and manifestations. For the HRA practitioner the concern is not whether an operator will make a slip, but whether that slip will lead to a specific consequence. The most concrete use of information processing models has been as part of the Human Cognitive Reliability (HCR) method which was described in Chapter 1 (Hannaman et al., 1984). Here the principles of the SRK-framework were used to determine the required level of information processing in a task, and that in turn was used to select a specific TRC curve to provide a probability value. Yet even here the information processing model was used in a rather general way, to describe the required level of information processing independently of the context. It was used as an intermediate step for making a prediction (or rather, for selecting a probability value) and not as the main principle.

For cognitive systems engineering almost the same comments can be made as at the end of the previous section. The focus is on the link between error modes (manifestations) and causes with the important proviso that this link is bi-directional. This is an open recognition of the fact that methods for prediction cannot be developed independently of methods for analysis, nor vice-versa. Predictions must necessarily be based on the accumulated experience from past events, and such experience can be obtained only by analysis. It follows that in order for this principle to work the analysis and the prediction should use the same fundamental concepts and, specifically, the same principles of classification. This will be the topic for the following chapters.

7. CHAPTER SUMMARY

Chapter 3 has shown that a number of classification schemes exist which can be used to guide empirical investigations of erroneous actions. The relative abundance of classification schemes indicates the amount of effort that has been put into research on this important question in the last decade or so. The extent to which the classification schemes adequately represent the strength of the research is, however, much less clear. The fact that so many schemes have been produced by investigators could be interpreted as evidence of significant gaps in our understanding of how human cognition plays a role in the occurrence of erroneous actions.

A key factor in classification is the extent to which categories of erroneous action can be related to a viable model of human cognition. Traditional HRA methods are particularly weak in this area because their analysis of human performance is based entirely upon the observable behaviours that "cause" a system to deviate away from a desired state. Such observations are clearly important in developing an understanding of why systems fail. Yet confining the focus of the analysis to the consideration of external manifestations of "errors" provides little information about how and/or why a particular action occurred. For this kind of explanation it is necessary to refer to a model of the human operator that will allow the investigator to make inferences about the possible psychological (or indeed sociological) causes that underpin the occurrence of the observed erroneous action.

Classification schemes based upon the information processing metaphor provide a more rounded explanation of "error" events. This is because they refer to models that describe how information is generally perceived, how decisions are made, and how actions are executed. (Strictly speaking, however, cognition is treated as an epiphenomenon of information processing. In a proper model of cognition this should not be the case.) While no-one would take issue with the idea that actions must involve cognitive activity in each of these three broad areas, it can be questioned whether the information processing viewpoint provides the most appropriate metaphor for discussing human thought processes. Since the late 1980s there has been growing feeling that the analogy may have been over-utilised to the detriment of alternative developments. It is a major problem for information processing models that aim to specify error modes in human performance that explanations of action failures are couched in terms of breakdown in the natural course of information processing. Such models are therefore of limited value for dealing with erroneous actions that have their origins in the environment, such as those which arise due to false or misleading signals. Put differently, information processing models tell us little about the **contextual** causes of actions, nor do they allow us to interpret why a particular course of action was selected for use by the human operator.

In contrast, classifications developed from a cognitive systems engineering perspective are predicated on the assumption that erroneous actions require explanations that refer to a contextual model of human cognition, in which a model of competent human performance is intrinsically linked to factors present in the environment. The event is therefore interpreted relative to factors that define the working environment at the time when the erroneous action occurred. In this respect the cognitive systems classifications view erroneous actions as a type of **mismatch** that occurs between cognition and context. Thus, classifications based on cognitive systems engineering represent a clear break with both traditional and information processing approaches.

The classification schemes have generally been proposed within the context of analysing accidents or "errors" that have occurred and providing explanations for them. This is a cause of concern because the focus of HRA is prediction rather than explanation. For a number of reasons the efforts of "error" psychology and HRA seem to have gone in opposite directions, with relatively little interaction and exchange of results. There is thus a real schism between HRA and psychology where neither the models, nor the methods are compatible. This has the unfortunate consequence that it is difficult to use the experiences from the qualitative analysis of causes of events to improve the quantitative prediction of probabilities of consequences. The conceptual foundation of HRA has to a large extent remained incomplete because it has been impossible to "translate" the details of human of information processing models to match the needs of HRA methods. In order for any progress to be made it is necessary to recognise that this schism exists and to develop models and methods that try to overcome it by addressing the needs of "error psychology" and HRA at the same time.

Chapter 4
A Conceptual Framework

We can never avoid the dilemma that all our knowledge is of the past, while all our decisions are about the future.
Gro Harlem Brundtland (Norwegian Politician, 1939 -).

1. INTRODUCTION

One of the central points made so far has been the importance of making a distinction between manifestations and causes, both in event analysis and in performance prediction. The previous chapter concluded that none of the reviewed classification schemes made this distinction consistently and that some even failed to make the distinction at all. Traditional human factors and ergonomics use a reasonably effective classification of "errors" in terms of their external manifestations, and approaches belonging to this tradition have developed a number of practical tools to predict the probability of simple error types. The analytic capability of traditional human factors approaches is, however, quite low and they contain few general principles that can be used to provide a description of failures in terms of cognitive functions. In this respect information processing models fare better because they contain a large number of general statements relating to "error" tendencies in human cognition. The validity of these statements is, however, unproved and the explanations are therefore mostly of a hypothetical nature. The same goes for performance predictions that are made in terms of theoretically defined categories rather than in terms of empirically established performance types. Finally, models based on cognitive systems engineering tend to perform quite well across the range of criteria, although their current ability to make predictions concerning likely failure modes still needs development. Models from cognitive systems engineering are particularly strong in terms of their technical content because they are based on viable and well-articulated models of human action. They offer an opportunity for the development of practical tools for the analysis of erroneous actions as well as performance prediction.

Chapter 3 also argued that there is a considerable gap between, on the one hand, the psychological study of the causes of erroneous actions and, on the other, the practice of HRA and performance prediction. This is a serious issue because it means that HRA as a practical discipline has little chance to take advantage of the findings from the theoretical study of erroneous actions - and *vice versa*. Although the problem of the missing distinction between manifestations and causes may at first seem unrelated to problem of the gap between theory and practice, they nevertheless have a common solution. That solution is to develop a consistent conceptual framework for performance analysis and prediction, which firmly integrates psychological theory with the practice of HRA. As I will argue this inevitably requires that the distinction between manifestation and cause is strictly maintained which means that in philosophical terms they must be logically independent categories. Otherwise the prediction of effects from causes is analytic, hence without practical value.

2. THE NEED TO PREDICT

Every system, no matter how simple it is, can fail. In systems where human actions are needed to ensure that it functions properly, actions may go wrong and unwanted consequences may occur. Although it is generally realised that the variability of human performance has positive as well as negative consequences, there has been an understandable tendency to focus on cases where things can go wrong since these may lead to a loss of material, money, and even human life - and, in general, to the loss of control of the system. Unexpected positive outcomes are gratefully accepted but are seen as serendipitous and therefore rarely give cause to further deliberations - such as how one should reinforce them. Unexpected negative outcomes, on the other hand, are treated as situations that must be avoided and over the years significant efforts have been put into that. It seems as if the assumption about causality is stronger for negative than for positive outcomes.

The concern for the occurrence of unwanted outcomes is considerable in industries and processes where the cost is high and / or where public opinion is important as, for instance, aviation and nuclear power production. In these cases one can therefore find many attempts to deal with the problem of human actions as a causal factor, usually referred to as "human error" or human erroneous actions. There have typically been two main concerns. The first is to develop classification systems or taxonomies that will enable the identification of the specific causes of unwanted consequences, in particular in terms of human actions or performance conditions. This is usually accompanied by the development of methods for event reporting and data analysis. The second concern is to develop methods to predict the possible occurrence of human erroneous actions, typically as a quantitative human reliability assessment (HRA). Both concerns have received much attention during last 30 years, and are still the focus of extensive interest and controversy (Dougherty, 1990). A number of books have - each in their own way - dealt with this issue in recent years, e.g. Dougherty & Fragola (1988), Gertman & Blackman (1994), Hollnagel (1993a), Kirwan (1994), Park (1987), Reason (1990 & 1997) and Woods et al. (1994).

2.1 Initiating Events And Response Potential

Wherever complex technological systems are needed, it is important to ensure their adequate functioning before they are put into operation. This means in particular that steps must be taken to prevent the system from failing under normal conditions and contingencies that are within the design base. (Such contingencies are usually referred to as design-base accidents.) Furthermore, if a failure should occur then the system should be capable of containing the possible adverse outcomes and restoring normal operation as soon as possible.

Consider, for instance, a tube break in a power plant or a petro-chemical process installation. It is first of all essential to ensure that a tube break only occurs infrequently, i.e., with a given low probability - or preferably not at all. If the tube break nevertheless should occur, then it is essential that the system is designed so that the event can be reliably detected and the effects can be quickly contained. Most systems are equipped with various safety mechanisms and barriers that will provide a first line of defence, but this will usually have to be followed up by operator actions. Such actions will typically require a clear **indication** that the event has happened followed by **procedures** or plans to limit the consequences, which typically involve uncontrolled releases of energy and matter. The actual carrying out of the actions depend on a close co-operation between the operators, the control and instrumentation, and the technological safety features. There are thus two main issues that face the analyst:

- To **predict** and **evaluate** the situations where something could possibly go wrong. This is an analysis of the possible **initiating events**.

- To **assess** the ability of the system to recover appropriately when something goes wrong, regardless of whether the cause was a mechanical or a human failure. This amounts to an analysis of the **potential to respond.**

In the context of PSA/HRA the analysis of initiating events is generally confined to technological failures, i.e., the T of the MTO triad (cf. Figure 8). The developments that are possible following the initiating event do, however, depend on both human actions and the socio-technical environment. The set of initiating events that are due to failures in the technology is therefore a subset of the total set of initiating events. The analysis of the potential to respond should not only include all possible initiating events, but should also consider how **performance conditions** may affect the ability to respond. Thus, the same conditions may affect the initiating events and the developments after that.

(As an aside, it may seem strange that a PSA normally begins by the technological failures given that the majority of causes are acknowledged to involve human actions, cf. Chapter 1. This is probably due more to the tradition within the field than the facts.)

In order to perform either type of analysis, it is necessary that the analyst can predict correctly what may happen, i.e., that it is possible to account for the propagation of events through the system. In the case of a purely technical system where humans are not involved in the actual operation - in other words a completely automated or completely mechanical system - such propagation can, in principle, be reliably predicted. This is because the system is made up of components whose nature is well known (since they have been explicitly designed) and whose connections are also known. The only acknowledged limitation is the possibility of so-called sneak paths, which are design errors that under certain conditions may lead to unwanted consequences even though no single component has failed (Burati & Godoy, 1982). The concept of sneak paths has also been applied to operators (e.g. Hahn et al., 1991), although an appropriate operator model should be able to account for this in other ways. In practice, of course, even relatively small systems are usually too complex for a deterministic analysis to be feasible, cf. the discussion in Chapter 2.

2.2 Prediction For Interactive Systems

In cases where the functioning of the system depends on the actions of one or more operators, the situation is more complex. The fact that operators are involved does not by itself mean that the resulting performance must be worse than for a fully automated system. Quite to the contrary, the performance may in many situations be improved, for instance during off-normal conditions. This is one essential reason why operators are retained in the system. In many cases the analysis of the response potential considers or even relies upon the human capacity to provide solutions that have not been pre-defined because the complexity or uncertainty was too great. Yet in the present discussion we are not concerned with the net effect of human actions and whether they are an advantage or a disadvantage but only with the problems of analysing the interaction and accounting for it.

The higher complexity of man-machine systems is due to two things. Firstly, operators not only react, but also do things proactively. Human actions are driven by intentions as well as by events. It cannot therefore simply be assumed that operators will respond to an event but otherwise remain inactive, i.e., that they can be described as simple stimulus-driven mechanisms. Quite to the contrary, operators are able to look ahead and "react" to things that they expect will happen but which have not yet occurred. What people do, therefore reflect their intentions and expectations as well as the actual events. Their actions will not only be reactions and people will usually deviate from the strict regime of design-based performance that the analyst would prefer. Even when actions are supposed to be in a sequence, operators may deviate from it due to a number of internal and external reasons.

Secondly, the state or the configuration of the system may change. This may, of course, also happen in non-interactive systems, but here even these changes are predictable - at least in principle. In contrast, human performance in interactive systems may change the configuration of the system as well as the operators' understanding of the system. (I shall, however, not consider the latter, except indirectly.) This means that the functional composition of the system changes although the physical composition remains the same. The interrelations of tasks, events, and components are not fixed and unchanging. The analyst therefore cannot afford to assume that the system will be the same in different branches of an event tree; doing an action correctly or not doing an action may easily lead to different system configurations (Cacciabue et al., 1992; 1996). It is therefore necessary to be able to describe how changes in the system can affect the way in which an event can propagate. This is usually referred to as the influence of the context on actions.

3. METHOD, CLASSIFICATION, MODEL

The development of a system to support the analysis of accidents and events must as a minimum include a **method** and a **classification scheme**. The purpose of the method is to provide an account of how the analysis shall be performed, preferably by describing each step of the analysis as well as how they are organised. The purpose of the classification scheme is to provide a consistent basis for describing details of the events and for identifying the possible causes. A system to support prediction must, of course, include the same elements. In addition, it is also necessary that the classification scheme explicitly refers to a **model** of the phenomena being studied. In the following each of these will be considered in more detail.

The need to have these three elements is not specific to accident analysis and performance prediction, or even to studies of human action. It is easy to find examples across the whole spectrum of science as, for instance, in classical botany or particle physics. The latter can also serve as a good example of the dependency between the model and the classification scheme. The changes in the understanding of the nature of the fundamental particles have led to changes in the categories for classification of the observed phenomena, as well as to changes in the methods.

3.1 Method

A method is defined as a regular or systematic way of accomplishing something. In event analysis the method describes how the analysis of actions should be performed in order to find the possible and probable causes, in particular how the concepts and categories used for explanation should be applied. In performance prediction the method describes how the consequences of an initiating event may be derived and how the performance conditions may affect the propagation of consequences. In both cases the method should explicitly describe how each step is to be carried out, as well as define the principles to determine when the analysis or prediction has come to an end, i.e., the **stop rule** or criterion.

The method must obviously refer to the classification scheme, i.e., the set of categories that is used to describe and represent the essential aspects of the event. The method should be clear so that the classification scheme is applied consistently and uniformly, thereby limiting the opportunities for subjective interpretations and variations. Error classification and performance prediction should depend not on personal insights and skills such as inspired guesses or a sixth sense, but should rely on generalised public knowledge and common sense. The method is important as a way of documenting how the analysis / prediction has been done and of describing the knowledge that was used to achieve the results. It also helps to ensure that the analysis / prediction is done in a systematic fashion so that it, if needed, can be repeated with the same results. This reduces the variability between analysts, hence improves the reliability.

3.2 Classification Scheme

A classification scheme, as an ordered set of categories, is necessary both to define the data that should be recorded and to describe the details of an event. Analyses very often start from existing event descriptions, e.g. reports from the field. Such event reports are, however, of a varying quality because the initial descriptions depend on local practices, i.e., the guidelines or procedures that have been established for a field or an application. In cases where the job is heavily regulated, for instance in aviation or nuclear power, there are specific and well defined reporting systems in place. In other cases, where there is less public concern for safety, reporting may be of a mixed quality. It is therefore often necessary to bring event descriptions on a **common form** before an analysis is attempted. It is in particular necessary to ensure that the information provided is as complete as possible. This can best be achieved by referring to a systematic classification scheme.

A consistent classification scheme is also necessary in order to analyse the event and identify the possible or probable causes. I have argued at length for the importance of having a complete and consistent classification scheme, and in particular for maintaining a separation between manifestations and causes - the phenotypes and the genotypes (Hollnagel, 1993b). If the two are mixed, as when we use our intuitive understanding of human behaviour to describe an action in terms of its causes, then it is difficult to guard the consistency and reliability of the analysis. (Many of the classification schemes discussed in Chapter 3 provided examples of that.) In addition, it becomes impossible either to reverse or revise the analysis. The term "erroneous action" is itself an expression of this principle. An erroneous action is defined as "an action that fails to produce the expected result and which therefore leads to an unwanted consequence" (Hollnagel, 1993a). In contrast, the term "human error" can be used to mean either the action, the causes of it, or the outcome.

A consistent classification is finally necessary to predict how an event may develop. The classification scheme describes the relations between causes and manifestations (effects), and thereby forms the basis for predicting the outcome of specific changes in the causes. This might also be achieved differently, e.g. by means of a functional model or a set of differential equations - even as an operator model! However, in the case of performance prediction and HRA there is as yet insufficient knowledge to provide such models - although not for the lack of trying. Most computer simulations of cognitive models suffer from the limitation that they cannot adequately describe the interaction between the operator and the process or between cognition and context, which makes them incapable of providing the predictions required for an HRA. (The notable exception is CES by Woods et al., 1988; the status of model based HRA will be described in Chapter 5.)

Performance predictions and event analyses have often started by using a simple set of categories, the best known being "omission - commission" and "slip - mistake - lapse", only later to realise that the categories are in need of extension. If it is the case that the basis for the initial categories is not known or has not been explicitly described, then it may be very difficult to extend them. This is clearly shown by the many problems the HRA community has in accommodating the concept of a "cognitive error", which refuses to fit neatly into any of the present schemes. CREAM uses a particular classification scheme, i.e., a specific set of categories organised in subsets or groups. The underlying argument for the need to have a classification scheme will, however, remain valid even if another set of categories is used. Although the classification scheme proposed here has shown itself to be a useful one, no claims are made as to its general superiority.

3.3 Model

In addition to having a method and a classification scheme, it is also necessary to have a **model**. A classification scheme must, by definition, refer to an underlying model or description of the domain. The model provides the principles according to which the classification scheme is organised. This is so whether we are talking about event analysis in the domain of human performance, or analysis of other phenomena; biological taxonomies are a prime example of that. In the present case, the underlying model must refer to the principles that govern human action and performance at the sharp end (cf. Chapter 2). Although cognition is an essential part of human action, the model should not be of cognition *per se*, but rather of cognition and context.

The model is a convenient way of referring to the set of axioms, assumptions, beliefs, and facts about human cognition that form the basis of how we view the world and the events that happen in it, in particular about how human cognition depends on the current context. It is also a useful reminder that we are not dealing with an objective reality. This is particularly important in the field of behavioural sciences - to which performance analysis clearly belongs - because differences in the basic view may easily be lost in theoretical elaboration. Thus, if discussions only take place on the level of the classification scheme, disputes about the meaning of terms and the proper way of applying them (i.e., the method) may be difficult to resolve. If, however, a clear reference can be made to the underlying model, it will be easier to determine what causes the differences, and possibly also to resolve them - or at least to acknowledge them fully.

The importance of having a model as a basis for the classification scheme can be seen by considering what happens if it is missing. In this case the application of the classification scheme becomes a syntactic exercise, i.e., everything can be - and must be - combined with everything else. There is no way of knowing which particular combinations of categories - causes and consequences - are meaningful and which are meaningless. Even if the classification scheme is small, the number of combinations can be very large. It is therefore be necessary to filter or reduce the combinations. To prevent this from being done arbitrarily, the combinations must be selected according to whether or not they make sense. But this obviously can only be done with reference to a model. Furthermore, to make a virtue out of necessity, the model should be as simple as possible, i.e., a minimal model (Hollnagel, 1993c). Otherwise it may contain a large number of concepts and principles with an uncertain epistemological status that may have to be revised in the light of future developments. By using a minimal model, the risk of revisions becoming necessary is reduced - but, of course, never completely eliminated.

3.4 The MCM Framework

We thus end up with a system that has three essential elements, which are related as shown in Figure 1. I shall refer to this as the MCM framework - for **M**ethod, **C**lassification scheme, and **M**odel - in analogy with the Root Cause Analysis (RCA) framework described by Cojazzi & Pinola (1994). The first element is a viable **model of human cognition.** This is necessary to link the description of specific system failures to the principles of the model - relative to the context in which the behaviour occurs. The second element is the **classification scheme** itself. Definitions of the categories of erroneous action embodied within the scheme should follow naturally from a consideration of the workings of the model of cognition, hence be a subset of the set of general actions. This requirement is essential if the assignment of causes for observed behaviour is to be justified on psychological grounds. Finally, the system must also incorporate a **method** that describes the links between the model of cognition and the classification of causes. The utility of analysis systems that lack a clear method is strictly limited. The absence of a method is a potential source of inconsistency when the classification scheme is used by different investigators, or applied by the same investigator working on different occasions, since it forces the investigator to "invent" a method *ad hoc*.

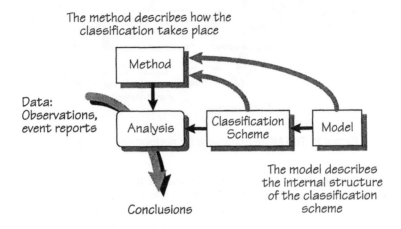

Figure 1: The relation between model, classification scheme, and method.

3.5 The Role Of Data

In addition to method, classification scheme, and model, a few words need be said about data. As indicated in Figure 1, data play a role as the input for the analysis. Such data do not magically appear by themselves, but must be specified and collected. The data therefore depend on the assumptions inherent in the MCM framework, as discussed in the following section. It is always important to remember that data, as factual information, are not just waiting to be discovered, but depend on what we look for. In addition data also have another important role. In cases where the purpose of the analysis is to make predictions of one kind or another - either quantitative or qualitative - data constitute the basis for the predictions. The analysis helps to describe the specific features of the situation or the action, but predictions can only be made if the specific features are seen in relation to the general characteristics, e.g. as frequencies, probabilities, levels, modes, etc. In this sense data constitute the basis for generating specific results from the analysis, as illustrated in Figure 2.

3.6 Data Analysis

As mentioned above, data types may vary from one source to another; the range may go from routine event reports and field studies to sessions from training simulators and protocols from controlled experiments. Similarly, the purposes of data collection and analysis may be quite different. The way in which raw data are analysed depends upon their type as well as the purpose of the activity. Observation and analysis of human performance data may conveniently be described by a series of steps, originally reported in Hollnagel et al. (1981). Each step transforms the data according to explicit rules:

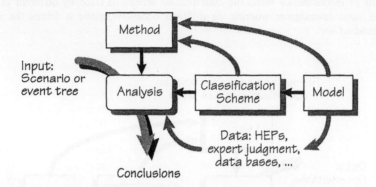

Figure 2: The MCM framework for performance prediction.

♦ **Raw data** - constitute the basis from which an analysis is made. Raw data can be regarded as **performance fragments**, in the sense that they do not provide a coherent or complete description of the performance, but rather serve as the necessary building blocks or fragments for such a description. Raw data can be defined as the elementary level of data for a given set of conditions. The level of raw data may thus vary from system to system and from situation to situation.

♦ **Intermediate data format** - represents the first stage of processing of the raw data. In this stage the raw data are combined to provide a coherent account of what actually occurred. It is thus a description of the **actual performance** using the terms and language from the raw data level rather than a refined, theoretically oriented language. The step from raw data to the intermediate data format is relatively simple since it basically involves a rearrangement and coding rather than an interpretation of the raw data. A typical example of that is a time-line description, which is used to combine various data sources.

♦ **Analysed event data** - where the intermediate data format is used as the basis for generating a description of the task or performance using formal terms and concepts. These concepts reflect the theoretical background of the analysis. The transformation changes the description of the actual performance to a **formal description** of the performance during the observed event.

The step from the intermediate data format to the analysed event data involves the use of the classification scheme, since the analysed event data are expressed in terms of the defined categories. The transformation is one from task terms to formal terms. The emphasis is also changed from providing a description of **what** happened to providing an explanation of **why** it happened, i.e., to finding the causes. This clearly involves an interpretation of the data.

♦ The **conceptual description** aims at presenting the common features from a number of events. By combining multiple formal descriptions of performances one may end up with a description of the **generic** or **prototypical performance**. The step from formal to prototypical performance is often quite elaborate and requires an analyst with considerable experience, in addition to various specialised translation aids. It also involves and depends on the use of the classification scheme.

- The **competence description** is the final stage of the data analysis, and combines the conceptual description with the theoretical background. The description of **competence** is largely synonymous with the model of cognition, i.e., it is the description of the behavioural repertoire of a person independent of a particular situation - though, of course, still restricted to a certain class of situations. The step from the conceptual description to the competence description may be quite elaborate and require that the analyst has considerable knowledge of the relevant theoretical areas as well as a considerable experience in using that knowledge. The analyst must be able to provide a description of the generic strategies, models, and performance criteria that lie behind the observed performance in terms that are task-independent.

The relations among the five steps described in the preceding can be shown as in Figure 3. The data analysis corresponding to the right side of Figure 3 exemplifies the bottom-up analysis of events to find the causes. The analysis moves from the level of the raw data to the level of a conceptual description that effectively is an explanation. The classification scheme is used throughout - and in particular in the translation from the intermediate data format, via the analysed event data, to the conceptual description. The model, i.e., the competence description, serves as the point of reference for the analysis.

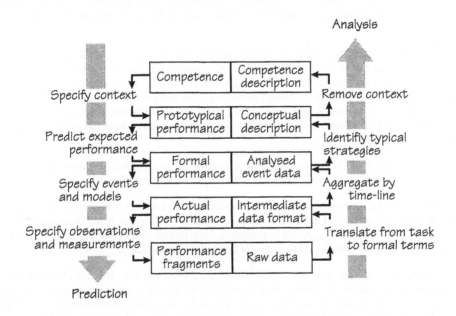

Figure 3: The dependence between data collection and data analysis.

In the same way the prediction corresponding to the left side of Figure 3 exemplifies the top-down process of performance prediction and reliability assessment. The prediction starts from the general description of performance and enhances that by providing details about the specific situation under investigation. The prediction usually stops on the level of formal performance, since the conditions normally are so variable that a description of the expected actual performance is of limited value. In any case, the concern is not so much for the expected actual performance as for the possible range or variation of performance.

4. MODELLING OF COGNITION

4.1 Modelling Traditions

The use of models of the human operator goes a long way back. Sheridan (1986) described a number of examples from the 1950s, and the field has been growing rapidly since then. It is well worth remembering, however, that the general acceptance in the 1960s and 1970s of the use of operator modelling was low, and that the notion of an operator model at that time was very different from what we tend to assume at present. It is a characteristic human tendency to believe that the way we look at things at the moment is not only the right way but also the way it always has been. Our present view is coloured by the substantial developments that took place during the 1980s in disciplines such as cognitive science and cognitive systems engineering, in AI and knowledge-based systems, and in human-computer interaction. In the 1990s it is quite common for conferences on engineering and process control to have special sessions on operator modelling, cognitive modelling, etc. - or even to arrange special conferences on the topic. Examples can easily be found in conferences held regularly by societies such as IFAC, IFIP, IEEE, etc. People with longer memories will nevertheless realise that such was not the case ten or fifteen years ago.

Practically all complex systems include and depend upon the interaction between operators and machines. Examples go from the exotic to the mundane: the space shuttle, nuclear power plants, petro-chemical production (and chemical industrial production in general), aeroplanes, trains, ships (passenger and cargo vessels alike), cars, computer aided telephony, automated teller machines, etc. In each case there is a real need to improve operator models, not only for the purposes of designing the systems and possibly increasing their level of automation, but also for the purpose of being able to assess the reliability of their functioning.

Complex interactive systems are characterised by a changed balance between the roles of operators and machines. Sheridan (1986) expressed this very precisely when he wrote that:

> "In one sense the success of man-machine modeling has contributed to putting the human operator out of business. As what pilots did in flying airplanes become better understood, that knowledge was immediately applied to autopilots. Now human pilots have less to do - at least with respect to those aspects of their behavior amenable to modeling."

The last sentence is important, because it recognises that there are limits to what can be modelled. Two conflicting conclusions can be drawn from that. Firstly, that we should improve the concepts and methods that are available for modelling, so that more and more can be automated. The underlying assumption is, of course, that a higher degree of automation will improve the safety and reliability of the system. This assumption is obviously debatable. Secondly, that we should carefully consider the consequences of automation, precisely because we can only automate some parts and must leave others untouched. The danger exists that automation is introduced for that which **can** be automated, rather than for that which **should** be automated.

With respect to safety and reliability analyses the attainment of complete automation would in one sense be ideal because it, paradoxically, would eliminate the need for operator modelling. If a system or sub-system can be completely automated, then the risk and reliability aspects can be treated entirely within conventional reliability methods as they have been developed for engineering systems. Even if the system includes human operators, the essence of complete automation is that the operators perform strictly within the boundaries defined by the engineering parts of the system, i.e., the operators' output must fall completely within the control systems' input and *vice versa* (cf. Hollnagel, 1995a). In these cases operators can rightly be treated as automata and as parts of the control system, hence be modelled and analysed using the same methods. Models with a psychological content are therefore not required. Since complete automation of complex systems is not achievable, neither in principle nor in practice, it follows that the need for operator modelling remains, and further that the modelling must have a psychological basis. (A corollary of this is that full automation is not desirable either.)

The issues of operator modelling in process control have been treated in several books (e.g. Rasmussen, 1986; Hoc et al., 1994), in handbooks (e.g. Helander, 1988) and in numerous papers and conference proceedings. Here I will try to look at operator modelling in the context of systems analysis - particularly in risk and reliability analysis - rather than at operator modelling in general. Operator models are obviously used for several different purposes, and the primary application appears to have been in design, notably the design of man-machine interaction and of human-computer interaction. (I consider the latter partly a subset of man-machine interaction and partly as separate from it. HCI has on the whole been vigorously pursued by people from the fields of computer science and psychology, with limited interest for industrial engineering applications.) The developments of operator models have, however, generally taken place outside the disciplines of safety and reliability analysis. This means that these disciplines in principle should have been able to use the developments from other areas and to learn from the experiences - and mistakes - that have made by these. As noted above there seems to have been little direct transfer in practice, although some of the terminology and some of the basic concepts have been eagerly adopted by the risk community.

4.2 Micro-And Macro Cognition

The sobering fact about theories and models of the **causes** for human erroneous actions is that there are hardly any. There are several models that describe specific, but hypothetical, information processing characteristics of the human mind but these usually lead to a set of categories that mix causes and manifestations and which therefore do not describe how human erroneous actions come about (cf. Chapter 3). There are also a number of theories and models that pertain to describe specific features of human cognition (the so-called mental mechanism) but they are generally concerned with narrow micro-phenomena. Micro-cognition is here used as a way of referring to the detailed theoretical accounts of how cognition takes place in the human mind. (This distinguishes it from the concern about cognition that is part of Artificial Intelligence. Here the focus is on the "mechanisms of intelligence" *per se*, rather than the way the human mind works.) Micro-cognition is concerned with the building of theories for specific phenomena, and with correlating details of the theories with available empirical and experimental evidence. Typical examples of micro-cognition are studies of human memory, of problem solving in confined environments (e.g. the Towers of Hanoi), of learning and forgetting in specific tasks, of language understanding, etc. Many of the problems that are investigated are "real", in the sense that they correspond to problems that one may find in real-life situations - at least by name. Yet when they are studied in terms of micro-cognition the emphasis is more on experimental control than on external validity, more on predictability within a narrow paradigm than on regularity across conditions, and more on models or theories that go in depth than in breadth. Micro-cognition relinquishes the coupling between the phenomenon and the real context to the advantage of the coupling with the underlying theory or model (Cacciabue & Hollnagel, 1995a).

It follows that theories and models of micro-cognition are of limited value for the analysis and understanding of human erroneous actions. The important point in the study of erroneous actions is to recognise them first and foremost as **actions**. As argued previously (see also Woods et al., 1994), there is little use for explanations of erroneous actions unless they can also be used to explain or describe non-erroneous actions - or in other words actions in general. As the following survey will show, specific theories or models of human "error mechanisms" are of limited value for practical applications, because they make the underlying assumption that actions that go wrong are fundamentally different from actions that succeed. This is a sharp contradiction of common sense; furthermore in a philosophical sense it implies a form of strict determinism.

As an alternative, macro-cognition refers to the study of the role of cognition in realistic tasks, i.e., in interacting with the environment. Macro-cognition only rarely looks at phenomena that take place exclusively within the human mind or without overt interaction. It is thus more concerned with human performance under actual working conditions than with controlled experiments, and is largely synonymous with the current trend towards natural decision making and the study of cognition in the wild - although the conceptual underpinnings of this trend remain to be clarified. Typical examples of macro-cognition are fault diagnosis, controlling an industrial process, landing an aircraft, writing a program, designing a house, planning a mission, etc. (e.g. Klein et al., 1993). Some phenomena may, in principle, belong to both categories. Examples are problem solving, decision making, communication, information retrieval, etc. When they are treated as macro-cognition the interest is, however, more on **how** they are performed and how well they serve to achieve their goals than on the details of **what** goes on in the mind while they are performed.

4.3 Cognitive Functions

The evidence for cognitive functions is based on a combination of introspection (privileged knowledge) and reasoning. We know from first hand experience that we observe, try to identify, plan, compare, decide, etc., because we know what these processes are and can apply the labels in a consistent and meaningful way to describe what goes on in our minds. We can also reason analytically that certain steps or pre-conditions are necessary to achieve certain goals. Introspection and reasoning obviously depend on each other, and must furthermore both refer to a common language that can be used to describe what we observe or reason about.

As an example, consider the typical description of decision making as a series of information processing steps, such as described by the step-ladder model (Chapter 3). It is usually assumed that this model is the result of induction, i.e., that it is the outcome of analysing a set of observations or the results from a set of experiments. It can, however, also be argued that the very same model description is the result of a deductive process (Hollnagel, 1984). Thus, in order to make a decision it is necessary to compare at least two alternatives and to choose whichever is subjectively (or objectively) the more attractive. In order to do that it is necessary to evaluate the alternatives, which in turn requires that the alternatives are identified. Each of these steps can be further analysed in a logical fashion according to a goals-means principle, where the corroborating evidence in all cases is our introspective and privileged knowledge.

The complexity of human performance and behaviour can be described in many different ways. Consider, for instance, a distinction between the following three levels of description (cf. Figure 4):

◆ **The level of (overt) behaviour** consists of the observable performance or actions of a person. The data includes movements, utterances, gestures, intentional actions, expressions, etc. In short, everything that we can observe and record. It is assumed that actions and behaviour are caused by cognition, but many of the things we do may also be caused by e.g. emotion. Although it may possibly suffice to describe how emotions affect cognition, and thereby remain on the *terra firma* of human information processing (Mandler, 1975), it should not be ruled out that a complementary theory about emotion and behaviour is necessary.

- **The level of cognition** is a description of the cognitive functions that are assumed to constitute the basis for our actions. The cognitive functions refer either to what we know from introspection (e.g. reasoning, problem solving) or to what we can infer with confidence from experiments (e.g. short-term memory, forgetting). The cognitive functions are therefore closely linked to the level of subjectively and objectively observable behaviour.

- **The level of brain functions** contains the set of hypothetical brain functions that correspond to established mental faculties or capacities, as cell assemblies rather than on the level of neurones. The main data are here from electrophysiological measurements and tomography, typically linked to specific tasks or modalities. This can be supplemented by descriptions on the level of molecular function that focus on properties of individual neurones, cell components, and even further down to the level of chromosomes and genes.

The three levels represent significantly different descriptions although they do not exhaust the ways in which a description can be given. It is important to recognise the existence of different levels of description because the complexity of human behaviour cannot be contained by a description on a single level. Thus a description of overt behaviour alone accounts only for what can be perceived, and does not provide any information about the underlying processes, cf. the dilemma of the Turing test (Turing, 1950). It must therefore be complemented by descriptions on other levels. Descriptions on any of the other levels must, however, refer to a specific range of observable behaviour in order to be informative. The description of the processes in a cell assembly may be relevant for a neuro-biologist, but is not sufficient for the study of the complexity of human behaviour. More to the point, a description on the level of information processing (as causes) is of limited value unless it can be set in relation to a description of overt performance (as manifestations).

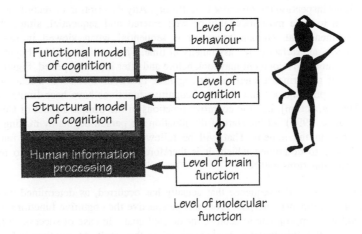

Figure 4: Three levels of description of human behaviour.

4.4 Structural Models

A **structural analysis of cognition** is based on the assumption that cognition can be explained by referring to a more elementary level, that is the level of the **human information processor**. This human information processor can, in turn, be described in terms of a set of well-defined structures and their functions corresponding to the strong assumption that the human mind is a physical symbol processing system (Newell & Simon, 1976; Newell, 1990). This is taken to be true whether the underlying metaphor is taken from the von Neumann machine or Parallel Distributed Processing (Norman, 1986). The structural approach starts from a set of hypotheses about what constitutes the basis of cognition, i.e., about what the fundamental elements or structures must be. While it is beyond doubt that cognition takes place in the brain and therefore in some sense is based on how the brain is structured and functions, the cognitive psychology that emerged in the 1960s found the distance between known descriptions of brain functions and descriptions of cognitive functions to be too large. Models and theories therefore made use of an intermediate level of human information processing. The information processing paradigm was used as a metaphor for a machine in the brain (sometimes called wetware), thus providing a possible link between the level of brain functions and the level of cognition (cf. Figure 4). It was assumed that human cognition **was** information processing, and therefore that models of cognition had to be information processing models. This assumption is, however, not necessary.

The notion of human information processing also carried with it two assumptions that unfortunately and inevitably limited the understanding of human actions. One was about the sequentiality of cognition; the other was about the existence of context free processes. While the information processing metaphor on the whole has been very useful and has led to many significant developments it clearly has its limits. Alternative ways of solving the problems should therefore be considered.

4.4.1 The Sequentiality Of Cognition

Analyses of human behaviour in general, and of cognition in particular, are necessarily based on descriptions of what **has** happened. Descriptions of human performance often appear simple and well ordered, but on closer inspection the orderliness is illusory. Any description or recording of events that is organised in relation to time must inevitably be well-ordered and sequential, simply because time is directed and one-dimensional. Human behaviour is sequential when viewed in retrospect, but the succession of events is an artefact of the asymmetry of time. This is illustrated by Figure 5 and Figure 6 that show how the "same" development may look before and after it has happened. Figure 5 shows all the possible sequences of performance elements or cognitive functions that could occur in the immediate future. In order to simplify the account the set of cognitive functions have been limited to [observation, identification, planning, action], i.e., the cognitive functions in the Simple Model of Cognition (SMoC; Hollnagel & Cacciabue, 1991). At the start of the situation the person begins by making an observation, e.g. noting a change in the system. That will be followed by another cognitive function, but it is impossible in advance to determine which. It is therefore not possible precisely to predict what the sequence of cognitive functions will be.

In contrast, Figure 6 shows the sequence that actually has occurred, as determined by e.g. concurrent verbal protocols or performance analysis. From this perspective the cognitive functions did take place in an orderly way and were appropriate to achieve the desired goal - in case of success, at least. Yet it was not possible to predict that this specific sequence would be the result. At any time before an action or a task is undertaken there may be a number of plausible or likely ways in which the sequence of events can develop. Afterwards, it is easy to see what happened, and to ascribe a rational process to explain the specific sequence of events. The point is, however, that it is a mistake to assume that there is an orderly underlying process or "mental mechanism" that produces the observed actions. In practice, people usually prepare themselves mentally for the most likely event sequences, whether they are planning to fly a mission over enemy territory (Amalberti & Deblon, 1992) or whether they are going for an important meeting with their boss. The actual sequence of actions is, however, the result of a complex coupling between the internal and external processes - between the person's control of the situation and the conditions that existed at the time (Hollnagel, 1996) - rather than of a built-in dependency among the cognitive functions.

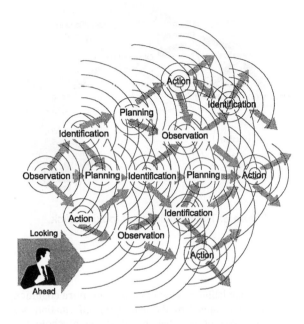

Figure 5: The indeterminacy of looking ahead.

Since the sequentiality of behaviour is an artefact of time, it is not necessary that models or explanations of cognition reproduce specific prototypical sequences. **Any** set of outputs that have been generated by a model will be sequential when viewed from the present, but the order of the observed events does not require a similar order in the functional principles of the model. In other words, models of cognition do not have to account for cognitive processes as a set of steps or stages that are executed one by one. In particular, the seemingly disorder or complexity of actions - jumps, disruptions, reversals, captures - does not need to be explained as **deviations** from an underlying sequence or from a well-ordered flow of cognition. This well-ordered flow is nevertheless an unavoidable consequence of human information processing models.

It is, indeed, noteworthy that the "disorder" of behaviour can be found even on the level of brain functions. For instance, recent research has shown that there are few stable patterns of activity within the olfactory system (Holmes, 1994). The same stimulus (a smell) may produce a response in one set of cells at one time, and a response in a different set of cells only ten minutes later. Thus it looks as if even neurones do not simply gather information from below and pass it up a well-defined pathway to the higher levels. Although the resulting function may be the same, there are no pre-defined pathways or sequences of elementary processes.

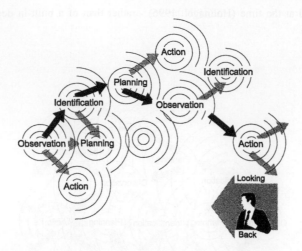

Figure 6: The orderliness of looking back.

4.4.2 Context Free Processes

Many structural approaches to the modelling of cognition focus on **pure** cognition, i.e., the unadulterated processes of mental activity that are implied by the underlying information processing structure. This trend has existed from the early days of information processing psychology, e.g. the EPAM model (Feigenbaum, 1961) and is particularly evident in the classical studies of cognition (e.g. Newell & Simon, 1972) and the choice of such experimental vehicles as cryptoarithmetic, mental puzzles, etc. The information processing metaphor favoured experimental paradigms and tasks that focused on human information processing.

This focus is understandable when seen in relation to the underlying premise. If cognition is information processing, then it makes sense to investigate the performance of that *in vitro*, so to speak, as a pure process. Information processing in the technological domain can successfully be described in a context free manner, e.g. as an adder or an AND-gate - or even as a Turing machine. But it does not make sense to speak of basic human information processes in the same manner. More specifically, **cognition is not an information processing epiphenomenon**. If experiments look for pure information processing and create conditions that favour specific manifestations, then it is not really surprising that the models are confirmed; staying within a well-defined and well-controlled domain ensures that the results usually comply with the assumptions. Yet whenever experiments and investigations are carried out on real-life tasks, it is inevitably found that the models are insufficient (e.g. Neisser, 1982).

It has become clear that cognition must be studied *in vivo*, i.e., as it occurs **within** a context. (Another term is that cognition must be studied in the wild, cf. Hutchins, 1995; Miller & Woods, 1996.) Even if we consider the solitary thinker or the experimental subject doing cryptoarithmetic, their cognition takes place in a context and is predicated on a set of assumptions about the world, the environment, the purpose, and the constraints (something that experimental psychologists traditionally have made much out of, e.g. Maier, 1930). Pure cognition only makes sense if we consider the level of the underlying structure,

---

e.g. processes on the level of brain cells. Yet in these cases the processes are not cognition as we normally use the term, since the gulf to the level of experience and consciousness has yet to be spanned.

The consequence of this for the modelling of cognition is that the basic approach must be **functional** rather than **structural**. Instead of starting by considering what the possible mechanisms of cognition may be, we should consider the **role** of cognition, i.e., what cognition does for the person. Cognition is assumed to be at the root of how people adapt to changing circumstances and cope with complexity. Cognition should therefore be described primarily in terms of what the actions accomplish, rather than in terms of the hypothetical mental "mechanisms" that may be involved. This functional approach has the added advantage of linking the study of cognition more closely to the practice of real-life situations.

4.5 A Simple Model of Cognition (SMoC)

The purpose of event analysis is to find and explain the path from the manifestations of erroneous actions (error modes) to the probable causes for why these erroneous actions occurred. In Western science at the end of the 20th Century any such explanation must include human cognition as an essential part, lest we regress to a behaviourist line of thinking. That is, we must be able to identify the essential cognitive functions that were involved in determining the action, as well as the external factors that had an influence on those functions. We must be able to understand why human cognition was "configured" or functioned in a particular way given the particular conditions. The question is whether this requires a detailed causal model of human cognition, as exemplified by the information processing models that were developed in the 1980s.

As argued above it is not necessary to postulate a pre-defined organisation of human and cognitive functions as the basis for a causal explanation. A pre-defined organisation - called a procedural prototype (Hollnagel, 1993d) - assumes that cognitive functions must occur in a specific order or in a limited set of orders, and in a sense tries to subordinate the empirical data under that organisation. Even worse, a procedural prototype leaves very little room for the influence of the context.

The logical alternative to a procedural prototype is to impose as little *a priori* structuring on cognition as possible. Instead, the actual ordering or organisation of cognitive functions in a situation, i.e., the **control**, is seen as determined by the context and the conditions. (The conditions are obviously as they are experienced by the person, i.e., subjective rather than objective. I shall, however, save a discussion of the philosophical implications of that to another time.) A model of this nature would include the accepted features of cognition such as limited capacity and the persistent or characteristic cognitive functions, but not as model primitives. Instead, a functional approach will try to explain the specific capacity limitations and the specific organisation of the cognitive functions (the sequence in which they actually occurred) as a function of the context, i.e., of the environment and the situation. In accounting for an event the theory would try to understand and explain how cognition was controlled under the given circumstances.

As an example, consider a very Simple Model of Cognition, called SMoC (Figure 7). The SMoC contains the essential elements of cognition and tries to organise them in a way that is generally applicable. Even the SMoC, however, imposes a sequence or organisation that may not always be correct. It also does very little for the aspect of limited capacity that is not part of the model in any sense. On the whole, the know restrictions of human cognition are not represented in SMoC, and to a considerable degree this is not problem at all. It turns out that in practice there is no need to model these restrictions, since they have little, if any effect on the reliability of performance. Instead, the conditions under which the performance takes place - the Common Performance Conditions - prevail.

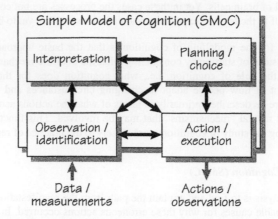

Figure 7: The Simple Model of Cognition (SMoC).

As I shall describe later, in Chapter 7, it is actually sufficient for an approach to HRA to assume that the cognitive functions are loosely structured parts of the underlying model without imposing an explicit organisation. Instead of referring to elementary cognitive functions a model of cognition can be based on a fundamental distinction between competence and control. The actual organisation of the cognitive functions is, in a sense, imposed by the analyst when the analysis is carried out. In the case of retrospective analyses the organisation is mainly determined by the context that is known or can be derived from the available empirical data. In the case of performance prediction, the organisation must be constructed by the analyst, e.g. in terms of the scenario and the likely (or expected) common performance conditions.

5. STANDARD CLASSIFICATION SCHEMES

It is by now commonly accepted that a theory or model for human erroneous actions must include three main sets of factors or influences, corresponding to the fact that all actions take place in a context and that this context can be described as a combination of **individual**, **technological** and **organisational** factors. In earlier chapters the context has also been referred to as the Man, Technology, Organisation (MTO) triad - in accordance with the Scandinavian tradition. (In the Scandinavian languages "M" stands for "Menneske" or "Människa", meaning human being, which grammatically is neuter.) These factors determine whether the action has any unwanted consequences, hence whether it is classified as an erroneous action.

♦   The first set of factors relates to the individual who carries out the action, and in particular to the characteristics of human cognition; they are here called **person related factors**.

♦   The second set of factors describes the technological characteristics of the system and in particular the various failure modes for the system, the sub-systems and the system components; they are here called **system related factors**.

♦   The third and final set of factors relates to the organisational context, e.g. established practices for communication and control, performance norms, and - in particular - the possibility of latent failures conditions and system resident pathogens (Reason, 1991); they are here called **organisation related factors**.

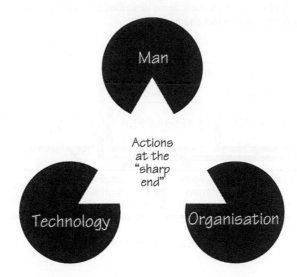

Figure 8: The relation between person-related, technology-related, and organisation-related factors.

The link among the three sets of factors can be shown in different ways, depending on how they are going to be applied. The relation among the three sets of factors is illustrated by Figure 8, which shows actions at the sharp end as depending on the reference backgrounds of human characteristics ("man"), system-related factors ("technology"), and the organisation. Figure 8 uses a well-known visual illusion to emphasise that if either of these reference backgrounds are missing, then the actions at the sharp end cannot be completely described. In cases where the purpose has been the analysis of human erroneous actions or the prediction of human performance, there has been a tendency to focus on person- and system-related factors and to give less weight to organisational factors. In cases where the purpose has been to understand the nature of organisational accidents, there has been a corresponding tendency to focus on organisational factors and pay less attention to actions at the sharp end and in particular to person-related factors. In both cases the description, and therefore the understanding, remains incomplete.

There is clearly not one specific view that is correct in an absolute sense. The choice of perspective depends on the purpose of the investigation. In each case human actions are a function of person related factors **and** system related factors **and** the organisational context. It is perfectly legitimate to put more emphasis on some of the aspects that determine performance, but it is a sign of misdirected zealousness to neglect any of them.

5.1 Factors Influencing Vulnerability To Error

Many empirical studies of human performance have treated human erroneous actions in terms of a simple cause-consequence model of erroneous behaviour (e.g. Otway & Misenta, 1980; Canning, 1976). However, several investigators have taken issue with such a view and proposed that, in addition to stimulus-response considerations, a number of factors are relevant to the classification of error events. Rasmussen et al. (1981), for example, has suggested a sevenfold classification of factors relevant to investigations of error, which is shown in summary form in Figure 9. The reader may want to compare this to the more recent NUPEC classification scheme described in Chapter 3, and also with the discussion of performance shaping factors later in this chapter.

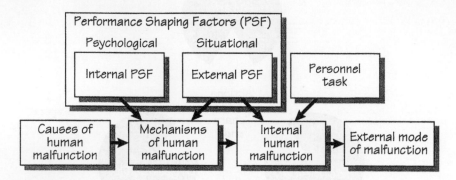

Figure 9: The CSNI classification scheme.

When Figure 9 is viewed from the bottom up, the boxes in the lower tier indicate that a sequence leading to an erroneous action typically begins with the **occurrence of an event** in the environment (called "cause of human malfunction") which activates a psychological "failure mechanism". This in turn invokes an "internal human malfunction" that may or may not manifest itself in the operating environment as an observable error ("external mode of malfunction"). In contrast to this, the boxes in the top row of Figure 9 show the number of factors that can contribute to increase the likelihood that an error mechanism will release a malfunction. According to the CSNI scheme these are primarily factors known to influence the adequacy of human performance (e.g., Swain, 1967; Embrey, 1980) and sociological considerations that define the interrelationships between the relevant actor(s) and the current situation. They are commonly known as Performance Influencing Factors (PIFs).

Another way to read Figure 9 is with regard to how the different factors are presumed to **interact,** thus affording alternative explanations of error events as shown in Figure 10. For example, the two boxes shown at the bottom right hand side of Figure 10 specify the factors that describe **what** error occurred - either in terms of external events (e.g. in an aviation context, the failure to set flaps) or in terms of the human activity which went wrong (e.g. pilot forgot to implement check-list item or failed to use an established procedure). The box labelled "personnel task" immediately above these describes **who** committed the error (e.g. control room operator, maintenance / flight-deck crew, dispatcher, etc.).

Conversely, the two boxes labelled "performance-shaping factors" essentially describe **why** a malfunction occurred. The "causes of human malfunction" category specifies **what** the events were that caused a process to deviate from the norm. Examples are task demands that exceed human performance capabilities, equipment malfunctions that produce misinformation and induce an erroneous belief among operators regarding actual system states, or simple external disturbances in the process. Also included here would be any off-normal event that acted to divert an individual's attention away from the task in hand.

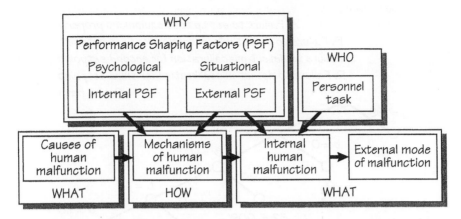

Figure 10: The CSNI classification scheme as describing interaction.

Internal and external PSFs overlap considerably with the causes' category in a conceptual sense. However, the main distinction being made here is with regard to the immediacy of effects. PSFs are not normally seen to be causal in a strict sense, but are generally thought of as factors that contribute to how erroneous actions manifest themselves. Thus, in the CSNI "model" the presence of any single factor should not by itself lead to a failure of information processing. If it did, it should rightly be treated as a proper cause. Rather, a combination of causal and contributory factors creates a situation that is conducive to certain kinds of human malfunction. For example, in a transport environment a high level of personal stress combined with bad weather in busy conditions might cause task demands to exceed the individual's capability to perform appropriately, and this could be a catalyst for the occurrence of an error of some type. Rouse (1983) has provided a further discussion of the distinction between causes and catalysts of error.

Finally, the box labelled "mechanisms of human malfunction" denotes **how** an error occurs with reference to the underlying mental process which in a specific instance acted as the main error mechanism. For example, if a person forgot a critical action this could be interpreted either in terms of an error of omission (an error category at the behavioural level) or, alternatively, in terms of a distraction brought about by competing memory demands (e.g. Baddeley, 1990).

The CSNI classification scheme provides a useful framework for investigating the causal pathways in the analysis of the erroneous actions taken by human operators. It is important to recognise, however, that the approach provides little guidance regarding assignment of operator errors to specific categories, nor to performance prediction. For these reasons the CSNI classification has not proved to be particularly useful for reliability analysis or accident investigation.

5.2 Classification In First-Generation HRA

Some of the early approaches to classification of erroneous actions came from the field of human reliability analysis, cf. Chapter 3. The objective of HRA is to characterise how actions of an operator may lead to system states that are in conflict with those which were expected or desirable. The primary tool used to describe performance deviations of this type was the PSA/HRA event tree or one of its derivatives. An example of an event tree is shown in Figure 11 (from Swain & Guttmann, 1983).

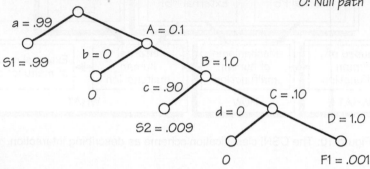

A: Failure to set up test equipment properly
B: Failure to detect miscalibration for first setpoint
C: Failure to detect miscalibration for second setpoint
D: Failure to detect miscalibration for third setpoint
O: Null path

a = .99   A = 0.1
S1 = .99   b = 0   B = 1.0
O   c = .90   C = .10
S2 = .009   d = 0   D = 1.0
O   F1 = .001

Figure 11: Example of an HRA event tree for a hypothetical calibration task.

There are three important points to make in relation to an event tree such as the one shown in Figure 11. Firstly, the event tree represents work activities solely in terms of a linear and immutable sequence of operations that unfolds in an orderly fashion. Thus, an event tree does not convey any information regarding possible covert activities, cognitive or otherwise, which may lead to erroneous actions or failed events. Secondly, the event tree does not allow the occurrence of extraneous actions, except as the generic failure to perform an action; it is therefore a reduced and basically closed representation. Thirdly, in an event tree each step of the linear sequence is treated as a binary choice node in which the only possible outcomes are success or failure. This means that an event tree cannot represent inefficient behaviour, i.e., behaviour that is neither incorrect nor completely adequate. This is an obvious limitation for a classification scheme developed for the purpose of assessing human reliability.

This last observation leads naturally to a fourth point that relates to the kinds of error classification scheme found in association with probabilistic risk assessments. There is a tendency for HRA analysts too readily to adopt the principles of the binary representation and thereby confine their attention to classification schemes such as the one shown in Figure 12. The binary representation is consistent with the background of the descriptions of the traditional human factors approach given in Chapter 3.

The most obvious shortcomings of classification schemes devised for the purpose of risk assessment is that they cannot be defended on psychological grounds because they do not refer to a viable model of human cognition. The extent to which the widespread distinction between errors of commission and omission represents a valid classification of operator actions is also questionable. In a relatively simplistic task, for example, making a distinction between what the operator **does** and **does not** do, can often serve as a useful first approximation for discussing system failures. Yet even the simple systematic discussion of fundamental phenotypes in Chapter 2 showed that an omission can have several different meanings.

In their favour first-generation HRA classification schemes do incorporate a rigorous application method that must be followed in making the analysis. This assignment of error types represents a natural extension of the characterisation of operator actions in terms of an event tree and for this reason HRA techniques have proved popular with many investigators. In relation to human performance in more complex environments the omission-commission distinction nevertheless has some weaknesses that limit the practicality of the scheme. The problems associated with a classification such as the one proposed by Swain & Guttmann (1983) have been discussed at length by both Singleton (1973) and Reason (1986). Both authors point out that such schemes typically confound two important variables: (1) whether actions are taken or not taken, and (2) whether the outcome is correct or erroneous. It is easy to envisage a situation where a specified action is omitted from a sequence but the system still behaves as desired, or, alternatively, where a prescribed intervention is made by the operator but the system still deviates from an expected course. In such cases the classification fails to establish whether the actions of the operator are to be considered erroneous. Criticisms such as these raise serious doubts about the general utility of schemes based upon the binary classification of erroneous actions in terms of omissions and commissions, given the inherent philosophical problems that they appear to embody.

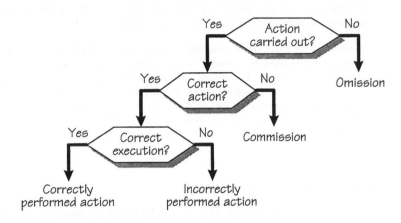

Figure 12: Binary classification of erroneous actions.

5.3 *Classification In Human Information Processing*

Each of the classifications belonging to the information processing approach described in Chapter 3 incorporates to a greater or lesser extent aspects of the three essential elements, i.e., a model, a method, and a classification scheme. In essence, the model(s) in the information processing approach have been variants of the step-ladder model, which quantitatively represented the state-of-the-art at the end of the 1980s. While the step-ladder model has proved to be a valuable tool for describing the ways that human information processing can go wrong, it has been less successful when used to specify categories of human erroneous actions. The discussion in Chapter 3 referred to three separate occasions where the same model had been used to guide the specification of error categories (Pedersen, 1985; Reason, 1987; and Rouse & Rouse, 1983). Nevertheless on each occasion a different classification scheme was the result. This finding suggests that classifications of erroneous actions do not necessarily follow from a consideration of the underlying model - to the extent that the underlying model has been applied consistently.

This problem is partly caused by the failure among the investigators to agree upon a set of principles that specifies how the step from the model to error assignment should be made - not in the analytical sense but in terms of using the model in practice. Of the three schemes mentioned above, for example, only the one proposed by Pedersen (1985) provided any guidance about error assignment. Here the process of error assignment involved evaluating success or failure during each stage of information-processing such that if the response to a question was negative, then an error was assigned to that particular stage of processing. Clearly, such descriptions can be valuable for providing general explanations of error causation, to the extent that the error types correspond to empirically verifiable categories. Even so, the results fall far short of an exhaustive classification of error causation in human cognition.

A second major problem with the classification schemes derived from the step-ladder model relates to the fact that this model essentially provides an explanation of the erroneous actions of an expert - i.e., erroneous actions that are based in the habit-strength of the person. The major concern is providing a description of erroneous actions made by skilled persons in familiar circumstances although these only represent a small proportion of the total corpus of possible erroneous actions. It could reasonably be argued that an analysis of the factors that underpin e.g. operator mistakes (defined as actions as planned, but where the plan is inadequate) is at least equally important. Yet erroneous actions falling into such categories lie outside the scope of either of the schemes outlined above.

The challenge to the adequacy of a classification scheme based upon an information processing analysis relates to the more fundamental problem of whether it is appropriate to base explanations of characteristic performance features on what amounts to "design-defects" of human information processing. For example, investigators working from the standpoint of the cognitive systems engineering perspective would argue that the processes that underpin erroneous actions on one occasion are the same as those which underpin accurate performance on another. This suggests that the search for the psychological causes of erroneous actions requires a much broader range of analysis than is implicit in the human information processing approach.

5.4 Classification In Cognitive Systems Engineering

So far there has been no explicit mentioning of a classification scheme based on the principles of cognitive systems engineering. The reason for that is simply that as yet none have been available. The main purpose of this book is to present such a classification scheme, together with the method and the model. The details of that will be provided from Chapter 7 and onwards. The preceding chapters, including this, all serve to lay the foundation for the approach, to explain the basic principles, and to show how it answers the criticisms that have been raised against the first-generation HRA methods.

In the wake of Dougherty's (1990) exposure of the shortcomings of first-generation HRA, there have been several suggestions for how improvements could be made, in particular how the issue of "cognitive errors" could be addressed. Some of these will be discussed in Chapter 5, but none of them are seen as belonging to the category of cognitive systems engineering approaches. The reason for that is quite simply that they have attempted to account for the impact of cognitive functions by augmenting either traditional human factors approaches or information processing approaches. Such attempts have, however, failed to address the roots of the problem as they have been discussed until now. To recapitulate, the problems with the existing approaches to HRA and error analysis are: (1) a failure to distinguish clearly between cause and manifestation (genotype and phenotype), (2) insufficient ability to describe how the context, e.g. as performance shaping factors or as the confluence of technology and organisation, wields an influence on performance, and (3) incomplete recognition of the three aspects of method, classification scheme, and model - to the extent that some approaches almost completely lack one or the other (cf. Chapter 5).

The Cognitive Reliability and Error Analysis Method (CREAM) presented in this book explicitly recognises these issues and proposes a solution to them. It is unlikely that CREAM is the solution to all the problems of first-generation HRA. But by making the foundation for the approach clear it should hopefully be easier to compensate for any shortcomings.

6. PERFORMANCE SHAPING FACTORS AND COMMON PERFORMANCE CONDITIONS

It is an axiom of the study of human behaviour that everything we do is influenced - but not completely determined- by the conditions that exist at the time. Despite this self-evident truth, much effort has been spent on describing the "mechanisms" of the mind as a pure mental information processing system that reacts or responds to the events that occur, but where the "mechanism" itself works in what amounts to splendid isolation. This tradition has been most evident in the case of information processing psychology, to the degree that it was hailed as a significant discovery that human cognition was situated, by which was meant that it took place in a context (e.g. Clancey, 1992). The same notion has prevailed in the practice of HRA - although implicitly rather than explicitly. This is demonstrated by the notion of Performance Shaping Factors (PSF) which was introduced above. (In the following, the term PSF will be used as a generic term to refer to Performance Shaping Factors, Performance Influencing Factors, etc.)

HRA has naturally been concerned with the variability of human performance, since unpredictable and uncontrolled variability may become part of the causal pathway that leads to incidents and accidents. Although there is a certain inherent variability to human performance, the main effect is generally assumed to come from external factors which directly or indirectly may change how actions are carried out. The inherent variability is possibly due to the nature of the physiological and neuro-physiological constitution of humans. Phenomena such as saturation, adaptation, habituation, etc., are basic features of the neuro-physiological system and play an important role for performance. Other causes may be found on a different level such as circadian rhythms, vigilance, fatigue, hunger, etc. It is relatively easy to produce a long list of factors that influence human performance, even without including the factors that are related to information processing, such as channel capacity, decay, noise, entropy, etc. All of these factors contribute to the variability of human performance, and may rightly be seen as internal rather than external to the human as an organism.

In addition to these internal factors there is a long list of factors that have to do with the external conditions of work that also have an effect on human performance. There is a basic agreement about a core set of these factors, although there are competing lists of various lengths. There is, furthermore, less agreement about **how** these factors exert their influence. If a PSF, such as noise level or insufficient training, has an effect on human performance then it is necessary also to provide an acceptable account of how this influence comes about. This requires a psychological explanation of human action - essentially, a theory of action - combined with an explanation of how performance conditions may influence action. Mainstream psychology has largely been unable to achieve this, and it is no surprise that HRA has failed as well.

6.1 Performance Shaping Factors In THERP

In the classical HRA approaches, such as the Technique for Human Error Rate Prediction (THERP, Swain & Guttman, 1985), the problem is bypassed in a rather simple and, in some sense, elegant way. The trick is to assume that if PSFs have an effect on performance, then this effect must be measurable in terms of the Human Error Probability (HEP). As described in Chapter 1 the solution is simply to calculate the effect on the HEP, and thereby completely avoid the thorny question of how this effect comes about. The net outcome is simply a quantitative change of probabilities and the method neither requires nor provides any suggestion for how this may take place in a psychological sense.

In THERP, the only description of the human component is that reproduced in Figure 13. The acknowledgement that something goes on in the mind is shown by the box labelled "mediating activities and processes". This statement is not meant as an undue critique of THERP. One must consider the level of development of cognitive psychology when THERP was first proposed and that THERP was never intended to be or to provide a scientifically adequate description of the human mind. (Neither, for that matter, is CREAM.) THERP was developed to answer a practical need, and it did so in a very efficient way. The efficiency was, however, partly obtained by neglecting some details that we now realise are essential for a good HRA approach. Yet hindsight should never serve as the basis for undue critique.

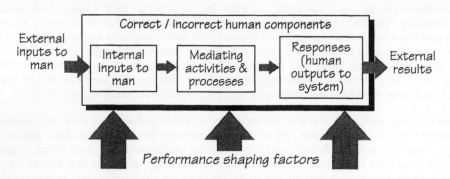

Figure 13: The operator model in THERP.

It follows from the above description that the effect of PSFs, in THERP and other first generation HRA approaches, occurs via the HEPs only. It should, however, rightly be via the "human components" of Figure 13, including the functions of input, mediation, and output. This will go further towards satisfying the psychological demands to an explanation, although it may not necessarily make the method of applying the PSFs any simpler. In fact, one may reasonably suspect that the method - if a method can be developed - will be considerably more complex, hence less attractive to the practitioner. We should, however, strive to find an acceptable compromise.

6.2 Classical Performance Shaping Factors

The information processing approach implied a more complex description of PSFs as illustrated by the CSNI taxonomy for error classification discussed above (Section 0). In the CSNI taxonomy a main distinction was made between PSFs and situation factors where the latter also influenced or shaped performance. In the general understanding, PSFs were nevertheless not considered to be causal in a strict sense, such as the "mechanisms of human malfunction".

Swain (1989) provided an authoritative overview of the PSFs that traditionally have been used in HRA. This classification made a distinction between **external** PSFs, **stressor** PSFs, and **internal** PSFs. External PSFs were subdivided into situation characteristics, job and task instructions, and task and equipment characteristics. Stressor PSFs were subdivided into psychological stressors and physiological stressors. There were no subdivisions for internal PSFs. The three sets of PSFs are reproduced in Table 1, Table 2, and Table 3 below.

Table 1: External performance shaping factors in THERP.

Situational characteristics (PSFs general to one or more jobs in a work situation)	Architectural features. Quality of environment: (temperature, humidity, air quality, and radiation, lighting, noise and vibration, degree of general cleanliness). Work hours / work breaks. Availability / adequacy of special equipment, tools and supplies.	Shift rotation. Staffing parameters: Organisational structure (authority, responsibility, communication channels). Actions by supervisors, co-workers, union representatives, and regulatory personnel. Rewards, recognitions, benefits.
Job and task instructions; single most important tool for most tasks	Procedures required (written or not written). Cautions and warnings.	Written or oral communications. Work methods. Plant policies (shop practises).
Task and equipment characteristics (PSFs specific to tasks in a job)	Perceptual requirements. Motor requirements (speed, strength, precision). Control-display relationships. Anticipatory requirements. Interpretation. Decision-making. Complexity (information load). Narrowness of task. Frequency and repetitiveness.	Task criticality. Long- and short-term memory. Calculation requirements. Feedback (knowledge of results). Dynamics vs. Step-by-step activities. Team structure and communication. Man-machine interface factors (design of prime / test / manufacturing equipment, job aids, tools, fixtures).

Table 2: Stressor performance shaping factors in THERP

STRESSOR PSFs		
Psychological stressors (PSFs which directly affect mental stress	Suddenness of onset. Duration of stress. Task speed. High jeopardy risk. Threats (of failure, loss of job). Monotonous, degrading or meaningless work. Long, uneventful vigilance periods.	Conflicts of motives about job performance. Reinforcement absent or negative. Sensory deprivation. Distractions (noise, glare, movement, flicker, colour). Inconsistent cueing.
Physiological stressors (PSFs which directly affect physical stress	Duration of stress Fatigue Pain or discomfort Hunger or thirst Temperature extremes Radiation G-force extremes	Atmospheric pressure extremes Oxygen insufficiency Vibration Movement constriction Lack of physical exercise Disruption of circadian rhythm

Table 3: Internal performance shaping factors in THERP

Organismic factors (characteristics of people resulting from internal & external influences)	Previous training/experience State of current practice or skill Personality and intelligence variables Motivation and attitudes Knowledge required (performance standards) Stress (mental or bodily tension)	Emotional state Sex differences Physical condition Attitudes based on influence of family and other outside persons or agencies Group identification

Although this classification includes all of the common PSFs it does not represent a very clear structure; there is in particular considerable overlap both between the main groups and between the entries within a group.

In the CSNI taxonomy, a smaller set of performance shaping or performance influencing factors were proposed (cf. Figure 9). The CSNI PSFs were divided into subjective goals and intentions, mental load / resources, and affective factors. The CSNI situation factors were likewise divided into task characteristics, physical environment, and work time characteristics. The complete list is shown in Table 4.

Table 4: Performance shaping factors in the CSNI taxonomy.

PSFs		
Subjective goals and intentions	Exaggeration of some task performance aspects. Inappropriate extension of task contents.	Task perceived as secondary. Conflicting goals. Other.
Mental load, resources	Inadequate ergonomic design of work place. Overlapping tasks. Inadequate general education.	Inadequate general task training and instruction. Other.
Affective factors	Social factors. Insufficient load, boredom.	Time pressure. Fear of failure. Other.
SITUATION FACTORS		
Task characteristics	Familiar task on schedule. Familiar task on demand.	Unfamiliar task on schedule. Unfamiliar task on demand. Other.
Physical environment	Noise. Uncomfortable temperature, humidity, pressure, smell, etc.	Light. Radiation. Other.
Work time characteristics	Day shift. Night shift. In beginning of shift.	In middle of shift. At end of shift. Not stated or applicable.

The CSNI taxonomy can be seen as an attempt to provide an explanation for the PSFs that had generally been used and were recognised by HRA, in particular by THERP. Otherwise there is little in common between the CSNI taxonomy and THERP. Other, and more detailed, attempts have been made to explain the interaction between working conditions and the resulting performance; some of these will be presented in Chapter 5. This interaction is clearly an issue of great importance to HRA, since it provides the rationale for much of the quantitative methodology.

Although the CSNI classification is better structured than the one from THERP - mainly because it was deliberately made simpler - it is still not clear why some categories are grouped together. Furthermore, the coupling to the underlying description of human performance (cf. Figure 10) is a little weak, perhaps because the model itself is not sufficiently argued. It is therefore necessary to consider an alternative and more principled approach.

6.3 Error Modes And Error Models

A good starting point for describing how performance conditions can affect performance is to consider the possible ways in which an action can go wrong - the so-called error modes. In the context of PSA, this has often been done using a binary classification of whether the action will succeed or fail. This binary classification is, unfortunately, deceptively simple and consequently misleading. Rather than discussing in terms of success or failure we must consider how an action can go wrong, depending on the performance conditions. This can be achieved without any attempts of proposing an information processing model of human performance, but rather by looking at the phenomenology or appearance of actions.

The PSFs basically recognise that there are some characteristic links between conditions and performance, and as long as these links are adequately described there is no serious need to explain them theoretically. In the same way it may be noted that there are a limited number of manifestations of incorrectly performed actions. The reason for this limitation is that actions must take place in a five dimensional time-space-energy continuum. Any action can consequently be described with regard to space, time, and force - and so can the deviations or incorrectly performed actions. The basic error modes are shown in Table 5. (In Table 5 "sequence has been put under "time", but it might equally well have been put under "space".)

Whereas the error modes can be used to characterise performance reliability - in the sense that a reliable performance will exhibit few of the error modes - it is also important to be able to account for how performance conditions may influence the error modes. This can be achieved by proposing a description of how various kinds of causes can lead to specific error modes.

Table 5: Systematic error modes.

Dimension		Type of deviation		
		Not enough	Too much	Other
Space	Direction			Wrong direction (up / down / left / right / front / back)
	Magnitude	Too short	Too far	
	Object			Neighbouring object / Similar object / Unrelated object
Time	Duration	Too short	Too long	
	Timing	Too early	Too late	Not at all (omission)
	Speed	Too slow	Too fast	Fluctuating
	Sequence	Stop before end point	Continue beyond end point	Reversal / Omission / Repetition / Intrusion
Force	Force	Too little / too soft	Too much / too hard	Unsteady

In relation to performance reliability it is necessary to consider how the classification of error modes and their causes can be achieved so that both the nature of cognition and the context dependence or performance conditions are addressed. This can be done by using the classification based on a distinction between error modes (phenotypes) and failure causes (genotypes), cf Chapter 1. As shown in Figure 14, the classification system can initially be divided into three main groups: (1) the consequences of the failure, (2) the error modes, and (3) the context that can be described in terms of causes and common performance conditions. Each main group can be subdivided further; an example being the error modes described in Table 5.

Context information can further be used to define sets of **possible** error modes and **probable** error causes. In any given context some error modes will be impossible due, for instance, to the physical and functional characteristics of the interface. For instance, if a piece of equipment such as a pump or a switch is controlled indirectly via icons on a screen it is impossible to use too much force. Similarly, some causes will be more likely than others due to working conditions, task demands, organisational support, etc.

It turns out that it is possible to define a relatively small set of Common Performance Conditions (CPC) that contains the general determinants of performance, hence the **common modes** for actions in a context. The CPCs that are included in the present version of CREAM are shown in Table 6. For practical reasons, the context is described in terms of a limited number of factors or dimensions; the proposed CPCs are intended to have a minimum degree of overlap, although they are not independent of each other. Table 6 also shows the basic qualitative descriptors that are suggested for each CPC.

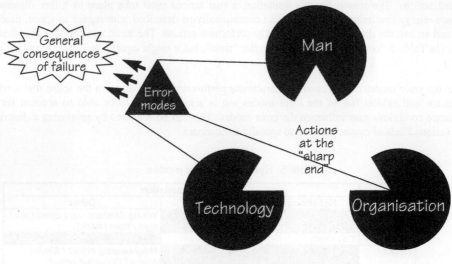

Figure 14: The CREAM classification scheme.

Of the nine CPCs many occur in most other attempts of listing or categorising PSFs. This is the case for available time, often called adequacy of time; adequacy of MMI and operational support, often called plant interface, indications or information available, or indication complexity; number of simultaneous goals, often called task complexity; availability of procedures and plans, often called procedural guidance or quality of procedures or procedure usability; and adequacy of training and experience.

Four of the nine CPCs are less common. Of these, the working conditions partly correspond to the notions of stress and workload, but is more oriented towards the description of the physical (and psychological) working conditions, rather than being a generalised term for the subjective working conditions (i.e., stress level). This CPC may also to some extent include the notion of significant preceding and concurrent actions, which can be found in several PSF systems. The adequacy of the organisation is rarely included in PSF lists, although this clearly is one of the more important determinants of performance that probably deserves to be described by more than one CPC. (This shortcoming has been addressed in more recent approaches, such as Barriere et al., 1994.). Although the time of day or the effects of circadian rhythms often is treated in a perfunctory manner, this CPC clearly cannot be overlooked. The variability of performance due to the time of day is one of the most robust phenomena in psychology, yet it is rarely explicitly included in HRA models and methods. Finally, the quality of crew collaboration is essential for the efficient use of the human resources. This has, for instance, been recognised in the field of aviation where with the introduction of principles for Crew Resource Management (CRM, e.g. Caro, 1988). It specifically addresses such important issues such as communication and delegation of responsibility.

Table 6: Common Performance Conditions.

CPC name	Level / descriptors
Adequacy of organisation	The quality of the roles and responsibilities of team members, additional support, communication systems, Safety Management System, instructions and guidelines for externally oriented activities, role of external agencies, etc.
	Very efficient / Efficient / Inefficient / Deficient
Working conditions	The nature of the physical working conditions such as ambient lighting, glare on screens, noise from alarms, interruptions from the task, etc.
	Advantageous / Compatible / Incompatible
Adequacy of MMI and operational support	The Man-Machine Interface in general, including the information available on control panels, computerised workstations, and operational support provided by specifically designed decision aids.
	Supportive / Adequate / Tolerable / Inappropriate
Availability of procedures / plans	Procedures and plans include operating and emergency procedures, familiar patterns of response heuristics, routines, etc.
	Appropriate / Acceptable / Inappropriate
Number of simultaneous goals	The number of tasks a person is required to pursue or attend to at the same time (i.e., evaluating the effects of actions, sampling new information, assessing multiple goals etc.).
	Fewer than capacity / Matching current capacity / More than capacity
Available time	The time available to carry out a task and corresponds to how well the task execution is synchronised to the process dynamics.
	Adequate / Temporarily inadequate / Continuously inadequate
Time of day (circadian rhythm)	The time of day (or night) describes the time at which the task is carried out, in particular whether or not the person is adjusted to the current time (circadian rhythm). Typical examples are the effects of shift work. It is a well-established fact that the time of day has an effect on the quality of work, and that performance is less efficient if the normal circadian rhythm is disrupted.
	Day-time (adjusted) / Night-time (unadjusted)
Adequacy of training and experience	The level and quality of training provided to operators as familiarisation to new technology, refreshing old skills, etc. It also refers to the level of operational experience.
	Adequate, high experience / Adequate, limited experience / Inadequate
Crew collaboration quality	The quality of the collaboration between crew members, including the overlap between the official and unofficial structure, the level of trust, and the general social climate among crew members.
	Very efficient / Efficient / Inefficient / Deficient

There is obviously a significant overlap between the CPCs and the traditional PSFs. This is because the set of possible conditions that may affect performance is limited. The difference between the CPCs and the PSFs is therefore not so much in the names and meaning of the categories that are used, but in **how** they are used. The main difference is that the CPCs are applied at an early stage of the analysis to characterise the context for the task as a whole, rather than as a simplified way of adjusting probability values for individual events. In this way the influence of CPCs becomes closely linked to the task analysis.

6.4 Specific Effects Of Performance Conditions

Whereas Table 6 describes the Common Performance Conditions and the typical "values" they can take, it is necessary to relate this to the potential effect on performance reliability. This can be done using the general principle that advantageous performance conditions may improve reliability, whereas disadvantageous conditions are likely to reduce reliability. If reliability is improved, then the operators are expected to fail less often in their tasks. Conversely, if the reliability is reduced, then operators will fail more often. The descriptions used in Table 6 are directly amenable to such an assignment, and the result is shown in Table 7.

The assignments shown in Table 7 are based on general human factors knowledge, specifically the experience from the HRA discipline. On the whole the assignments correspond directly to the descriptors used for the CPCs, in terms of their semantic contents. The assignments imply a bi-modal distribution where the extremes are improved and reduced reliability respectively. The middle region corresponds to the category of "no significant effect". This means that it is impossible to predict whether the effect will be positive or negative, and furthermore that the effect in general will be relatively small.

Table 7: CPCs and performance reliability.

CPC name	Level / descriptors	Expected effect on performance reliability
Adequacy of organisation	Very efficient	Improved
	Efficient	Not significant
	Inefficient	Reduced
	Deficient	Reduced
Working conditions	Advantageous	Improved
	Compatible	Not significant
	Incompatible	Reduced
Adequacy of MMI and operational support	Supportive	Improved
	Adequate	Not significant
	Tolerable	Not significant
	Inappropriate	Reduced
Availability of procedures / plans	Appropriate	Improved
	Acceptable	Not significant
	Inappropriate	Reduced
Number of simultaneous goals	Fewer than capacity	Not significant
	Matching current capacity	Not significant
	More than capacity	Reduced
Available time	Adequate	Improved
	Temporarily inadequate	Not significant
	Continuously inadequate	Reduced
Time of day (circadian rhythm)	Day-time (adjusted)	Not significant
	Night-time (unadjusted)	Reduced
Adequacy of training and experience	Adequate, high experience.	Improved
	Adequate, limited experience.	Not significant
	Inadequate.	Reduced
Crew collaboration quality	Very efficient	Improved
	Efficient	Not significant
	Inefficient	Not significant
	Deficient	Reduced

Finally, this use of the CPCs and error modes is essentially model free. Rather than postulating a strong information processing model of human cognition and human action, the approach tries to identify the salient or dominant (or requisite) features of performance, as links in the space of Man-Technology-Organisation (MTO) rather than as paths in an information processing model. This requires the analyses to be made in the MTO space as well, something that existing approaches have only done to a limited degree.

6.5 Dependency Of Performance Conditions

One of the basic, but tacit, assumptions of the use of PSFs/PIFs is that they are independent, as expressed in the conventional formula:

$$\log f = \sum_{k=1}^{N} PSF_k * W_k + C$$

where the total effect of the PSFs is calculated as the sum of the (weighted) PSFs. Yet even a cursory investigation of the meaning of the commonly used PSFs shows that they cannot be independent. For instance, stress (as an internal PSF) cannot be independent of complexity or information load (as a task and equipment characteristic). If the information load is high it is reasonable to expect that it will lead to an increase in stress. In order to determine the effects of CPCs on performance it is therefore necessary to describe the dependency between performance conditions. In other words, it is necessary to develop a model that describes how the performance conditions affect each other, in addition to describing how they affect performance.

A possible reason why such a model is not an inseparable part of the first-generation HRA approaches is that the performance shaping factors typically have been developed in response to a practical need. It is always possible for a given situation to identify some factors that either have had, or can be assumed to have, a major influence on performance. For a representative range of situations, a small number of factors will occur repeatedly, and it is these that form the core of the performance shaping factors. Yet inventing performance shaping factors in this way is similar to curve fitting or parameter fitting (Kantowitz & Fujita, 1990). With a little imagination it is always possible to come up with a set that adequately describes or explains the specific situation. The drawback of this approach is that it does not refer to a firm conceptual basis or a model.

One approach to establishing the required model is to consider each of the CPCs in turn with respect to how it may influence the others. The following is an illustration of what this might produce, referring to the general understanding of human performance rather than to any specific domains or types of situations.

Adequacy of organisation. This CPC describes the quality of the roles and responsibilities of team members, additional support, communication systems, Safety Management System, instructions and guidelines for externally oriented activities, role of external agencies, etc. The adequacy of the organisation clearly has an effect on the general *working conditions*. The better the organisation is, the better the working conditions will be and *vice versa*. There is a similar relation to the *adequacy of MMI and operational support*, to the *availability of procedures / plans*, to the *adequacy of training and experience,* and to the *crew collaboration quality*. In all four cases, and perhaps particularly in the latter three, the adequacy of the organisation will be a determining factor. To some extent one could also argue that the adequacy of the organisation has an impact on the time of day, in the sense that a good organisation will be able to minimise the negative consequences of changing work hours. However, this is probably a rather indirect effect, hence need not be considered.

Working conditions. This CPC describes the nature of the physical working conditions such as ambient lighting, glare on screens, noise from alarms, interruptions from the task, etc. The working conditions have a direct impact on the *number of simultaneous goals* that the user must attend to in the sense that improved working conditions may lead to a reduction in the number of goals. Conversely, improved working conditions can be assumed to increase the *available time*. Inadequate working conditions may also to a more limited extent reduce the availability of procedures / plans, although this may also be seen as a more direct result of the adequacy of the organisation.

Adequacy of MMI and operational support. This CPC describes the Man-Machine Interface in general, including the information available on control panels, computerised workstations, and operational support provided by specifically designed decision aids. The adequacy of MMI and operational support has a direct effect on the *working conditions*; if the MMI and operational support are inadequate, the working conditions will clearly suffer. More importantly, there is also a significant effect on the *number of simultaneous goals* and the *available time*. If the MMI is improved, the number of goals will be reduced, while at the same time the available time will be increased.

Availability of procedures / plans. Procedures and plans include operating and emergency procedures, familiar patterns of response heuristics, routines, etc. As for the adequacy of MMI and operational support this CPC has the same kind of impact on *number of simultaneous goals* and *available time*. If the availability of procedures and plans is high, the number of goals will be smaller and the available time will be larger.

Number of simultaneous goals. This CPC describes the number of tasks a person is required to pursue or attend to at the same time (i.e., evaluating the effects of actions, sampling new information, assessing multiple goals etc.). It has a direct impact on *available time* in the sense that the two CPCs are inversely related. Thus, if there are more goals to attend to, the available time will decrease - and vice versa.

Available time. This CPC describes the time available to carry out a task and corresponds to how well the task execution is synchronised to the process dynamics. It has an impact on one aspect of the general *working conditions* with respect to task interruptions, in the sense that the working conditions will deteriorate as the available time decreases. In combination, the number of simultaneous goals and available time correspond to the classical notion of workload.

Time of day. The time of day (or night) describes the time at which the task is carried out, in particular whether or not the person is adjusted to the current time (circadian rhythm). Typical examples are the effects of shift work. It is a well-established fact that the time of day has an effect on the quality of work, and that performance is less efficient if the normal circadian rhythm is disrupted. Concretely, this can be expressed as an effect on *working conditions*, as well as an effect on *available time*, in the sense that less efficient performance means that the available time potentially is reduced.

Adequacy of training and experience. This CPC describes the level and quality of training provided to operators as familiarisation to new technology, refreshing old skills, etc. It also refers to the level of operational experience. The adequacy of training and experience has an impact on *working conditions*, *available time*, and *crew collaboration quality*. The better the training and experience are, the better the working conditions will be, the more time there will be available, and the crew will be able better to work together.

Crew collaboration quality. The quality of the collaboration among members of the crew is important for normal as well as accident situations. In a crew where members work well together tasks will be performed efficiently, the need for unnecessary communication will be reduced, incorrect actions will often be noticed and recovered in time, and responsibilities and resources will be shared in a flexible manner. Crew collaboration quality has an impact on the *available time*, since efficient communication and the efficient use of resources will avoid misunderstandings and reduce delays.

	Adequacy of organisation	Working conditions	Adequacy of MI and operational support	Availability of procedures/plans	Number of simultaneous goals	Available time	Time of day	Adequacy of training and experience	Crew collaboration quality
Adequacy of organisation	■								
Working conditions	+	■	+				+	+	+
Adequacy of MI and operational support	+		■						
Availability of procedures/plans	+			■					
Number of simultaneous goals		-	-	-	■				
Available time		+	+	+	-	■		+	+
Time of day							■		
Adequacy of training and experience	+							■	
Crew collaboration quality	+							+	■

Figure 15: Dependencies between Common Performance Conditions.

The dependencies described in the proceeding can be summarised as shown in Figure 15. This shows how the CPCs in the left hand column are affected by the CPCs defined in the upper row. Clearly, a CPC cannot depend on itself, hence the diagonal is blackened out. The dark grey cells indicate a dependency between two CPCs. Thus, for the second column both "number of simultaneous goals" and "available time" are assumed to depend on the "working conditions". If the working conditions improve, then the number of simultaneous goals, i.e., the number of tasks that the operator has to attend to at the same time, is assumed to be reduced as indicated by the "-" sign. Similarly, if the working conditions are improved, then the available time is also assumed to improve as indicated by the "+" sign. Both of these are assumed to be direct dependencies.

Another way of showing the dependencies between the CPCs are by means of a diagram, as in Figure 16. Here the dependencies are shown by means of arrows, where an arrow indicates that one CPC influences another. White arrows denote a direct influence (increase-increase, decrease-decrease) while black arrows denote an inverse influence (increase-decrease, decrease-increase). For instance, "adequacy of organisation" has an influence on both "working conditions", adequacy of training and experience", availability of procedures/plans", and " adequacy of MMI". As can be seen from Figure 16, as well as from Figure 15, "adequacy of organisation" does not depend on any of the other CPCs. In relation to the performance in a task "adequacy of organisation" is therefore a background variable, which can be assumed to be constant for the duration of the situation. Similarly the main dependent CPCs are "available time" and "number of goals", corresponding to the recognised importance of the level of work load.

7. CHAPTER SUMMARY

The *raison d'être* for HRA is the need to be able to predict the likely performance of a system. Predictions for interactive systems (joint cognitive systems) are difficult to make because the possible future developments depend on the complex interactions between technology, people, and organisations. The predictions concern both the consequences of specific initiating events, and the joint system's ability to respond correctly to abnormal conditions.

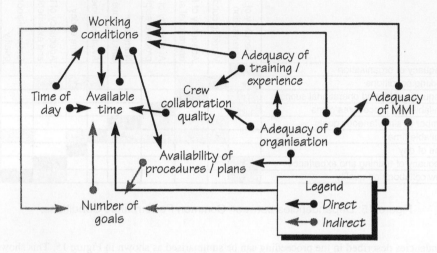

Figure 16: Links between Common Performance Conditions.

In order for predictions to be reliable, they must use a well-defined method. In order for them to be valid, they must use an adequate classification scheme and a sound model of the constituent phenomena. Together this can be described as a Method-Classification-Model (MCM) framework. Chapter 4 introduced the MCM framework and provided a description of two of the parts: the model and the classification scheme. The method will be described separately in later chapters.

The modelling of cognition has been a focus of research since the early 1960s, although the notion of an operator model only became widely accepted in the 1980s. For many years the dominating approach has been towards the modelling of micro-cognition, i.e., the details of how cognition happens in the mind. This has led to an emphasis on structural models of cognition as an epiphenomenon of context free information processing in the mind. An alternative is to model macro-cognition, i.e., how cognition functions in realistic situations of work. This puts the emphasis on what purposes cognition serves, hence tends to favour a functional approach to modelling, such as the Simple Model of Cognition (SMoC). The cognitive modelling approach that is part of CREAM is described in more detail in Chapter 6.

Classification schemes are needed both to describe the different error manifestations and to account for the factors that may influence the vulnerability to error. A classification scheme must therefore be based on an underlying model of the causal nature of actions - although this often occurs implicitly rather than explicitly. A good classification scheme requires both that the categories are meaningful and unambiguous, and that there is a clear procedure for how different error modes and causes can be assigned to the various categories.

Classifications in first-generation HRA are based on a binary principle that is used to develop detailed event categorisations; the most common example of that is the omission-commission distinction. Their advantage is a rigorous application method; their disadvantage is the artificial simplicity of the categories, which makes them inadequate for realistic descriptions. In terms of the MCM triad there are clear methods, limited classification schemes and no modelling worth mentioning. Classifications in the information processing tradition rely heavily on the features of detailed information processing models, but the step from model to application is often insufficiently described. In addition, the classification schemes typically fail to maintain a clear distinction between causes (genotypes) and manifestations (phenotypes). In terms of the MCM triad the methods are not well defined, the classification schemes can be detailed but are often quite diverse, and the models are elaborate but with unknown validity. Cognitive systems engineering has so far provided few examples of classification schemes. The following chapters, and in particular Chapter 6, will provide a description of what such a classification system may look like.

Every HRA approach relies on the assumption that performance is affected by the conditions, traditionally described as performance shaping factors. In first-generation HRA approaches the PSFs are used directly to modify the HEP, and there is little justification for this in terms of an underlying performance model. Cognitive systems engineering also assumes that performance is influenced by the performance conditions, but tries more explicitly to account for how the common performance conditions depend on each other and how they have an overall influence on performance reliability. Chapter 4 introduced the notion of the common performance conditions, while the details of their application will be given in Chapter 9.

Chapter 5

HRA - The First Generation

Though this be madness, yet there is method in't.
William Shakespeare, Hamlet [II. ii. (211)]

1. RELIABILITY AND SAFETY ANALYSIS OF DYNAMIC PROCESS SYSTEMS

The purpose of this chapter is to provide a summary of first-generation HRA methods by using the triad of method, classification scheme, and model (the MCM framework), developed in Chapter 4. The historical / conceptual background of the methods will be pointed out, since this is useful in understanding how they have developed. The chapter will also outline the present ideas about a second generation HRA, such as they are. The main argument is that the first-generation HRA methods are unsatisfactory because they have been moulded on the principles of PSA without fully considering what that implies. The proper approach to develop a second generation HRA must accordingly include taking a step back to see HRA in its historical context. We will then be in a better position to understand the real needs of HRA, and to understand why HRA has developed into what it is today.

Over the years several surveys of HRA approaches have been made. In a few cases explicit criteria were established to rate the various approaches, hence to point out "best" HRA approach. These criteria have typically included cost, ease of application, ease of analysis, availability of data, reliability, and face validity. Clearly, criteria such as these are important in the final choice of a specific HRA approach. However, in the present work the main concern is to illustrate the assumptions that each approach makes about performance reliability. It is therefore necessary to propose another set of criteria.

In general, the purpose of an analysis is to find and describe the elements of that which is being analysed and account for how they are related - in terms of functions, causes, dependencies, etc. This means that an analysis must refer to a consistent **classification scheme** that is relevant for the domain under investigation. The analysis must further employ a **method** as a systematic way of finding the elements and relating them to the constituents of the classification scheme. In HRA the topic of study is human actions - in particular erroneous actions - and the aim is to describe what can cause an action to go wrong. Both the classification scheme and the method must therefore be adequate for this purpose.

It can further be argued that the classification scheme needs to refer to a set of concepts for the domain in question, i.e., a set of supporting principles. This is often called a model of the domain. In relation to HRA and the analysis of human erroneous actions the classification scheme must be based upon a viable **model of human actions** rather than just erroneous actions - which at the present stage of development corresponds to a model of human cognition. The model will guide the definition of specific system failures based on the characteristics of human cognition relative to the context in which the behaviour occurs.

The need for models in safety and reliability analysis is obviously different from the need for models in design. The important questions in design are whether the intended system will match the capabilities of the operator or user, whether automation and task allocation will be at an appropriate level, etc., cf. Section 3.4 below. Models are therefore used to support design specifications, either implicitly as good human factors practise or explicitly as simulation models. Operator models may also be used to identify places where the design does not fit with the model, which presumably are places where things can go wrong - that is, provided the design is not changed.

The concerns of safety and reliability analyses are whether an operator is likely to make an incorrect action and which type of action it will be. In the simple sense this requires a prediction of when the operator is likely to fail - which in the information processing approach meant seeing the operator as a "fallible machine". In a more generic sense this means an identification ahead of time of situations where the combination of the information and conditions with the operator's current state and capabilities lead to the choice of a wrong action or to the incorrect performance of an action.

Design models are not appropriate for safety and reliability analyses because they are developed for a different purpose and because they are prone to consider the static rather than dynamic aspects of the system. For instance, a set of good human factors guidelines will tell the designer about the size of letters, the choice of colours, the shape of controls, information density, etc. Yet these aspects are not equally applicable to analyses of actual use. If there is a discrepancy between what is needed by the operator and what is provided by the design, then things are likely to go wrong. Unfortunately, it cannot safely be assumed that the operator will perform flawlessly if the design is appropriate, since experience shows that human performance is variable. We therefore need models that can tell us when this is likely to be the case, considering the effects of the environment and the situation.

This purpose requires that the dynamics of the joint man-machine system are adequately described. If we consider a dynamic system, we are interested in the causes of events, i.e., the origin of the dynamics. To the extent that the joint system performs as expected and remains within the envelope of design performance, there may be no need to go beyond a technological description. Operators act as regulators (and as more complex regulators than those which can be designed) but since they perform within the prescribed band there is no need to model them differently from how a technical regulator is modelled. However, when human performance begins to play a separate role, there is a need to understand it in its own terms. This can happen either when human performance or human actions are seen as the root cause(s) of unwanted system events, but also when human actions are seen as the events that save the system, i.e., as the ultimate safeguard. In the latter cases the variety of the human operators fortunately exceeds the variety of the control systems.

The focus in safety and reliability analyses has predominantly been on the negative sides of performance, i.e., the human operator as an unreliable element. This has led to the explicit goal of removing the human by increasing automation or at least of improving human performance and increasing system reliability by reducing as far as possible the responsibilities of the operator. Yet this solution misses the point entirely. Rather than trying to improve the reliability of human performance by restraining it, one should try to increase the role of the human by designing a natural working environment that encourages operators to make use of their unique capabilities. This notion of human centred automation is reflected in the slowly changing concepts of the human controller.

As far as the classification scheme is concerned, the definition of the categories of effects or manifestations (phenotypes) and causes (genotypes) within the scheme should follow naturally from a consideration of the workings of the model. This requirement is essential if assignments of erroneous actions are to be justified on psychological grounds. As far as the method is concerned, the importance of that is self-evident. The utility of analytical approaches that lack a method component will be limited due to potential sources of inconsistency when the classification scheme is applied repeatedly.

The detailed relations among the three main elements were discussed in Chapter 4 and shall therefore not be repeated here. Instead the characterisation of the typical HRA approaches will used the principles of the MCM framework, looking at the way in which a method, a classification scheme, and a model have been described for each approach. Following the discussion of Chapter 4, an evaluation will also be made of whether the approaches describe how the influence of the PSFs comes about. Considering these evaluations, conclusions about the current treatment of human erroneous actions in PSA methods will be drawn.

2. FIRST-GENERATION HRA APPROACHES

According to the current lore there are around 35-40 clearly distinguishable HRA approaches. It is safe to assume, however, that some of them are variations of the same approach, and that the number of significantly different HRA approaches therefore is smaller. Several studies have in the past provided surveys of extant HRA approaches. The coverage of six of these studies is summarised in Table 1 below. For each approach, the main reference is indicated. (The approaches are ordered alphabetically by the first author.)

In addition to the surveys listed above, a number of books and reports have described HRA approaches for specific purposes other than to provide an overview of the state-of-the art. Thus Dougherty & Fragola (1988), in their book on human reliability analysis, presented an integrated systems engineering approach. The HRA approach they described was centred on THERP, although additional discussion about the use of Time-Reliability Correlation (TRC) was provided. Another example is Park (1987), who in his book on human reliability, looked more into the nature of human error and its prevention. As part of that, descriptions were given of THERP, OAT, and various fault-tree methods.

The approaches listed in Table 1 do not constitute an exhaustive list, but represent the ones that have received sustained attention during the 1980s. It has a geographical bias since most of the methods have been developed and used in the US. There are several utilities in Europe that have developed proprietary HRA methods, and the same may be the case in other parts of the world. Even if information about these methods was available it could, for obvious reasons, not be included here. Despite this bias, the table clearly shows that there is a group of HRA approaches that are commonly considered. In addition, there are a number of specific HRA approaches that fall outside the mainstream, usually because they are only part approaches or proposals for describing specific aspects. If we make the "arbitrary", but not entirely unreasonable, decision to consider only HRA approaches that have been included in at least four of the six surveys in Table 1, the result is a list of selected HRA approaches shown in Table 2. (The reader will notice that AIPA also is included in Table 2 despite it being included in only one survey. The reason for that is that AIPA is one of the early HRA methods. The order of Table 2 has been changed from alphabetical to chronological.)

The entries of Table 1 provide the year for the first acknowledged reference of each method (at least as far as can be judged from the available material). If these indications are used as an estimate of when the commonly known HRA methods were developed, an interesting perspective on the history of HRA emerges, which makes clear that the history is relatively short. (It is, of course, reasonable to assume that most methods have been developed in the years immediately preceding their publication, so that their actual time of origin is one or two years earlier than the time of the first reference. Since this, however, will be the case for all methods, we can disregard it.) If the "year of birth" for each of the methods is seen in relation to a time-line, the result is something like Figure 1. This shows that there were a few early developments in the last half of the 1970s, and that most of the methods were developed in the first half of the 1980s. It is also interesting to note that the development slowly stopped by the end of the decade. There have been some new methods proposed in the 1990s, partly in response to the call from Dougherty (1990), but nothing like on the scale of the 1980s. This distribution is clearly not independent of the occurrence of the accident at Three Mile Island of March 28, 1979.

Table 1: Summary of HRA surveys.

Method	Lucas (1988)	Haney et al. (1989)	Swain (1989)	Spurgin & Moieni (1991)	Gertman & Blackman (1994)	Kirwan (1994)
Time-dependent Accident Sequence Analysis (Apostolakis, 1984).	♦		♦			
Simulator data (Beare et al., 1983).	♦					
Expert estimation (Comer et al., 1984).	♦		♦	♦	♦	
HAP - Human Action Probabilities (Dougherty, 1981)	♦					
SHERPA - Systematic Human Error Reduction and Prediction Approach (Embrey, 1986).	♦					
ORCA - Operator Reliability Calculation and Assessment (Dougherty & Fragola, 1988).					♦	
SLIM/MAUD - Success Likelihood Index Method / Multi-Attribute Utility Decomposition (Embrey et al., 1984).	♦	♦	♦	♦	♦	
AIPA - Accident Investigation and Progression Analysis (Fleming et al., 1975).			♦			
Fullwood's method (Fullwood & Gilbert, 1976).	♦					
TRC - Time-Reliability Correlation (Hall et al., 1982).					♦	
HCR - Human Cognitive Reliability (Hannaman et al., 1984).	♦	♦	♦	♦	♦	
JHEDI - Justification of Human Error Data Information (Kirwan, 1990)						♦
SHARP - Systematic Human Action Reliability Procedure (Hannaman & Spurgin, 1984).	♦	♦				
Variation diagrams (Leplat & Rasmussen, 1987)	♦					
Tree of causes (Meric et al., 1976)	♦					
Murphy Diagrams (Pew et al., 1981)	♦					
STAHR - Socio-Technical Assessment of Human Reliability (Phillips et al., 1983)	♦	♦		♦	♦	
CM - Confusion matrix (Potash et al., 1981)	♦	♦	♦		♦	
Human Problem Solving (Rouse, 1983)	♦					
MSFM - Multiple-Sequential Failure Model (Samanta et al., 1985)	♦	♦				
MAPPS - Maintenance Personnel Performance Simulation (Siegel et al., 1984).	♦	♦		♦	♦	
Licensee Event Reports (Speaker et al., 1982).	♦					
ASEP - Accident Sequence Evaluation Procedure (Swain, 1987; Gore et al., 1995).				♦	♦	♦
THERP - Technique for Human Error Rate Prediction (Swain & Guttmann, 1983).	♦	♦	♦	♦	♦	
SRM - Sandia recovery model (Weston et al., 1987).	♦			♦	♦	
HEART - Human Error Assessment and Reduction Technique (Williams, 1988)						♦
Speed-accuracy trade-off (Woods et al., 1984).		♦				
SAINT - Systems Analysis of Integrated Networks of Tasks (Wortman et al., 1978).				♦		♦
OAT - Operator Action Tree (Wreathall, 1982)	♦	♦	♦	♦		

Table 2: Selected HRA approaches.

	Lucas (1988)	Haney et al. (1989)	Swain (1989)	Spurgin & Moieni (1991)	Gertman & Blackman (1994)	Kirwan (1994)
AIPA - Accident Investigation and Progression Analysis (Fleming et al., 1975).			♦			
CM - Confusion matrix (Potash et al., 1981)	♦	♦	♦		♦	
OAT - Operator Action Tree (Wreathall, 1982)	♦	♦	♦	♦		
STAHR - Socio-Technical Assessment of Human Reliability (Phillips et al., 1983)		♦	♦		♦	♦
THERP - Technique for Human Error Rate Prediction (Swain & Guttmann, 1983).	♦	♦	♦		♦	♦
Expert estimation (Comer et al., 1984).		♦		♦	♦	♦
SLIM/MAUD - Success Likelihood Index Method / Multi-Attribute Utility Decomposition (Embrey et al., 1984).	♦	♦	♦	♦	♦	♦
HCR - Human Cognitive Reliability (Hannaman et al., 1984).		♦	♦	♦	♦	♦
MAPPS - Maintenance Personnel Performance Simulation (Siegel et al., 1984).	♦	♦			♦	♦

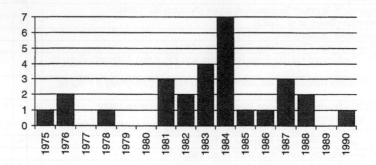

Figure 1: Distribution of HRA approaches according to year of publication.

2.1 Accident Investigation And Progression Analysis (AIPA)

The Accident Initiation and Progression Analysis (AIPA) was developed in the mid-1970s as a method to estimate the probability of operator responses in the operation of large High Temperature Gas-Cooled Reactors (Fleming et al., 1975). The operator's actions were described in terms of whether or not the required responses were made, i.e., the classification scheme was only success or failure. The purpose of AIPA was to determine the probability of whether the action would be carried out, described in terms of the mean time to operator response. This, of course, is an early version of the Time-Reliability Correlation. The method was used to estimate the probabilities for each action, based on expert judgements.

2.1.1 Method

According to Swain (1989) the AIPA method included the following basic modelling assumptions:

- The operator had a probability of zero for making any response instantaneously - defined as the interval from 0.2 to 40 seconds after an event.

- Given enough time, the operator would eventually take some action, which most likely would not increase the potential event consequences.

- If the operator found that the first actions were insufficient, he would then take further corrective action until a mitigating condition was achieved. (This seems to be a very optimistic point of view!)

2.1.2 Classification Scheme

The classification scheme for AIPA is the fundamental distinction between success and failure, defined as whether the required response was made. This is consistent with the purpose of finding the mean time for an operator response.

2.1.3 Model

AIPA cannot be said to include an operator model as such. The operator is basically seen as a black box that emits a - possibly successful - response according to an unknown "mechanism". The characteristics of the "mechanism" are approximated by the TRC, but no attempt is made to speculate about how the response actually comes about. AIPA did not explicitly address the issue of PSFs.

2.1.4 Conclusion

In relation to the MCM framework, AIPA is clearly not very sophisticated neither in terms of the classification scheme nor of the operator model. This is hardly surprising considering that it was developed in the early 1970s. Even though AIPA may no longer be in use it is relevant to include it in this survey because it very clearly exhibits the fundamental features of first-generation HRA approaches. As such it is highly representative of the trends that can be found in practically all other methods developed later.

2.2 Confusion Matrix

The confusion matrix technique is used to estimate the probability of an operator not correctly diagnosing an initiating event. The confusion matrix normally includes all actions taken during an event sequence that supplement automatic system response, but excludes actions that fall outside the accident-response envelope, such as "errors" made in test and maintenance activities.

2.2.1 Method

A confusion matrix is concerned with the initiating events for accidents that have similar signatures. The rationale is that accidents that look the same may confuse the operators, hence provoke an incorrect diagnosis. A confusion matrix is constructed by listing similar initiating events on the vertical and horizontal axes of the matrix. The similarity of symptoms and their expected frequency is analysed, and this is used to rank each initiator with respect to how likely it is to be confused with the actual event (i.e., another initiator). Once the ranking is completed, probabilities based on e.g. simulator data or other sources are assigned for each misdiagnosis, and modified to reflect the quality of the control room design and operator training.

Two matrices are constructed in this manner, one for each of two different phases of the accident sequence. For example, one might be for the early phase covering the first 15 minutes after the first indication of the accident, and the second for a subsequent phase. Finally, a frequency analysis is carried out for each initiating event and estimated total probabilities for misdiagnosis are calculated.

2.2.2 Classification Scheme

The classification scheme for the confusion matrix is very simple, being basically reduced to the case of incorrect diagnosis. In terms of the omission-commission classification this would correspond to a commission, but in practice only the single event of misdiagnosis is considered. Thus, the confusion matrix does not explicitly use a classification scheme.

2.2.3 Model

It follows from the above, that the confusion matrix does not use a model either. One of the steps in the method is to estimate the probability that the crew either misdiagnoses the event or fails to act appropriately. This estimation is, however, done without considering the influence of cognition, indeed without having any explicit model of the operator. Neither does the confusion matrix explicitly address the role of the PSFs, although they indirectly must be taken into account when considering the likelihood of confusing two initiating events.

2.2.4 Conclusion

The confusion matrix does not provide any explicit treatment of human erroneous actions, apart from the over manifestations of misdiagnosis. It nevertheless differs from many of the first-generation approaches by including an overall situation assessment, although this is done implicitly through the ranking of each situation. Another positive aspect is that it recognises that the situation may change as the event develops, and that the confusion matrix therefore may not be the same at different points in time. Basically it remains, however, a statistical approach.

2.3 Operator Action Tree (OAT)

The Operator Action Tree System (OAT or OATS) was developed by John Wreathall in the early 1980s and has been described by Wreathall (1982). The Operator Action Tree approach to HRA is based on the premise that the response to an event can be described as consisting of three stages: (1) observing or noting the event, (2) diagnosing or thinking about it, and (3) responding to it. It is further assumed that errors that may occur during the third phase, i.e., carrying out the necessary response actions, are not the most important. The primary concern should rather be on the errors that may occur during the second stage, the diagnosis (cf. the Confusion Matrix approach, Section 2.2). The OAT approach therefore concentrates on the probability that the operator correctly diagnoses accident and identifies the responses that are necessary in terms of system operations.

2.3.1 Method

The method associated with the OAT contains the following five steps (Hall et al., 1982):

1.   Identify the relevant plant safety functions from system event trees.

2. Identify the event specific actions required to achieve the plant safety functions.

3. Identify the displays that present relevant alarm indications and the time available for the operators to take appropriate mitigating actions.

4. Represent the error in the fault trees or event trees of the PSA.

5. Estimate the probabilities of the errors. Once the thinking interval has been established, the nominal error probability is calculated from the time-reliability relationship. (The thinking interval is defined as the difference between the time when an appropriate indication is given and the time that the corresponding action must be initiated.)

2.3.2 Classification Scheme

The OAT relies on a representation of the appropriate actions as a binary tree, the Operator Action Tree. Each action is considered with regard to whether it is correctly or incorrectly performed, i.e., a classification in terms of success or failure. This is the basic classification that can be found in practically all first-generation HRA approaches.

2.3.3 Model

The OAT is ostensibly aimed at the cognitive functions that are related to diagnosis. It would therefore be reasonable if the OAT referred to a model capable of accounting for diagnosis as a cognitive function. Although the OAT does distinguish among three phases of a response and thereby can be said to embody a rudimentary operator model, the error probability is derived from the Time-Reliability Correlation, which usually serves in lieu of an operator model. In principle, the OAT could be said to make use of a high-level information processing model where only the probability of response is considered. In practice, this is equivalent to not having a detailed model at all. As for most of the other approaches, the role of the PSFs is not explicitly addressed.

2.3.4 Conclusion

The OAT considers the probability of failure in diagnosing an event, i.e., the classical case of response or non-response. Because of this, the OAT cannot be said to provide an adequate treatment of human erroneous actions.

It nevertheless differs from the majority of the first-generation HRA approaches by maintaining a distinction among three phases of the response: observation, diagnosis, and response. This amounts to a simple process model and acknowledges that the response is based on a development that includes various activities by the operator, rather than being generated *ex nihilo*. Unfortunately the OAT approach only looked at the second of these phases, and thus missed the chance of providing the first cognitive HRA model.

2.4 Socio-Technical Assessment Of Human Reliability (STAHR)

The STAHR differ many ways from other first-generation HRA approaches. It represents a consensus-based psychological scaling method to assess human reliability in complex technical systems, and consists of a technical and a social component. The technical component is an influence diagram that shows the network of causes and effects that links the factors to the outcome of the situation, cf. Figure 2. The social component refers to the elicitation, by group consensus, of expert judgements of the conditional probabilities of the various factors shown in the influence diagram as well as of their respective weight of evidence.

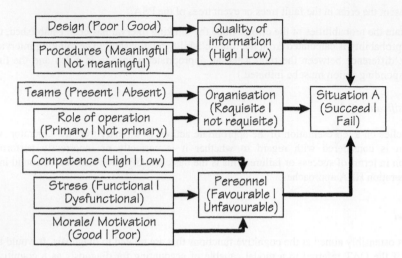

Figure 2: STAHR influence diagram.

2.4.1 Method

Compared to other first-generation HRA approaches, the method of STAHR is quite complex as illustrated by the following summary of the main steps. The method refers explicitly to the influence diagram.

1. Describe all the relevant conditioning events.

2. Define the target event.

3. Choose a middle-level event (in the influence diagram) and assess the weight of evidence for each of the bottom-level influences leading to it. (The diagram in Figure 2 has been rotated 90°. The bottom-level therefore corresponds to the leftmost column, while the middle-level corresponds to the middle column.)

4. Assess the weight of evidence for this middle-level influence conditional on the bottom-level influence.

5. Repeat steps 3 and 4 for the remaining middle- and bottom-level influences.

6. Assess probabilities of the target event conditional on the middle-level influence.

7. Calculate the unconditional probability of the target event and the unconditional weight of evidence of the middle-level influence.

8. Compare these results to the holistic judgements of the assessors; revise the assessments as necessary to reduce discrepancies between holistic judgements and model results.

9.   Iterate through the above steps as necessary until the assessors have finished refining their judgements.

10. Perform sensitivity analyses on any remaining group disagreements. If there are no disagreements of any consequence, then report point estimates. If disagreements are substantial, then report ranges.

2.4.2 Classification Scheme

Human erroneous actions play a smaller part in STAHR than in the other first-generation HRA approaches, since it does not depend so heavily on the PSA event tree. STAHR basically provides a method for obtaining and combining probability estimates. There is therefore little need for an explicit classification scheme.

2.4.3 Model

It follows from the above that STAHR does not make use of an explicit model of the operator, nor of an explicit model of human erroneous actions. The influence diagram is a model, of a sort, that describes the relation between the various factors that may influence an outcome. The basis for this description is, however, not explicitly defined. PSFs are explicitly addressed in the sense that the leftmost column describes the main performance determinants.

2.4.4 Conclusion

STAHR, as an expert estimation approach, does not provide any explicit treatment of human erroneous actions. Since it is not based on a the principle of decomposition, there is also less of a need to consider PSFs as a separate measure.

2.5 Technique For Human Error Rate Prediction (THERP)

The Technique for Human Error Rate Prediction (THERP) is probably the best known of the first-generation HRA methods. According to Swain (1989) the development of THERP began already in 1961, but the main work was done during the 1970s, resulting in the so-called THERP handbook, of which the final version is Swain & Guttmann (1983). Later a simpler variation of THERP was developed as ASEP, the Accident Sequence Evaluation Procedure (Swain, 1987; Gore et al., 1995).

The aim of THERP is to calculate the probability of successful performance of the activities necessary for the accomplishment of a task. The calculations are based on pre-defined error rates (the so-called Human Error Probabilities or HEPs), and success is defined as the 1's complement to the probability of making an error. THERP involves performing a task analysis to provide a description of the performance characteristics of the human tasks being analysed. The results of the task analysis are represented graphically in an HRA event tree that is a formal representation of the required sequence of actions, cf. the example in Chapter 4). The nominal probability estimates from the analysis of the HRA event tree are modified for the effects of sequence-specific PSFs, which may include factors such as dependence between and within operators, stress levels, experience, quality of information provided, display types, etc.

2.5.1 Method

The basis for applying THERP is a schematic representation of human actions and related system events, the so-called HRA event tree. The method consists of the following six steps:

1. Define the system failures that may be influenced by human errors and for which probabilities are to be estimated.

2. Identify, list, and analyse human operations performed and their relationships to system tasks and functions of interest.

3. Estimate the relevant human error probabilities.

4. Determine the effects of the human errors on the system failure events of interest.

5. Recommend changes to the system in order to reduce system failure rate to an acceptable level.

6. Review the consequences of proposed changes with respect to availability, reliability, and cost-benefit.

2.5.2 Classification Scheme

The classification of erroneous actions was originally made in terms of the now commonplace distinction between omissions and commissions. The classification was later extended to include a more detailed breakdown of incorrectly performed actions, using some of the standard error modes. The main principle of the classification scheme is, however, still a series of binary choices starting from the need to act. THERP, however, clearly makes use of a classification scheme, whether it is simple or more elaborate.

2.5.3 Model

In the early versions of THERP the operator model was a variation of the well-known Stimulus-Organism-Response model. This has later been extended with a more detailed description of the cognitive functions that are assumed to take place between perception and action, using a variation of Rasmussen's (1986) step-ladder model. The model is, however, not defined as an explicit feature of THERP, but can rather be implied from how the classification scheme is used.

The effect of the PSFs is provided by modifying the nominal HEP rates. This is a purely quantitative approach and does not propose any explanation of how the PSFs exert their influence.

2.5.4 Conclusion

THERP has been described as one of the few complete HRA methods, in the sense that it describes both how events should be modelled (in terms of the event tree) and how they should be quantified. The dominance of the HRA event tree, however, means that the classification scheme and the model necessarily remain limited, since the event tree can only account for binary choices (success-failure). It is thus difficult in THERP to introduce more complex error modes, and the full consequences of even the SRK-framework cannot easily be implemented.

A final feature of THERP is the use of performance shaping factors to complement the task analysis. While the task analysis identifies the important individual tasks that are to be analysed, the PSFs describe the general conditions that may influence the performance of the tasks, and which therefore may change the probability estimates. The use of this technique to account for non-specific influences is found in most first-generation HRA methods (including AIPA). The separate use of performance shaping factor is relevant for an evaluation of the operator model, since it suggests that the model by itself is context independent.

2.6 Expert Estimation

This HRA approach uses expert judgement to estimate human error probabilities. There are several ways in which this can be accomplished, via e.g. a paired comparisons procedure, a ranking / rating procedure, by direct numerical estimates, by indirect numerical estimation, and through a multi-attribute utility procedure.

The experts base their estimates of human error probabilities on their own background and experience. Thus, no on-site data collection is required. It is an essential part of this approach that the experts can either develop a common task description or work from a task description supplied to them. The face validity of the resulting human error probabilities is based on this common frame of reference as well as the experts' experience in performing or in assessing the performance of similar tasks.

2.6.1 Method

As an example of a method, the procedure for using a numerical estimation technique described in NUREG/CR-2743 (Seaver & Stillwell, 1983) is summarised below:

1.   Assemble approximately six judges who will participate in the estimation.

2.   Train the judges in the use of basic probability and statistics.

3.   Train the judges according to a model for human error probabilities, if appropriate.

4.   Clearly define and outline all aspects of the events under evaluation.

5.   Provide experts with a logarithmically spaced scale of odds. The scale should be detailed enough so that experts can indicate differences in odds between events presented during the discussion and should also capture the range of odds associated with the events to be evaluated.

6.   Obtain odds judgements from the log scale instrument.

7.   Aggregate individual judgements to provide a single odds judgement. Create a table of odds provided by the experts.

8.   Transform the experts' odds estimates into HEPs, using a pre-defined procedure

9.   Use one of several accepted methods to estimate reliability, e.g. ANOVA.

2.6.2 Classification Scheme

The expert estimation approach does not make use of an explicit classification scheme. Instead the human error probabilities are estimated directly for specific actions. Presumably this implies, as a minimum, the category of omission and possibly also the category of commission - since failing to do something could be due either to an omission or because something else was done. These concerns are, however, not explicitly addressed by the method.

2.6.3 Model

As was the case for the confusion matrix, the expert estimation does not make use of any kind of operator model. Step 3 of the method refers to a model, but it is not clear what this entails. It nevertheless does not appear to be a model as the term is used in the MCM framework. Since the HEPs are calculated directly, the influence of the PSFs is presumably included in the experts' judgements.

2.6.4 Conclusion

The expert estimation does not provide any explicit treatment of human erroneous actions. It relies on the intuitive understanding of the experts. In that sense it has a holistic rather than a decompositional flavour.

2.7 Success Likelihood Index Method / Multi-Attribute Utility Decomposition (SLIM/MAUD)

SLIM-MAUD is an approach to estimate HEPs by means of an interactive computer-based procedure, MAUD, which serves to elicit and organise the assessments of experts within the framework of SLIM. The approach is based on the assumption that the failure probability associated with task performance is based on a combination of PSFs that include the characteristics of the individual, the environment, and the task. A second assumption is that experts can estimate these failure rates or select reasonable anchor values and allow the software to perform the estimate for them. Anchor values used as upper and lower bounds on failure rates for a particular task are provided by the analysts. The SLIM approach uses ratings of the relative goodness of PSFs in a particular situation weighted by their relative impact as determined by a panel of experts. The end result of a SLIM-MAUD session is a probability estimate that can be used in HRA fault and event trees.

2.7.1 Method

The method of SLIM-MAUD is highly developed and described in great detail in the existing documentation. It can be summarised as follows:

1. Identify experts competent to review the task and its requirements of personnel.

2. Have the experts discuss the task and define the ways in which error could occur.

3. Use the materials available in step 2, to identify the PSFs that influence the various potential error modes identified.

4. Document the definitions for the various PSFs.

5. Determine the importance weights for each of the PSFs.

6. Rate each task by assigning a numerical value on a scale of 0.0 to 100.0. Whereas the weights determined in step 5 reflect the relative importance of the PSFs for the tasks, step 6 is the expert opinion regarding the situation as it exists.

7. Calculate the Success Likelihood Index (SLI) by summing the products of each of the ratings by its associated weight.

8. Transform the SLI to probabilities by selecting anchor values and use a calibration equation.

9. Generate uncertainty bounds by having the experts perform direct estimation of upper and lower bounds.

2.7.2 Classification Scheme

Swain (1989) made the point that SLIM-MAUD, like the confusion matrix and STAHR, was based on expert judgement rather than being a complete HRA approach. SLIM-MAUD explicitly refers to a characterisation of error modes, as shown in steps 2 & 3 of the method. The classification of the error modes is, however, made by the experts and is therefore not explicitly described by the approach. In terms of the categories used in this survey, SLIM-MAUD does not have a classification scheme over and above what can be found in the other first-generation HRA approaches.

2.7.3 Model

It follows from the above that SLIM-MAUD does not make use of an explicit model of the operator, nor an explicit model of human erroneous actions. Similar to the expert judgement and confusion matrix approaches, SLIM-MAUD relies on the implicit model used by the experts. The fact that they may agree on the surface characteristics of the situation (cf. step 2) does not, however, constitutes evidence that there is a common understanding of what an erroneous action is, neither of what the causal links may be.

SLIM-MAUD explicitly considers the influence of the PSFs, but only in a quantitative fashion. Since there is no explicit operator model there clearly can be no explicit description of how the PSFs exert their influence.

2.7.4 Conclusion

SLIM-MAUD, as an expert estimation approach, does not provide any explicit, systematic treatment of human erroneous actions. In particular, it does not provide an explicit account of how the PSFs affect the final probabilities, except in a mathematical sense.

2.8 Human Cognitive Reliability (HCR)

The HCR approach serves the purpose of quantifying the time-dependent non-response probability of control room operators responding to accident situations. The name of the HCR approach alludes to three different types of "cognitive processing", corresponding to the skill-based, rule-based, knowledge-based (SRK) classification made popular by Rasmussen (1986). The HCR approach explicitly assumes that the three different types of "cognitive processing" have different median response times.

The basis for the HCR approach is actually a normalised time-reliability curve, where the shape is determined by the dominant cognitive process associated with the task being performed. The analyst determines the type of cognitive process, estimates the median response time and the time window, and uses the HCR model to quantify the non-response probability.

2.8.1 Method

The HCR method can be described as having the following six steps:

1. Identify the actions that must be analysed by the HRA, using e.g. a task analysis method.
2. Classify the types of cognitive processing required by the actions. This classification in itself uses a sequence of binary choices, resulting in a classification in terms of skill-based, rule-based and knowledge-based actions.

3. Determine the median response time ($T_{1/2}$) of a crew to perform the required tasks.

4. Adjust the median response time to account for performance influencing factors.

5. For each action, determine the system time window (T_{sw}) in which action must be taken.

6. Finally, divide the T_{sw} with $T_{1/2}$ to obtain a normalised time value. On the basis of this, the probability of non-response is found using a set of time-reliability curves.

The median response time is obtained from simulator measurements, task analyses, or expert judgement. The effects on crew performance of operation-induced stress, control room equipment arrangement, etc., are accounted for by modifying the median time to perform the task.

2.8.2 Classification Scheme

The HCR uses the classification common to operator action trees and intrinsic to the notion of the Time-Reliability Correlation: success and failure (or correct response and incorrect response). The HCR extends this by adding a third category of "no response". The "no response" can also be seen as a result of "slow cognition", i.e., when the operator uses more time than allowed. Since the HCR was developed around 1984, it was too early to make use of the more detailed proposals for error classifications that have since been developed from the SRK framework, e.g. Reason (1990).

2.8.3 Model

It is clear from this description of the method that reference is made to a specific operator model, i.e., the skill-based, rule-based, knowledge-based framework developed by Rasmussen (1986) as an elaboration of the step-ladder model. This model is used in two different ways. Firstly, to classify the "cognitive processing" required by the actions. Secondly, to distinguish among three time-reliability curves that provide different values for the probability of non-response for each type of "cognitive processing". As noted above, it is not used as a basis for the classification scheme as such.

In the HCR approach, the median response time is adjusted by the effect of the PSFs, This is, however, done in a purely quantitative fashion, without considering the possible information processing mechanisms that the SRK framework might imply.

2.8.4 Conclusion

The HCR is probably the first-generation HRA approach that most explicitly refers to a cognitive model. The method is clearly developed with reference to the SRK framework. Since it was developed before a detailed error model based on the SRK framework was available, the treatment of human erroneous actions in the HCR remains on the same level as in many other first-generation methods, i.e., a basic distinction between success and failure - and in this case also no-response.

In the HCR, the PSFs are applied **after** the type of processing has been determined (step 4). This seems to imply that the type of processing depends on context free task characteristics, hence is independent of the actual performance conditions! Furthermore, there is no explicit account of how the PSFs exert their influence on performance.

2.9 Maintenance Personnel Performance Simulation (MAPPS)

The MAPPS computer model was developed to provide a tool for analysing maintenance activities in nuclear power plants. A principal focus of the model is to provide maintenance-oriented human performance reliability data for PRA purposes. The influence of select PSFs is also included, such as environmental, motivational, task, and organisational variables. The output from MAPPS provides information about predicted errors, personnel requirements, personnel stress and fatigue, performance time, and required ability levels for any corrective or preventive maintenance actions. Gertman & Blackman (1994) notes that extensive use of the model has not occurred because of the difficulty in accessing the mainframe version of the simulation and the limited distribution of the personal computer version.

2.9.1 Method

Unlike the other HRA approaches described here, MAPPS (and the PC version MicroMAPPS) is by itself a simulation of the "event tree". From the users' perspective, the method mainly consists of providing the required information as input to the simulation, rather than going through the actual steps of the calculations. The input includes a definition of the sub-task, selected from a pre-defined list of generic sub-tasks, and specification of the sub-task parameters.

2.9.2 Classification Scheme

Since MAPPS does not examine an event tree or considers the human erroneous actions as such, there is no explicit classification scheme involved. MAPPS simulates whether the tasks can be successfully performed, so by implication it can be assumed that at least a distinction between success and failure is included in the modelling.

2.9.3 Model

MAPPS is unique in being able to simulate the performance of a 2-8 person crew, rather than the performance of a single operator. It follows, that the modelling included in MAPPS must relate to crew performance rather than to individual erroneous actions. Basically, performance is based on a calculation of the intellectual and perceptual-motor requirements of a task, and the ability of the crew to perform that task, including the effects of time stress. The model is stochastic, and can therefore not be considered as an operator model in the usual sense.

2.9.4 Conclusion

MAPPS contains neither a classification scheme nor a model in the usual sense. Since the main effort of the user is to provide the required input for the simulation, it is impossible to say anything specific about the way in which MAPPS treats human erroneous actions. It is, however, assumed that this does not play a great part in MAPPS.

3. CONCLUSIONS

The first parts of this chapter have provided a survey of nine of first-generation HRA approaches, which were selected because they were among the most commonly cited - although they have not been used equally often. Although the selection was to some extent arbitrary the methods do include the main first-generation HRA approaches in terms of usage and also provide a good illustration of their breadth. The description of the selected approaches in particular looked at:

- whether an approach included an explicit classification scheme for human erroneous actions, as a way of identifying the basic elements of the fault tree;

- whether the classification scheme was based on an identifiable model of operator action and cognition, or at least included explicit assumptions about the nature of human behaviour;

- whether the model included or supported an account of how the PSFs influenced performance; and

- whether the approach included an explicit method, and whether the method explicitly made reference to the classification scheme.

The characteristics of the nine HRA approaches can be summarised as shown in Table 3.

Table 3: Summary of first-generation HRA approaches.

HRA Approach	Method description	Classification scheme	Operator model	PSF effects
AIPA	Not very detailed.	Success / failure.	Black box (TRC)	Not addressed
CM - Confusion matrix	Basic principles described.	Incorrect diagnosis.	None	Not addressed
OAT	Detailed procedure provided.	Incorrect diagnosis (success, failure).	Simple stage model, but focus on TRC.	Not addressed
STAHR	Basic principles described.	None.	None.	Implied by the influence diagram.
THERP	Detailed procedure provided.	Omission, commission. Later versions add more detail.	S-O-R. Later versions refer to SRK.	Only addressed quantitatively
Expert estimation	Basic principles described.	Not explicitly defined.	None	Only addressed implicitly and quantitatively
SLIM / MAUD	Detailed procedure described.	Not explicitly defined.	None	Only addressed quantitatively
HCR	Detailed procedure provided.	Success, failure, no response.	SRK, but only for selection of TRC curves.	Only addressed quantitatively
MAPPS	Not explicit, embedded in simulation.	Not explicitly defined.	Not explicitly defined. Possibly a model of crew performance.	Not explicitly described.

3.1 Method Description

In terms of method description all of the surveyed approaches are fairly explicit, with the possible exception of MAPPS. This is due to the fact that the core of MAPPS is a simulation, and that the essence of the HRA analysis therefore is encapsulated in the simulation. The responsibilities of the user are thus not to perform the analysis as such, but rather to prepare the input for MAPPS.

It is not really too surprising that the methods are well described, since the HRA approaches have all been developed in answer to practical needs. There seems, on the whole, to be no recognised HRA approaches which have developed from a theoretical or "academic" basis alone. The strong link to the practical need is probably also the reason why few of them show any significant connection between the method and the classification scheme - and, indeed, why few of them include a well-defined classification scheme.

The nine HRA approaches can be described as falling into two groups. (The above distinctions can, of course, also be extended to cover the methods from Table 1 that have not been reviewed here.) One group contains HRA approaches that have an important element of expert estimation or judgement. This group includes the Confusion Matrix, Expert Estimation, SLIM/MAUD, and STAHR. In these cases guidelines are given for how experts are to be involved in the HRA, and in some cases very detailed procedures are laid out (e.g. SLIM/MAUD). It nevertheless remains that the specific method cannot be automated completely, not even in principle, due to the significant involvement of human expertise. The other group contains HRA approaches where expert judgements are not required as an explicit part of the methods. In this group it will therefore, in principle, be possible to automate the analysis, for instance as a software system. This group includes AIPA, HCR, OAT, and THERP. For the last three the method is described in great detail, and the dependence on input from human experts is reduced. In practice, of course, none of the methods in the two groups can be completely automated, although software support can be - and has been - developed to facilitate their application. Again, MAPPS is the exception, since it takes a radically different approach.

3.2 Classification Schemes

In terms of the classification scheme, practically all of the HRA approaches described here are disadvantaged by the fact that they were developed in the early or middle 1980s, before the more detailed error classification schemes became widely known (e.g. Reason, 1990; Hollnagel, 1991). Furthermore, as described above, the development of HRA approaches took place in one community while the development of error classification schemes took place in another (e.g. Rasmussen et al., 1981) with little interaction between the two. The main distinction used by HRA seems to be between correct and faulty actions - or between success and failure. This dichotomy is at the root of all of the approaches, and is practically endemic to the concept of the binary representation used in the event trees. The notion of success and failure is also an integral part of the Time-Reliability Correlation, which describes the probability of a non-response in relation to elapsed time.

The concepts of omission and commission are usually attributed to Alan Swain and THERP. The omission-commission pair is an elaboration of the simple success-failure distinction because it introduces the notion of possible causes, and it has been developed further to distinguish between specific manifestations of omissions and commissions, cf. Chapter 1. It is therefore a little surprising that the HRA methods do not easily incorporate this development. Even in THERP, the basic representation is in terms of a binary event tree, where the nodes represent distinct events to which a human error probability can be assigned. In practice, it is next to impossible to introduce omission-commission on this level of description.

This state of affairs is recognised by many as unsatisfactory, particular in relation to the growing concern for "cognitive errors". One way in which this concern has been addressed is by considering the occurrence of specific types of "errors", e.g. as errors of diagnosis. errors of recovery, errors of planning or even errors of intention (cf. INTENT; Gertman et al., 1992). However, even in these cases the cognitive function - for instance, diagnosis - is considered as a category of action with the traditional binary outcome. These extensions to the classification scheme do therefore not have any impact on the fundamental approach, which is to consider unitary actions that can either succeed or fail. (As argued above, activities such as diagnosis and planning are by themselves very complex, and therefore defy the use of a single term such as "cognitive error".) Some of the HRA methods have tried to be more discriminative. In COGENT, presented as a cognitive event tree (Gertman & Blackman, 1994), the event tree nodes are described using a combination of the SRK framework and the slip-mistake classification. A different approach was seen in the HCR that included the notion of "no response". The HCR also used the elaborate structure of the SRK framework to distinguish between different types of required activity, but still described each of them in terms of the established dichotomy - or possibly a triad. It appears as if first-generation HRA approaches are entrenched in the limited universe of the binary event tree. Attempts of enriching the classification scheme therefore seem doomed to fail unless the basic notion of an event tree can somehow be changed.

3.3 Operator Models

In terms of an operator model most first-generation HRA methods simply do not have one. It can, of course, be argued that the use of the success-failure or even the omission-commission distinctions implies a certain type of model. The important point is nevertheless that there only are very few cases where a model has been explicitly described and acknowledged. This is, perhaps, most easily seen in relation to the use of PSFs. As argued above, the way in which PSFs exert their effect on performance should be described or explained by the operator model. None of the first-generation HRA approaches, however, do that. Instead, the influence of the PSFs is usually accomplished by simply multiplying the basic HEPs with a weighted sum of the PSFs, cf. the discussion in Chapter 1.

The descriptions of the first-generation HRA approaches clearly demonstrate that operator modelling, as it has been used over the years, has been based on rather simple concepts. As noted above, there presently seems to be two mutually exclusive ways of considering the operator. One is in terms of the Time-Reliability Correlation, which at best can be seen as a black box model. The TRC is not concerned with what goes on between event and response, but only with the probability that the response takes place within a certain time interval. Many of the main proponents of this approach clearly recognise that the TRC is not a model, but rather a coarse, although usually efficient, substitute for a more refined description. In theoretical terms, the TRC can be seen as a mathematical representation of a supposedly dominant performance characteristic. As such, it serves a very useful purpose, although it is problematic as a basis for developing more sophisticated methods and classification schemes.

The alternative to the TRC is in most cases a variation of the SRK framework developed by Rasmussen (1986). This view of human performance has proved fertile as a basis for developing more refined error classification schemes, and was for a long time recognised as a sort of "market standard". In terms of providing an adequate basis for the modelling of human cognition, it nevertheless has significant shortcomings and several alternative approaches have been developed (Bainbridge, 1993; Hollnagel, 1993d). One expression of the shortcomings is the imprecision of the terms. Thus, the notion of a skill-based error may have two different meanings. It can refer to a possible cause, i.e., that the cause of an observed consequence was an error that occurred in a skill - which in turn may have causes related to specific cognitive functions. It can also refer to the manifestation itself, since skills are defined as a specific class of actions with specific performance characteristics. Another problem is that it is difficult for a strict information processing model to account for the two main characteristics of human cognition: intentionality and context dependence.

The step-ladder model has in itself produced a number of proposals for error analysis such as CSNI (Rasmussen et al., 1981) or GEMS (Reason, 1990) but interestingly enough no specific suggestions for performance prediction, such as one could use in a PSA. Neither type of model goes into much detail about how the PSFs exert their influence.

3.4 Design And Performance Analysis

In contrast to this situation, there has been a quite strong development of user models in the field of design, specifically the design of human-computer interaction, as well as in the general developments of cognitive science and engineering. Here a development can be found which goes from classical behavioural models of the S-O-R type (Gagné, 1965; McCormick & Tiffin, 1974; Payne & Altman, 1962; Rook, 1962; Swain, 1963) over strict control theory models (Stassen, 1988; Sheridan, 1982; Johannsen, 1990) to information processing models (Lindsay & Norman, 1976) and finally to cognitive models (Bainbridge, 1991; Hollnagel & Woods, 1983; Woods, 1986; Woods et al., 1994).

As mentioned before there has, however been little collaboration or exchange of ideas between the two fields. In particular, safety and reliability analyses have apparently not been much influenced by the developments that have taken place in the field of design - or perhaps the transfer of concepts has not been possible. There could be several reasons for this. One is a difference in aims or purposes between the two fields. In safety and reliability analyses the aim is clearly to be able to **predict** events in a precise, usually quantitative sense, in order to fulfil the needs of technical analyses such as PSA. In the design field the needs have rather been to explain and understand the systems better, and to provide design guidelines, i.e., to establish the basis for design and improve the foundation for design decisions and specifications. (Ultimately, design depends on the ability to predict correctly the consequences of specific choices.) This has traditionally been done in a qualitative rather than a quantitative way. In fact, there has been a strong opposition to quantification and to engineering interests. Perhaps there has been too little awareness in the PSA/HRA field on what has happened in the design field, and *vice versa*. Academic achievements in cognitive modelling have probably seemed too esoteric for reliability engineers, and the problems of reliability engineers have seemed too mundane for the academic researchers. Whatever the reason may be there has been a growing discrepancy between the often sophisticated models used for design, and the simpler models used for safety and reliability analyses. To put it differently, there have been few suggestions for how the more elaborate design models can be used for safety and reliability analysis.

Spurgin & Moieni (1991) express a widely held opinion when they concluded an overview of HRA methods by stating that "(S)ince the availability of data on human performance is always likely to be a problem, more work needs to be done on the proper and systematic use of expert judgment methods". This conclusion was, however, made without considering the assumptions inherent in the various methods, in particular the assumptions about the operator model. It is hardly surprising that efforts to collect data based on a time-reliability "model" will only produce modest results when the standard of comparison is actual operator performance. If the model that implicitly controls the data collection is unrealistically simple, then the results will necessarily be of limited value. As the arguments above have shown, the models must be able to reflect human cognition in an adequate way. The paramount characteristic of human cognition is that it is **embedded** in a socio-technical context. Hence operator models must be able to model embedded cognition - or in other words, they must be able to model the influence of the situation and the context on the operator. If HRA is going to meet the growing demands from end users, it is necessary that second-generation HRA methods take heed of the significant developments in the modelling of cognition that have occurred in other fields.

In conclusion, the survey of first-generation HRA approaches shows that neither of them have a well-developed theoretical basis, in terms of a detailed model of human cognition and a corresponding classification scheme for erroneous actions. In particular, none of them provide an adequate theoretical treatment of "cognitive errors". Many of them do not even have sufficient categories to describe "cognitive errors" well. This does not mean that they are without any practical value. They have all been developed to solve a specific need, and several of the approaches have been and are still being widely used. The consequence is rather that there is no obvious way of "grafting" the notion of a "cognitive error" to any of the existing approaches. Instead, the conclusions of this survey are in full agreement with Dougherty's (1990) call for a "second coming" of HRA models.

4. HRA AND COGNITION: EXTENSIONS

The introduction of the notion of a first-generation HRA approach created a new problem, namely to define what constitutes a second-generation approach. The easy solution is refer to the chronology. In this case any HRA approach developed after, say, 1991 will be a second-generation approach. Another solution is to grab the bull by the horns and provide a definition of the qualities that would clearly separate first and second-generation approaches. A natural starting point would be the lists of criticisms of the first-generation HRA approaches, but it would be excessively optimistic to hope that they could all be solved by a second-generation approach. For instance, the problem of inadequate data is likely to remain for some time. As argued in the first chapters, one may also consider whether the criticisms are the right ones, in the sense that they identify the real problems of first-generation HRA.

The remaining sections of this chapter describe some HRA approaches that clearly differ from the first-generation approaches encountered so far. One of these, the CES, actually predates Dougherty's (1990) criticisms and can therefore, with some justification, be seen as a pragmatic articulation of these. Others have been developed in the wake of the criticisms, but are still mainly modifications of first-generation HRA approaches, in the sense that their basic structure has not been substantially changed. The approaches described below do not represent a complete survey, but nevertheless catch the representative attempts.

4.1 Cognitive Environment Simulator (CES)

The Cognitive Environment Simulator is not an HRA approach in the same way as the ones described above. It was developed as an "analytic computer simulation tool that can be used to explore human intention formation in the same way that reactor codes are used to model thermodynamic processes in the plant" (Woods et al., 1988, p. 170). Thus, instead of defining the possible error modes theoretically, the CES would generate them via a simulation, taking the characteristics of the situation into account. CES was complemented by a Cognitive Reliability Assessment Technique (CREATE) which specified how the CES could be used to quantify the human contribution to risk in a PSA (Woods et al., 1987).

4.1.1 Method

CES is a simulation tool that can be used to determine how an operator will respond in a situation, and there is therefore no manual method as such, i.e., no step-by-step procedure that an analyst should follow.

The CES takes as input a time series of values that describe the plant state. The values are generated by a process simulation. The values are provided to the CES via a virtual display board that represents the potentially observable plant behaviour. The CES uses this information to produce an intention to act for a given situation. This intention to act is, via a human mediator, returned to the process simulator, which then generates a new set of values, which are read by the CES, etc. In this way the possibility of incorrectly executed actions is introduced, although this must be specified explicitly as part of the scenario.

The CES performed three kinds of activities during a session (cf. Woods, et al., 1988):

- It monitored the state of the plant via the virtual display board.

- It generated explanations to account for the observations, in particular when something unexpected happened. These explanations made use of a detailed knowledge-base of the process.

- CES finally selected the appropriate responses (intentions to act), either actions to correct or cope with a system abnormality, monitoring to follow the execution of intentions, or adaptation of pre-planned responses to unusual circumstances.

4.1.2 Classification Scheme

Since the CES generates the intentions to act, rather than the actions themselves, it does not have an explicit classification of error modes or erroneous actions. It obviously can generate intentions that are erroneous, but the distinction is more subtle than the binary classification normally used in first-generation HRA approaches. The CES does not provide probability estimates of specific actions, and there is therefore no need to categorise actions as failures or successes. The CES rather generates in an analytical fashion the actions that an operator is likely to take under different operating conditions. The classification of the actions is best given in terms of possible problem solving strategies. The quality of the intentions depends on the strategies used by the model, and in that sense the classification scheme and the model are inseparable.

4.1.3 Model

CES makes use of a very detailed operator model, which is an adaptation of an Artificial Intelligence model of a limited resource problem solver (EAGOL). The model can use several problem solving strategies with suggestive picturesque names such as *vagabonding* (focus gambling), *Hamlet* (plethoric deliberation), *garden path* (excessive persistence in sticking to a solution), *inspector plodder* (meticulous, but slow), and *expert focuser* (efficient and flexible problem identification). The model is clearly defined and quite different from the types of models described in the preceding. It is, perhaps, closest to being an information processing model, but is more sophisticated due to its basis in Artificial Intelligence. It can therefore not easily be matched to any of the categories that have been used so far.

4.1.4 Conclusion

The CES stands apart as a unique attempt to develop an alternative to the existing HRA approaches. In doing so it was certainly way ahead of its time. Unfortunately, it differed so much from the established practice that it was not easily taken into use. Although an attempt was made to reconcile CES with the needs of the PSA, it was not an approach that could be applied without considerable effort. It also relied on the availability of quite powerful computers, which at the time were not available to the typical HRA practitioner. Finally, because the CES depended so heavily on the operator model and the strategies, it brought the problem of model validation to the fore.

4.2 INTENT

The background for the development of INTENT was the recognition that THERP only treats a few "errors of commission", namely "errors of selection" and "errors of execution". It was felt that there was a need to enlarge the scope to cover other types of commission errors, notably "errors of intention". On the surface, INTENT therefore addressed the same issues as the CES, although the approaches differ so much that the two cannot really be compared. INTENT, which by the way is not an acronym, has been developed and described by Gertman et al. (1990; 1992).

4.2.1 Method

INTENT only describes the steps needed to determine and quantify the probabilities for "errors of intention". The basic method is otherwise the same as for THERP (Section 2.5).

1. **Compile errors of intention.** This was done by looking for errors of intention in two data sources: the Nuclear Computerised Library for Assessing Reactor Reliability (NUCLARR) and the Licensee Event Reports (LER). The categories defined by INTENT included action consequence, response set, attitudes, and resource dependencies. Altogether these categories defined twenty different errors of intention.

2. **Quantify errors of intention.** This was achieved by a direct estimation method, which included determining the HEP upper and lower bounds as well as the PSFs and their associated weights. Eleven different PSFs were considered, and expert estimates were given of the influence of each PSF on the twenty errors of intention.

3. **Determine composite PSFs.** Each PSF was rated on a site and scenario specific basis on a scale from 1 to 5, where a low value corresponded to an unfavourable rating. The composite PSF for each error type used the common principles of multiplying and summing (cf. Chapter 1).

4. **Determine site specific HEPS for intention.** Finally, the site specific HEPs were calculated using a specially developed equation.

Only step 1 is specific for INTENT, and even this is generic in the sense that "errors of intention" can be replaced by another error type.

4.2.2 Classification Scheme

Although INTENT makes use of twenty different errors of intention grouped into four major categories, there is no systematic classification scheme to support this. In particular, the errors of intention are not set in relation to other types of classification, even though they include "errors" such as misdiagnosis,

---

inadequate communication, and several types of violation. The notion of "error of intention" thus represents a perspective, rather than a completely new set of error modes / causes.

4.2.3 Model

The emphasis on errors of intention clearly shows that some kind of model of cognition is implied. INTENT makes some reference to the GEMS model (Reason, 1990) but does not use it consistently. Since INTENT is meant to be an extension of THERP it presumably makes use of the same basic operator model.

4.2.4 Conclusion

INTENT is a closely focused proposal to enhance THERP by providing the capability to treat "errors of intention" as an additional type of commission errors. As such it stays firmly within the mould of first-generation HRA approaches, and therefore does not provide answers to any of the main criticisms.

4.3 Cognitive Event Tree System (COGENT)

Like INTENT, COGENT was also developed as an extension of an existing HRA approach (Gertman, 1992; 1993). In this case the purpose was to improve the HRA event tree representation, as used by THERP, to show various types of "cognitive errors". The limitation of the normal HRA event tree is that it can only represent the outcome of binary events, and that these typically are confined to the success-failure categories where the failures are of the omission-commission type. COGENT proposed an extension to the event tree representation by which several types of "cognitive errors" could be included.

4.3.1 Method

COGENT basically follows the principles of THERP in the construction of the event tree (Section 2.5). The innovation is that the nodes and leafs of the event tree are characterised by means of a small set of "cognitive error" types (cf. below). However, the method by which this can be done is not described, except as follows:

> "Determining the appropriate cognitive category (from those listed above) requires the HRA analyst to be familiar with control room equipment, procedures, crew knowledge and experience, and to have an understanding of plant response to off-normal or emergency events and how the unfolding of a particular scenario will most likely be presented to and responded to by a control room crew."
> (Gertman, 1992, p. 6)

In other words, the assignment depends almost exclusively on the expertise of the assessor.

4.3.2 Classification Scheme

The classification scheme used by COGENT is a combination of the SRK framework (Rasmussen, 1986) and the distinction between slips, lapses, and mistakes (Norman, 1984; Reason, 1990). The result is a set of eight failure types called skill-based slip, skill-based lapse, skill-based simple mistake, rule-based slip, rule-based lapse, rule-based mistake, knowledge-based lapse and knowledge-based mistake.

This set is superficially similar to the classification proposed by Reason (1990), but not nearly as well thought-through and argued. An interesting feature is the distinction between "small" mistakes and "normal" mistakes. A "small" mistake is seen as being due to inattention or an interruption, rather than due to a wrong intention. Although there may be good practical reasons for making this kind of distinction, it does not reflect a recognisable principle. On the whole, the classification of "cognitive error" types used by COGENT appears to express a pragmatic rather than a theoretical point of view.

4.3.3 Model

It follows from the above that COGENT does not refer to an explicit operator model, other than the allusion to the works of Rasmussen, Norman, and Reason. Since COGENT is an extension of the THERPian event tree, it can be assumed that the operator model is similar to the one implied by THERP.

4.3.4 Conclusion

COGENT and INTENT are alike in their attempt to enhance the typical first-generation HRA approaches, viz. THERP. The contribution provided by COGENT is, however, more cosmetic than the one provided by INTENT. The purpose of COGENT is to enable the representation of cognitive error types in the event tree. This does not lead to a change in the event tree representation as such, but only to the proposal for some new categories or labels for the nodes. The method by which this is to be done is, however, not described as part of COGENT.

4.4 EPRI Project On Methods For Addressing Human Error In Safety Analysis

This project, which has not yet acquired an acronym, was initiated directly as a response to the noted shortcomings of first-generation HRA approaches (Parry & Mosleh, 1996). The purpose was to develop an improved HRA method, which would overcome some of the problems in existing HRA approaches and also be applicable as part of a dynamic PSA. The latter, developed under the name of the Accident Dynamics Simulator, will not be described here.

4.4.1 Method

It was a premise for this project that an improved HRA approach should be applicable to the established ways of performing a PSA. The structure of the method is therefore quite similar to the first-generation methods described above, but the difference lies in the emphasis given to the specific steps. The main innovation is related to the following points.

♦ **Identification of error expressions**. The project recognised the need to use a set of error modes that was more complete that the traditional omission-commission distinctions (see Section 4.4.2). It was pointed out that several different error expressions might arise from the same error mode. Thus, a misdiagnosis can lead to several different consequences that constitute different error expressions in the PSA event tree.

♦ **Characterisation of PIFs/PSFs**. It was also clearly recognised that the PSFs (called Performance Influencing Factors or PIFs by the project) should be described on a level corresponding to the model of "error production". Rather than use the standard PSFs in the usual fashion, the PSFs should be described on a level relevant for the error modes, so that the effect of the PSFs could be directly included in the qualitative analysis.

Instead of describing a method in detail, the project provided a specification for an improved HRA method. In addition to the issues mentioned above, it was emphasised that the method should be applicable by the PSA analyst, without requiring detailed knowledge of human factors or HRA. Although this is understandable, it is probably hard to accomplish.

4.4.2 Classification Scheme

The classification scheme made a concerted effort of presenting a detailed set of error modes, and also relate these to possible error causes. The classification scheme was therefore firmly based on an operator model. The error modes were grouped according to the structure of the model, cf. below.

4.4.3 Model

The operator model used in this project was a simple information processing model with three modules called Information, Decision, and Action. An interesting development was a graphical notation used to explain the possible combinations of information processes that might correspond to typical error modes. The operator model had the virtue of being quite simple, but it is significant that the project used it explicitly to account not only for the error modes but also for the PSFs.

4.4.4 Conclusion

This project represents a deliberate attempt to develop an HRA approach that overcomes some of the main criticisms. At the present state of development a detailed method has not yet been formulated. It appears as if many of the developments aim directly at the inclusion in a dynamic simulation tool, which obviously reduce the need for a manual analysis method.

4.5 Human Interaction Timeline (HITLINE)

HITLINE is a proposal for an HRA approach that is based on a simulation of how the system develops in time (Macwan & Mosleh, 1994). It is thus a computerised approach rather than a manual HRA approach. The simulation uses the plant logic model, i.e., the PSA event tree representation, rather than a joint simulation of an operator model together with a process model. HITLINE specifically provides a way of incorporating the errors of commission.

4.5.1 Method

The method has three major steps:

- **Screening**. The purpose of the screening is to identify the set of initial conditions that will give rise to a specific HITLINE from the underlying PSA event tree. The initial conditions refer to expected failures of hardware, instrumentation, and operators.

- **Analysis of interaction between operator and plant**. The interactions between operator and plant are considered in the light of possible error modes and PSFs. An important part of HITLINE is the assumption that not all error modes are equally likely, but that they depend on the conditions. This leads to an identification of the likely error modes given the prevailing conditions, which clearly is an improvement over first-generation HRA methods.

- **Incorporation of interaction into the plant logic model.** The last step is to include the interaction and the likely error modes in the plant logic model. In the case where the outcome of a failure goes beyond the binary representation of the event tree, several HITLINEs can be constructed. This is particularly the case for failed operator actions, cf. Section 4.5.2 below.

The analysis is basically a forward propagation through each HITLINE. This simulation may be accomplished using different available tools, for instance DYLAM (Macwan et al., 1996). The purpose of the analysis is to generate automatically the actual event tree using the pre-defined probabilities for the various events, rather than do it by hand.

4.5.2 Classification Scheme

HITLINE uses an extended description of errors of commission, which applies the notion of intentional errors to make additional distinctions e.g. between intentional and unintentional commissions. Additional error types, based on empirical studies, are also proposed (Macwan et al., 1994). Examples of these are shortcut, delays, incorrect procedure selection, etc. The classification scheme is clearly more detailed than in first-generation HRA approaches, but there does not appear to be a systematic basis for the extensions.

4.5.3 Model

HITLINE does not refer to a specific operator model. The classification scheme suggests that some version of an information processing model is implicitly used.

4.5.4 Conclusion

HITLINE is an interesting proposal for extending the first-generation HRA approaches by increasing the complexity of the event tree modelling. This is accomplished partly by using an enlarged set of error types, and also by using multiple event trees corresponding to the different combinations of the initial conditions. In practice, this also requires a need for a computerised analysis, since the complexity easily becomes too demanding for a manual approach. HITLINE also addresses how the PSFs may influence the error modes in a qualitative rather than a quantitative sense. The quantification aspects of HITLINE are, however, quite traditional, possibly out of a need to interface with the established PSA approaches.

4.6 A Technique For Human Error Analysis (ATHEANA)

The purpose of ATHEANA (Cooper et al., 1996) was to develop an HRA approach that could improve the ability of PRAs/PSAs to identify important human-system interactions, represent the most important severe accident sequences, and provide recommendations for improving human performance based on an analysis of the possible causes. Although ATHEANA has been developed so that it can be used as part of the established PRA/PSA approach, it does not uncritically accept the PRA event tree as the only basis for an analysis, but provides a possibility of enhancing the initial PRA model, cf. below.

4.6.1 Method

The method used by ATHEANA is clearly described, and can be summarised by the steps shown in Figure 3 (from Cooper et al. 1996, p. 3-4).

The method begins by identifying the possible Human Failure Events (HFE) that are described by the PRA event tree. These are further characterised in terms of unsafe acts (slips, lapses, mistakes, and circumventions) referring to the set defined by Reason (1990). The next step is to consider the Error Forcing Contexts (EFC), which are defined as the combined effects of performance shaping factors and plant conditions that make human erroneous actions likely. This is an important extension of the traditional concept of PSFs, and acknowledges that human actions to a significant degree are determined by the context. The EFCs are provided as verbal descriptions, rather than as a set of pre-defined categories.

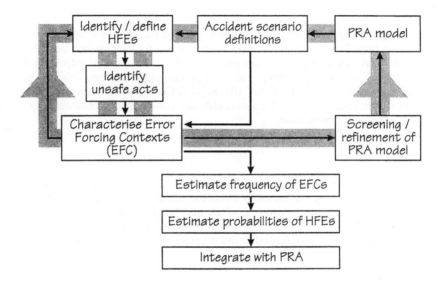

Figure 3: ATHEANA method

The ATHENA method contains two important loops. The first is from the characterisation of the EFCs to the identification of the HFEs. This recognises that an improved description of the context may enable a better identification of HFEs, and that this in turn may amend the description of the context. The second is from the characterisation of the EFCs to the PRA model. This suggests that the outcome of the qualitative part of the HRA may be used to modify the underlying PRA model, for instance by pointing to conditions or human-system interactions that have been missed in the first place.

The final quantification steps are disappointingly traditional, as expressed by the equation:

$$P(E|S) = \sum_{\substack{unsafe \\ act_i}} \sum_{\substack{error \\ forcing \\ context_j}} P_{ij}(S)$$

where P(E|S) is the probability of the HFE in scenario S and $P_{ij}(S)$ is the probability of unsafe action$_i$ resulting from EFC$_j$ in scenario S.

4.6.2 Classification Scheme

ATHEANA uses a classification scheme in two different ways. Firstly, it conforms with the PRA/PSA tradition in distinguishing between omissions and commissions as basic HFEs. Secondly, it uses Reason's (1990) characterisation of unsafe acts as a further refinement of the basic HFEs. Although ATHEANA acknowledges several of the recent developments in cognitive psychology and cognitive engineering, it does not go very far in proposing a classification system. It is possible that the need to interface to PRA/PSA has provided an obstacle for that.

4.6.3 Model

Even though ATHEANA, perhaps deliberately, remains with the traditional classifications of error types, it does argue that a better operator model is needed. In accounting for the links between EFCs and HFEs, reference is made to an information processing model with the following four stages: (1) detection, (2) situation assessment, (3) response planning, and (4) response implementation. This is a generic model that, incidentally, is quite similar to the SMoC (Hollnagel & Cacciabue, 1991; see also Chapter 6). The model could have been used as the basis for a more elaborate classification scheme, but this has not been done at the present stage of development.

4.6.4 Conclusion

ATHEANA is an HRA approach that has been developed in the mid-1990s and which only recently has been published. Although it very much remains within the mould of first-generation HRA approaches, it does propose an iterative qualitative analysis, which has the possibility of providing a significantly improved basis for quantification. At present, work is underway to develop an ATHEANA application tool and to strengthen the steps of the method. It will therefore be interesting to see how it develops in the future.

4.7 Conclusions

The characteristics of the six "newer" HRA approaches are summarised in Table 4, using the same categories as for the survey of the first-generation HRA approaches. Taken as a whole, the six approaches do not have many things in common, except trying to improve on some of the recognised shortcomings of existing approaches. The alternatives range from clearly focused extensions to the development of a radically different alternative. It is nevertheless possible to identify some themes that can be found in several of the approaches.

Enhanced PSA event trees. On the whole, the proposed alternatives are still linked PSA and several of them explicitly recognise the need to be able to interface to PSA as it is currently practised. The rigidity of the PSA event tree is nevertheless seen as something of a problem, and several of the alternatives propose ways of relaxing this constraint. In ATHEANA this is done by describing a feedback loop from the qualitative part of the HRA to the underlying event tree. In HITLINE, the solution is to use a number of instances or versions of the event tree, and to combine the results in the end. CES dispenses with the event tree altogether, and the EPRI project has the long term goal of developing a simulation based HRA, which also overcomes the limitations of a fixed event tree.

Diversified error modes. All the six alternatives propose a way of extending the traditional description of error modes. Common to them all is the acknowledgement of the importance of "cognitive errors" and / or the need to expand the notion of errors of commission. The solutions are, however, quite different and vary from a depth-first development of a single category (for example INTENT) to a breadth-first extension (EPRI project). In relation to a main theme of this book, none of the proposals maintain a clear distinction between phenotypes (manifestations) and genotypes (causes).

Expanded treatment of PSFs. Three of the six alternatives also provide a qualitative, rather than a quantitative, description of how the PSFs may affect performance. (INTENT and COGENT are extensions of THERP, and CES takes a completely different approach.) All though they do it in different ways, it is significant that they allow for the influence of the PSFs at an early stage of the analysis, that is, as part of the context rather than as an adjustment of the HEPs.

Table 4: Summary of alternative HRA approaches

HRA Approach	Method description	Classification scheme	Operator model	PSF effects
CES	Highly developed simulation tool	Intentions, problem solving strategies	Detailed model (EAGOL)	Implicit (as part of the approach)
INTENT	THERP with extensions	Errors of intentions, but not completely consistent	Same as THERP. Some reference to GEMS, but not really used.	Same as THERP
COGENT	Depends on expertise of analyst.	Cognitive errors (SRK + mistakes / slips / lapses)	General information processing model	Same as THERP
EPRI project	Method requirements specified, but details not given.	Elaborated error modes + causes.	Basic information processing model (three stages)	Integrated via operator model
HITLINE	Well described	Extended, but not systematic	No explicit model	Integrated via classification scheme
ATHEANA	Well described	Minor extension of basic schemes	Basic information processing model (four stages).	Integrated with classification as Error Forcing Contexts.

Improved operator models. Three of the six alternatives also consider the operator model explicitly. Again, INTENT and COGENT basically follow THERP, while HITLINE is not explicit about this issue. CES, of course, is the most radical by using a sophisticated AI-based model. In the other cases use is made of a basic multi-stage information processing model. As argued in previous chapters, HRA does not really have a need for a very detailed model, but it is important that there is some kind of foundation for the classification scheme. Examples of this are provided by several of the alternatives.

There are several reasons why the alternatives are so diverse. One is the lack of a commonly agreed conceptual basis for HRA. Another is the differences in motivation and objectives, going from an extension of THERP to completely new solutions. Yet another is the scope of the individual projects. Some represent well-funded research projects of considerable scope, while others are based on the dedicated efforts of a small group of researchers. Some are of a commercial nature as funded research contracts, while others are closer to basic research. Altogether these developments nevertheless show a clear recognition of what the basic problems are. The remaining chapters of this book will describe an attempt to develop an HRA approach that tries to address all four issues at the same time.

5. CHAPTER SUMMARY

In the past several surveys have been made of the existing approaches to HRA, and a considerable number of approaches have been described. The purpose of most of these surveys has been to identify one or more HRA approaches that could be considered as a best choice. The criteria have, however, often been rather *ad hoc*, and the results have therefore been open to debate.

In this chapter a characterisation has been given of nine different first-generation approaches to HRA, chosen because they represented some of the most frequently used techniques. The purpose of the characterisation was not to identify which approaches were better (or worse) than the others, but rather to characterise them in terms of a common frame of reference. That frame of reference was the MCM framework described in Chapter 4. Each of the nine HRA approaches was thus characterised with respect to the classification scheme it employed, the explicitness of the method, and the characteristics of the underlying operator model.

In terms of method description all of the surveyed approaches were fairly explicit, mainly because they have all been developed in answer to practical needs. As a matter of fact none of the recognised HRA approaches seems to have been developed from a theoretical or "academic" basis alone. This, however, also means that few of them show any significant connection between the method and the classification scheme; indeed, very few of them include a well-defined classification scheme.

In terms of the classification scheme, practically all the nine approaches were developed in the early or middle 1980s, before the more detailed error classification schemes became widely known. The main distinction therefore seems to be between correct and faulty actions - or between success and failure. This dichotomy is practically endemic to the concept of the binary representation used in the event trees. This state of affairs is quite unsatisfactory, particular in the growing concern for "cognitive errors". Yet such extensions to the classification schemes have little impact on the fundamental approach, which is to consider unitary actions that can either succeed or fail. First-generation HRA approaches appear to be entrenched in the limited universe of the binary event tree and attempts of enriching the classification scheme are likely to fail unless the basic notion of an event tree can somehow be changed.

In terms of an operator model there only are very few cases where the model has been explicitly described and acknowledged. This is, perhaps, most easily seen in relation to the way in which the operator model describes or explains how the PSFs exert their effect on performance. None of the first-generation HRA approaches can do that adequately. Instead, the influence of the PSFs is usually accomplished by simply multiplying the basic HEPs with a weighted sum of the PSFs, cf. the discussion in Chapter 1.

To complement this survey, six proposals for second-generation HRA approaches were also considered using the same principles (although one of them was developed well before Dougherty's criticism). Taken as a whole, they have little in common, except trying to improve on some of the recognised shortcomings of existing approaches. This is done in various ways by: (1) enhancement of the PSA event trees, for instance by using feedback from the HRA to modify the event trees; (2) diversification of error modes in order to enable a treatment of "cognitive errors"; (3) expanded treatment of PSFs by including them at an earlier stage of the analysis, and by considering qualitative as well as quantitative effects; and (4) improved operator models, by importing ideas from multi-stage information processing models. The proposed alternatives differ considerably from each other. This is due both to the lack of a commonly agreed conceptual basis for HRA as well as to different purposes for their development. These efforts nevertheless represent a clear recognition of what the basic problems are and indicate that the solution must be an approach that tries to address all four issues at the same time.

Chapter 6

CREAM - A Second Generation HRA Method

1. PRINCIPLES OF CREAM

The preceding chapters have tried to clarify the basic principles that must be applied when dealing with human reliability and human erroneous actions, whether for analysis or prediction. These principles have been used to characterise the current state of affairs in HRA and to explain how the first-generation HRA approaches have reached their current form. One important conclusion was that all the main first-generation HRA approaches have been developed to answer practical needs - although with varying degrees of success. Few, if any, HRA approaches have been developed from a theoretical or "academic" basis alone. The practical needs have put the emphasis on the HRA methods as useful ways of achieving the practical goals. Although a well-defined method is a *sine qua non* for an HRA approach, the method should nevertheless have a conceptual basis. It is nice if a method it works, but it is not enough. In order to have confidence in the results we need to know also **why** the method works. The lack of a theoretical foundation is therefore a weakness and not a virtue.

The strong link to practical needs is probably also the reason why few of the HRA approaches show any significant connection between the method and the classification scheme - and, indeed, why few of them have a really well defined classification scheme. The main distinction made by the first-generation HRA approaches seems to be between correct and faulty actions - or between success and failure. This dichotomy is at the root of all of the approaches, and is inseparable from the notion of the binary representation used in the PSA event trees. Two basic requirements to a second-generation HRA approach are therefore that it uses enhanced PSA event trees and that it extends the traditional description of error modes beyond the binary categorisation of success-failure and omission-commission. A further requirement is that a second-generation HRA approach must account explicitly for how the performance conditions affect performance which in turns leads to a requirement for a more realistic type of operator model.

In accordance with the conclusions of the preceding chapters, CREAM has been developed from the principled analysis of existing approaches and therefore explicitly contains a method, a classification scheme, and a model. Of these, the classification scheme and the method are the most important and the two are intrinsically linked. The underlying model serves mainly as a basis for relating some of the groups of the classification scheme. In other words, CREAM has not been developed from the underlying model of cognition, but simply uses it as a convenient way to organise some of the categories that describe possible causes and effects in human action. The primary purpose of CREAM is to offer a practical approach to both performance analysis and prediction. In accordance with the minimal modelling manifesto (Hollnagel, 1993c), the model that is part of CREAM is therefore as simple as possible, and is expressed in terms of its functions rather than in terms of its structure.

The present chapter will present the classification scheme and the model. The description of the method is provided in the two following chapters. In Chapter 7 the focus is on the retrospective application of CREAM, i.e., the method for event or accident analysis. In Chapter 8 the focus is on the predictive application of CREAM - the qualitative and quantitative methods for human performance prediction that are at the core of HRA.

1.1 Method Principles

The main principles of the method associated with CREAM is that it is fully bi-directional. This means that the same principles can be applied for retrospective analysis - in the search for causes - and in performance prediction. As argued in Chapter 3 this is important in order to be able to use the results from event analyses to improve performance prediction - which in practice is the easiest way of establishing the appropriateness of the categories and concepts. The method is furthermore recursive, rather than strictly sequential. This is a consequence of the organisation of the classification scheme, and will be explained in detail below. Finally, the method contains a clear stop-rule, i.e., there are well-defined conditions that determine when an analysis or a prediction has come to the end. This is important to ensure that the use of the method is consistent and uniform across applications and users. It is also necessary because the method is recursive; otherwise an analysis or prediction could go on forever.

1.2 Model Fundamentals

It is essential to acknowledge that any description of human actions must recognise that they occur in a context. It is therefore also essential that any model that is used as a basis for describing human performance and human actions is capable of accounting for how the context influences actions. In this respect most of the traditional models fail (cf. Chapter 4). Thus neither traditional human factors models nor information processing models are able to account adequately for how context and actions are coupled and mutually dependent.

I have previously (Hollnagel, 1993d) presented an approach to the modelling of cognition that was developed to overcome the limitations of information processing models and to describe how performance depends on context. The basic principle in this approach was a description of competence and control as separate aspects of performance. The competence describes what a person is capable of doing, while the control describes how the competence is realised, i.e., a person's level of control over the situation. The level of control clearly depends on the situation itself, hence on the context. This makes immediate sense in relation to a discussion of performance reliability. It stands to reason that if there is better control of the actions, then it is less likely that any given action will fail. Complete control, however, does not exclude that an action can be incorrectly performed. There is an underlying (or residual) variability of human performance that cannot be eliminated. Accepting that, a higher degree of control should still lead to a better performance, regardless of how "better" may de defined. A supporting argument is that a higher degree of control means that the person has a better chance of detecting incorrectly performed actions, hence on the whole that performance reliability will be improved (cf. Doireau et al., 1997).

2. MODELS OF COGNITION

The model is necessary to define the relationship between components of the classification scheme, in particular the ways in which actions are typically produced, hence the ways in which erroneous actions may come about. The model must be rich enough to describe a set of cognitive functions that can be used to explain human erroneous actions. Yet it must not be so detailed that it introduces categories or concepts that do not have a practical value, i.e., which cannot be related to observations and which cannot be linked to remedial actions.

As an example, consider the case of short-term memory. Many models of cognition refer to short-term memory, and there is clearly no reason to dispute the effects of something like a short-term memory in human cognition. (It may, however, be debated whether short-term memory is an actual physical structure in the brain or rather a persistent functional characteristic, i.e., a functional structure.) It would nevertheless be of limited value if an analysis identified short-term memory *per se* as the main cause of an erroneous action. There are several good reasons for that. Firstly, the status of short-term memory as a concrete entity is debatable. Secondly, there is little one can do about the fundamental characteristics of short-term memory. It would, indeed, be much more useful if the analysis ended up by identifying the set of conditions that **in combination** with the functional characteristics of a short-term memory could explain the erroneous action, hence the event. Finally, short-term memory is a "passive" element in the sense that it does not contribute to actions, i.e., it is not a necessary part of explanations of how actions are produced.

2.1 A Simple Model Of Cognition

Earlier versions of the CREAM classification scheme made use of a simplified model of cognition called **SMoC** - which just means **S**imple **M**odel **o**f **C**ognition (Hollnagel & Cacciabue, 1991). As discussed in Chapter 4, the purpose of SMoC was to describe the basic features of human cognition, and although it implied a typical path from observation over interpretation and planning to execution, the pathways in the model were not limited to that. The small set of cognitive functions in SMoC reflects a general consensus on the characteristics of human cognition as it has developed since the 1950s, and the basic functions were found in other HRA approaches such as ATHEANA.

The two fundamental features of the SMoC were (1) the distinction between observation and inference and (2) the cyclical nature of human cognition. The former emphasised the need to distinguish clearly between what can be **observed** and what can be **inferred** from the observations. Strictly speaking, and leaving out the thorny issue of introspection, what can be observed is overt behaviour, which match the two categories of **observation** and **action execution** (perception is taken to be the actual cognitive process, while observation is the surface manifestation of it). The remaining cognitive functions can only be inferred from observations of what the person does. The cyclical nature of human cognition (cf. Neisser, 1976) means that cognitive functions unfold in a context of past events - as well as anticipated future events. Action execution, for instance, can be preceded (or caused) by planning, by interpretation, or by observation. Observation in turn can follow as the consequence of an action, as well as of an external stimulus or event. The cyclical rendering of cognition serves to emphasise the multiple ways in which observable actions can depend on both unobservable cognitive functions and other events. A cyclical model, such as the SMoC, can therefore generate any sequential model, including the well-known step-ladder model (Rasmussen, 1986), if the appropriate number and types of functions are included.

2.2 Competence And Control

The model that is used as a basis for CREAM is a further development of the SMoC, called the Contextual Control Model (COCOM). A detailed treatment of COCOM as a candidate model of cognition for other purposes, e.g. simulation of cognition, is not provided here but can be found in other places (Hollnagel, 1997).

Cognition is not only an issue of processing input and producing a reaction, but also an issue of the continuous revision and review of goals and intentions (Bainbridge, 1993), i.e., a "loop" on the level of interpretation and planning. It is reasonable to assume that this occurs in parallel with whatever else may happen, while still in some way being determined by that. Cognition should therefore not be described as a sequence of steps, but rather as a controlled use of the available competence (skills, procedures, knowledge) and resources. This, of course, has significant implications for attempts to develop detailed models of cognition. Since the purpose of the operator model in CREAM is to support a classification scheme and an associated method, the important implications are with regard to how the analysis is carried out. A strictly sequential model of cognition would correspond to a strictly hierarchical ordering of the concepts and causes, hence also to a well-defined path or set of paths through the classification scheme (which in this case even might be called a taxonomy). A non-sequential model of cognition means that the path through the classification scheme is guided by the possible causal links between the various cognitive functions, as these unfold in a particular context. These links can not be defined *a priori*, but must reflect the prevailing conditions as they are known or assumed by the analysis or prediction. The basic assumption is that human performance is an outcome of the **controlled** use of competence adapted to the requirements of the situation, rather than the result of pre-determined sequences of responses to events. This corresponds to a fundamental principle of cognitive systems engineering that human action is **intentional** as well as reactive. The method must reflect that assumption.

The non-sequential nature of cognition could be accounted for simply by weakening or removing the links between the cognitive functions in the SMoC. This would, however, lead to an "anarchic" or "anomic" type of model with no obvious links between the cognitive functions. An alternative is to use a modelling approach where competence and control are described on equal terms. **Competence** can be defined in terms of a relatively small range of cognitive functions that appear, to a greater or lesser extent, in most contemporary attempts to model the essential characteristics of human cognition. In addition, competence includes the person's skills and knowledge that may have been compiled into familiar procedures and response patterns (action templates). **Control** can be described by referring to a continuum, going from a situation where a person has little or no control over events to conditions where events are under complete control, and by emphasising characteristic modes of control along the continuum. Hollnagel (1993a) suggested, as a minimum, the following four control modes: (1) scrambled control, (2) opportunistic control, (3) tactical control, and (4) strategic control. Altogether this leads to the Contextual Control Model (COCOM), which is shown in Figure 1.

The basic difference between COCOM and SMoC is that the links between the cognitive functions have been relinquished which means that there are no pre-defined cause-effect relations. In an earlier version of the analysis method (Cacciabue et al., 1993), it was assumed that causes should be traced backwards through the chain of planning ⇐ interpretation ⇐ observation. This assumption imposed a constraint on the classification scheme. On further analysis it turned out that this constraint was unnecessary and that its removal would improve the range of possible cause-effect links (Hollnagel & Marsden, 1995). Instead, the structuring of actions is provided by the notion of control. While there is no *a priori* sequence in which the functions must be configured, there will in each particular case be an **actual** configuration (cf. the discussion of the asymmetry of time in Chapter 4). At the present stage of development, the control modes described below are not actually used for incident analysis; instead, the analysis is guided by the way in which groups of causes are associated, as described below. The control modes are, however, important for performance prediction, as it will be described in Chapter 9. The control modes are also important for the further development of a genuine second-generation HRA approach. Firstly, they provide a way to include the influence of external conditions, which differs from the traditional Performance Shaping Factors. Secondly, they open the possibility of linking the classification scheme and the method with semi-dynamic or dynamic models of cognition. This is of particular interest for attempts to base HRA more explicitly on models of cognition.

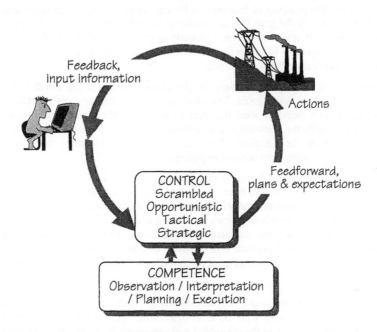

Figure 1: The Contextual Control Model of Cognition.

2.3 Four Control Modes

Control is necessary to organise the actions within the person's time horizon. Effective control is practically synonymous with the ability to plan future actions. The level of control is influenced by the context as it is experienced by the person (e.g. the cognitive goals described by Bainbridge, 1991), by knowledge or experience of dependencies between actions (pre-conditions, goals-means dependencies), and by expectations about how the situation is going to develop, in particular about which resources are and will be available to the person. In the Contextual Control Model (COCOM), a distinction is made among four characteristic control modes:

♦ In **scrambled control** the choice of next action is in practice unpredictable or haphazard. Scrambled control characterises a situation where there is little or no thinking involved in choosing what to do. This is typically the case when the task demands are very high, when the situation is unfamiliar and changes in unexpected ways, when thinking is paralysed and there accordingly is a complete loss of situation awareness. The extreme case of scrambled control is the state of momentary panic.

- In **opportunistic control** the next action is determined by the salient features of the current context rather than on more stable intentions or goals. The person does very little planning or anticipation, perhaps because the context is not clearly understood or because time is too constrained. In these situations the person will often be driven either by the perceptually dominant features of the interface or by those which due to experience or habit are the most frequently used, corresponding to the similarity matching and frequency gambling heuristics described by Reason (1990). The result is often functional fixation (De Keyser, et al., 1988).

- In **tactical control** performance is based on planning, hence more or less follows a known procedure or rule. The planning is, however, of limited scope and the needs taken into account may sometimes be *ad hoc*. If the plan is a frequently used one, performance corresponding to tactical control may seem as if it was based on a procedural prototype - corresponding to e.g. rule-based behaviour. Yet the regularity is due to the similarity of the context or performance conditions, rather than to the inherent "nature" of performance.

- In **strategic control** the person considers the global context, thus using a wider time horizon and looking ahead at higher level goals. The strategic mode provides a more efficient and robust performance, and may therefore seem the ideal to strive for. The attainment of strategic control is obviously influenced by the knowledge and skills of the person, i.e., the level of competence. In the strategic control mode the functional dependencies between task steps (pre-conditions) assume importance as they are taken into account in planning.

There are two important aspects of the modelling of control in the COCOM. One has to do with the conditions under which a person changes from one control mode to another (keeping in mind that the four modes only are points in a continuum); the other has to do with the characteristic performance in a given control mode - i.e., what determines how actions are chosen and carried out. The first aspect is important for the question of whether the operator maintains or loses control. The second aspect is important for the prediction of how the operator will perform and, possibly, of how lost control can be regained. It is also important for the prediction of the expected level of performance reliability, hence the probability that an action failure may occur.

The latter can be illustrated by the simple diagram shown in Figure 2. This indicates how performance reliability can be expected to co-vary with the level of control a person has over the situation. In cases where there is little control, such as in the scrambled and opportunistic modes, the performance reliability is low because the likelihood of making a failure is high. Note that this covers both action failures and more cognitive oriented failures, such as incorrect observations or diagnoses. When the level of control increases, in the tactical and strategic control modes, performance reliability correspondingly goes up. The reason why it is shown to taper off in the strategic control mode is that there may be too much thinking about what to do (planning, decision making), hence a less than optimal proportion of smooth, routine performance. (The notion of the control modes furthermore has a normative aspect in the sense that the tactical control mode is the optimum one). The relationship shown in Figure 2 is, however, for the purpose of illustration only, and the axes should be seen as ordinal scales. In reality the relationship is more complex, since - among other things - the available time plays a significant role. This will be described in further detail in Chapter 9.

For the purpose of CREAM there is no need to go into the technical details of how COCOM models operator cognition. The classification scheme described in this chapter only uses the overall distinction between the various types of competence, described as generic cognitive functions. The method for performance prediction, which will be described in Chapters 8 & 9, only uses the high level characterisation of the control modes. In accordance with the principle of minimal modelling, the features represented in Figure 1 will therefore suffice. Going into further detail does not really contribute to the understanding of how COCOM can be used in human reliability analysis.

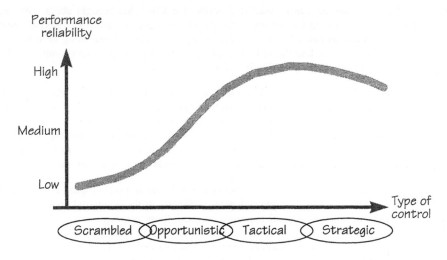

Figure 2: Proposed relation between control mode and reliability.

3. BASIC PRINCIPLES OF THE CLASSIFICATION SCHEME

In order to be of practical value, a classification scheme must obviously contain a large number of details. The classification scheme should be able to describe most, if not all, possible manifestations of erroneous actions as well as the majority of possible causes. It may be argued that a classification such as "omission - commission - extraneous action" covers all possible types of erroneous actions. However, the degree of resolution is not very high and it may require considerable skill of the analyst to assign an action to either of these categories in a consistent manner. On the other hand, if the classification scheme is very detailed the categories may be too narrow, and the number of categories may be so large that it becomes tedious or even difficult to make a classification. Clearly, the optimum is a level of detail that on the one hand is detailed enough to prevent valuable information from being lost, and on the other hand is sufficiently simple to make the assignment of events to categories manageable. This optimum cannot be defined analytically but must be based on practical experience. It is therefore important that clear principles are given for how the classification scheme can be modified, either making it simpler or more detailed.

3.1 Causes And Effects

On the highest level the classification scheme makes a distinction between effects (phenotypes or manifestations) and causes (genotypes). The effects refer to what is observable in the given system. This includes overt human actions as well as system events, such as indicated malfunctions, releases of matter and energy, changes in speed and direction, etc. It is obviously important that a classification scheme is capable of describing the possible effects. In the case of the retrospective analysis, the effects are the starting point for the analysis. In the case of performance prediction, the effects are the outcome of the analysed sequence - and typically represent something that should be avoided or prevented.

The causes are the categories that can be used to describe that which has brought about - or can bring about - the effect(s). In relation to human action the causes are typically covert or internal rather than overt or external, and must therefore be determined through a process of reasoning or inference. The classification scheme serves to define the links between possible causes and effects, and it is therefore very important that the organisation of the classification scheme is well defined.

While still remaining at the overall level, it is possible to distinguish between three major categories of causes which, according to the previous description of the MTO triad are the individual, technological and organisational phenotypes: (1) causes that have to do with the person, i.e., individual factors; (2) causes that have to do with the technological system; and (3) causes that have to do with the organisation or environment. It is nevertheless important to keep in mind that these categories reflect a convenient distinction for the sake of the method, rather than a clear ontological - or even pragmatic - demarcation.

The first category contains the genotypes that are associated with human psychological characteristics, for instance relating to cognition, to psycho-physiological variables, to the emotional state, to personality traits, etc. Depending on the psychological approach adopted, these genotypes can either be limited to those causes that are assumed to have an immediate and clear link to behaviour in a situation (such as most of the cognitive functions) or be extended to include factors that are more remote, e.g. personality traits. It may be useful to refer to the classical distinction between **proximal** and **distal** variables (Brunswick, 1956). Proximal variables or proximal genotypes are those that have a direct influence on the person's behaviour, while distal variables or distal genotypes are those that have an indirect influence. (The use of the terms proximal and distal variables does not completely solve the problem, since the terms require their own definition of what is meant by direct.) In practice we are only interested in the proximal genotypes, i.e., the possible causes for which a direct link can be established to the event characteristics.

The second category consists of the genotypes that are associated with the technological system, in particular to the state of the system and to state changes. This category includes everything that has to do with the state of components, failure of components and subsystems, state transitions and changes, etc. The category also includes everything that has to do with the man-machine interaction, the man-machine interface (information presentation and control), etc. A further distinction can be made between possible causes that have to do with technological hardware, and causes that have to do with software. This might lead to a definition of an even more detailed set of classification groups, but will not be pursued further here.

The third category contains the genotypes that characterise the organisation, the work environment and the interaction between people. Examples could be permanent features of the system (whereas temporary features would be included in the second category), aspects of the organisation (the local or the global organisation), and environmental conditions such as noise, temperature, etc. The third category might be seen as a garbage can for genotypes that do not belong in either the first or the second categories. In reality it is much more than that. The possible organisational / environmental causes are important in their own right, and human erroneous actions can only be explained fully by referring to the combination of the three major categories, as shown in Figure 3.

The genotypes of the three major categories can be further expanded to support a more detailed analysis, as shown by Figure 4. In the person related genotypes a further differentiation is made between specific cognitive functions - which in turn refer to the underlying cognitive model - and general functions that can be either temporary or permanent. In the technology or system related genotypes, a distinction is made between equipment, procedures, and interfaces - and the latter can further being divided into temporary and permanent causes. Finally, in the organisation related genotypes major subgroups are communication, organisation, training, ambient conditions and working conditions. The details of each classification group will be described in the following.

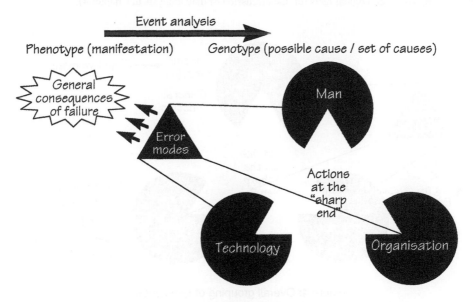

Figure 3: High level differentiation of genotypes.

In addition to the different genotypes, the phenotypes distinguish between the error modes and the category of general error consequences, i.e., the effects of the event in the system being analysed. The error modes denote observable characteristics of the person's actions while the error consequence describes what the effects are in the system. The consequences will in many cases be the starting point for the analysis; similarly a performance prediction may lead to a specification of the characteristic consequences, although this is not necessary. A typical analysis would begin either by a consideration of the error consequences (as would be appropriate for an incident analysis) or from the perspective of probable error modes (as when sources of potential risk are assessed).

In terms of context dependence, the error consequences are clearly completely application specific. Thus, the consequence of forgetting an action will depend on which system or which task is being considered, for instance whether it is a blast furnace or an aircraft. The main phenotypes and genotypes are application dependent, i.e., the details of the classification scheme may vary according to the application (this will be discussed in greater detail in the following). It specifically means that it is not feasible to develop a completely general or context independent classification scheme. The categories must always be specific either to a particular application or to a type of applications - for instance aviation or nuclear.

Finally, a subset of the genotypes will be relatively application independent. This is, for instance, the case for the genotypes that are part of the cognitive model, such as the basic cognitive functions. The cognitive functions refer to a general theory about the foundation for human action and will therefore be potentially applicable to all situations and for all applications. The cognitive functions may express themselves differently depending on the application and the context, but they will in principle always be there. For instance, diagnosis will be different for a physician, a mechanic, and a power plant operator; but the common element is the abstract cognitive function called diagnosis. Cognitive functions can therefore be described in more general terms, and do in fact provide the link between the classification scheme and the generally accepted psychological facts (cf. the discussion of data analysis in Chapter 4).

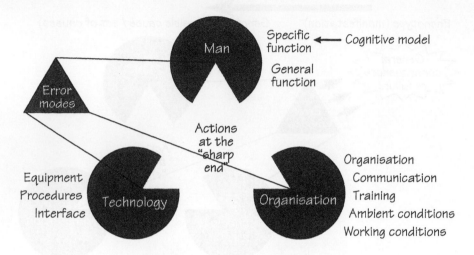

Figure 4: Overall grouping of genotypes.

3.2 A Note On Terminology

As Woods et al. (1994) so cogently have argued, the term "cause" is fundamentally misleading because it implies the existence of a well-defined relationship to the effect. In the context of accident or event analysis it is important to realise that "causes" are the outcome of a judgement made with the help of hindsight, corresponding to the definition proposed in Chapter 1 that "a 'human error' is the *post hoc* attribution of a cause to an observed outcome, where the cause refers to a human action or performance characteristic". In the context of performance prediction causes are less contentious, since they can be defined as the initiating events or starting points for the propagation of consequences through the system.

The purpose of the classification scheme is to provide a way of identifying significant links between "causes" and "effects". As long as the "causes" and "effects" refer to the beginning and end points in the description of the event, it is quite reasonable to call them causes and effects. Yet any description, whether for analysis or prediction, will contain a number of intermediate steps that provide the "explanation" of why the event developed - or would develop - in a specific way. As an example, consider the path from "causes of human malfunction" to "external mode of malfunction" in Figure 5 below. In between the starting and end points are "mechanisms of human malfunction" and "internal mode of malfunction". It is perfectly legitimate to consider these stages as intermediate "causes" and / or "effects", but it is confusing if the very same terms are used.

Although this problem must be faced by every kind of classification system, it is particularly cumbersome for CREAM because the way of developing "paths" between the categories of the classification scheme is essential for the associated method. In order to prevent any unnecessary confusion I shall therefore use the terms antecedent and consequence to separate aspects of the classification groups. The antecedent is that which gives rise to a specific consequent or consequents, given the premises of the classification scheme. (This has the additional benefit that antecedent refers to what has gone before and implies a relationship with what ensues - although not necessarily a causal one. Using the concepts from Figure 5 as an example, the "mechanisms of human malfunction" can be described as the consequent of "causes of human malfunction" and the antecedent of "internal mode of malfunction". The terms cause and effect will be used specifically to denote the end points of the event being considered. The cause can thus be either the "root" cause or the initiating event, while the effect corresponds to the error mode or phenotype.

4. CLASSIFICATION GROUPS

The classification scheme used in CREAM does not form a hierarchy of classes and subclasses. There are two reasons for this. Firstly, there is insufficient knowledge about the causes of human actions to produce a consistent hierarchical classification. This goes for human actions seen in isolation, and even more so if human actions are seen as part of a socio-technical context. Secondly, a hierarchical classification system forces the analysis to become strictly sequential, i.e., to move from one end of the hierarchy to the other - from top to bottom or from bottom to top, depending on where the analysis starts. This means that the depth of the analysis is pre-defined, or at least that there is a maximum (and usually also a minimum) depth of the analysis corresponding to the number of levels in the hierarchy. Similarly, the transitions between categories are defined in advance because they must follow the order laid down by the hierarchy. A hierarchical classification system must therefore either be correct for all applications or domains, or be limited to a specific range of applications. The first is not the case, at least since there is no empirical evidence for that. (It may also be impossible in principle although this may be difficult to prove.) If the classification system is limited to a specific range it means that there must be more than one classification system possible. But since a hierarchical classification system must reflect an underlying ordering principle, in this case a theory about the causal nature of human action, it is difficult to see how there could be several different hierarchical classification systems. This dilemma is nicely resolved if alternatives other than hierarchical classifications are considered.

As an example of the problems in hierarchical classification consider the CSNI classification scheme discussed in Chapter 4 and reproduced in outline below (Figure 5). For the retrospective analysis, the main path goes from the "external mode of malfunction" via "internal human malfunction" and "mechanism of human malfunction" to "causes of human malfunction". For performance prediction, the reverse path should be followed. (Note that the classification scheme is not strictly hierarchical due to the links from "personnel task" to "internal human malfunction" and "external mode of malfunction" respectively.) In neither case is it possible to deviate from the pre-defined path, nor is it possible to increase the depth of the analysis or to consider a combination of cause-effect relationships. This lack of flexibility is typical for most classification schemes. The clear structure of the scheme makes the classification simple to do, but it does have the unwanted consequence that the classification scheme is the same - and the paths through it are the same - regardless of the circumstances and the situation. A hierarchical classification scheme therefore does not allow for the influence of the context, except in a quite narrowly defined way.

An alternative to having a strictly hierarchical classification scheme is to make use of a number of more flexibly connected **classification groups**. This is the alternative used by CREAM. The definition of the groups is based on two principles. Firstly, that the groups shall reflect the principles of differentiation or specialisation illustrated by Figure 4. Secondly, that the number of groups shall be so large that all reasonable sets or clusters of antecedents and causes can be recognised. This is important in order to maintain sufficient overlap between the classification scheme proposed here and already existing classification schemes. The number of groups shall, however, not be so large that it becomes difficult to use them in an event analysis. In practice, an analysis must provide an acceptable trade-off between the amount of work required and the level of detail of the outcome. If there are too many classification groups, the task of managing them and maintaining an overview of the analysis may become prohibitively large. If there are too few classification groups - say, only omission and commission or the categories of the CSNI scheme - it very easy to perform the analysis, but the value of the outcome may be limited. Considering the overview provided by Figure 4, we can define the set of classification groups described below. Only one of these refers to phenotypes, namely the error modes.

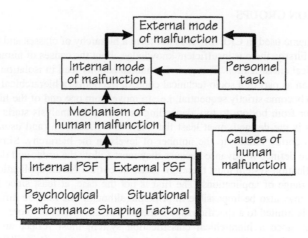

Figure 5: CSNI classification scheme as a hierarchy.

◆   **Error modes.** The error modes describe the manifestations on the level of observable behaviour. The term error mode denotes the particular form in which an erroneous action can show itself. Error modes can refer either to the phenomenal (in the philosophical sense) features of the time-space continuum or to the systematic phenotypes (Hollnagel, 1993b).

The main category "person related genotypes" contains the groups of antecedents that can be associated with the person or the user. The main category is not itself a group in the classification scheme but only a convenient label; the antecedents are ordered in the following three groups:

◆   **Specific cognitive functions.** The specific cognitive functions describe the functions that are assumed to constitute the basis for meaningful human action. The cognitive functions must reflect the principles of the underlying model of cognition. In CREAM these principles are taken from the Contextual Control Model (COCOM) as described above.

◆   **General person related functions (temporary).** Temporary person related functions are typically psycho-physical or physiological states - or even emotional states - that are characteristic of the person at a given time, but which are not constantly present. Classical examples from this classification group are circadian rhythm or time pressure.

- **General person related functions (permanent).** Permanent person related functions are constant characteristics of an individual, as for instance colour blindness or cognitive biases in thinking and judgement.

The main category "technology related genotypes" contains the groups of antecedents that can be associated with the technological system. The main category is not itself a group in the classification scheme but only a convenient label; the antecedents are ordered in the following four groups:

- **Equipment.** This group refers to the purely technological elements, such as mechanical or electronic components (including software), sub-systems, control systems, etc. There may be a potential overlap with antecedents in the "temporary interface" group.

- **Procedures.** This classification group refers to the system specific procedures or prescriptions for how a task shall be performed. It may conceivable overlap with the "organisation" group.

- **Interface (temporary).** This group describes antecedents related to temporary conditions of the man-machine interaction, such as failure of information presentation, limited access to controls, etc.

- **Interface (permanent).** This group describes antecedents related to permanent features of the man-machine interface, typically design flaws or oversights.

The final main category "organisation related genotypes" contains the groups of antecedents that can be associated with the organisational environment, such as the working conditions in general. The main category is not itself a group in the classification scheme but only a convenient label; the antecedents are ordered in the following five groups:

- **Communication.** This classification group refers to everything that has to do with the communication between operators, or between an operator and the technological system. There is a potential overlap with the "temporary interface" group, and with the "ambient conditions" group.

- **Organisation.** This classification group refers to antecedents that have to do with the organisation in a large sense, such as safety climate, social climate, reporting procedures, lines of command and responsibility, quality control policy, etc.

- **Training**. This classification group refers to antecedents that have to do with the preparedness of the operator, in particular the steps taken to ensure that operators have the competence (skills and knowledge) that are required to perform the task.

- **Ambient conditions.** This classification group refers to antecedents that characterise the ambient conditions, such as temperature, time of day (or night), noise, etc. These are all factors that have an impact on the well-being of the operator, hence on performance efficiency.

- **Working conditions**. This classification group refers to antecedents that characterise the working conditions, such as task demands, workplace design, team support, and working hours. The working conditions differ from the ambient conditions in the sense that the latter refer to factors that can be measured in a standardised fashion, while the former refer to factors that depend on the operator's perception of them (i.e., it will not be enough simply to measure them). Changes in working conditions will also be faster than changes in ambient conditions.

4.1 Details Of Classification Groups

In the following sections the details of the classification groups are described. This description makes a distinction between **general** consequents and **specific** consequents - or in the case of error modes, general effects and specific effects. In each case the categories describe the consequents that are characteristic for a group, i.e., the way in which the "effects" of that group - as an antecedent - can be detected or analytically identified. For the time being, the difference between general and specific can be seen as corresponding to the amount of information that is available for a situation. If sufficiently detailed information is available, the specific consequents should be used; otherwise the general consequents should be used. Later in this chapter it will be described how the distinction between the general and the specific has consequences for how event analysis and performance prediction take place, and how it constitutes the basis for defining links between the classification groups.

4.1.1 Error Modes (Basic Phenotypes)

The error modes are the categories that describe how an incorrect or erroneous action can manifest itself, i.e., the possible phenotypes. The term error mode is used in analogy with the term failure mode as it is applied to the technological parts of the system. Since actions must take place in a four-dimensional time-space continuum, it is possible to define an exhaustive set of error modes. (The same is not the case for the antecedents or phenotypes.) The possible error modes are illustrated in Figure 6, which corresponds to the description given in Chapter 4.

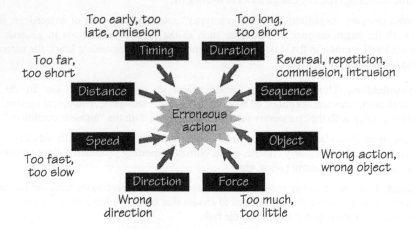

Figure 6: Dimensions of error modes.

Rather than define eight different classification sub-groups for the error modes, it is practical to divide them into the four sub-groups described below.

Action at the wrong time. This sub-group includes the error modes of timing and duration. The content of this sub-group is shown in Table 1. Note that the error modes refer to a single action, rather than to the temporal relation between two or more actions. In the latter case a more elaborate classification of temporal relations has been provided by Allen (1983).

Table 1: Categories for "action at wrong time".

General effects	Specific effects	Definition / explanation
Timing	Too early	An action started too early, before a signal was given or the required conditions had been established. (Premature action)
	Too late	An action started too late. (Delayed action)
	Omission	An action that was not done at all (within the time interval allowed).
Duration	Too long	An action that continued beyond the point when it should have stopped.
	Too short	An action that was stopped before it should have been.

Action of wrong type. This sub-group includes the physical characteristics of force, distance, speed, and direction. The content of this sub-group is shown in Table 2. Force denotes the power or effort that is applied to an action. If too much force is used equipment may malfunction or break; if insufficient power is used, the action may not have any effect. Distance denotes the extent in space of an action; it is synonymous with magnitude and may be linear or angular. Speed denotes how quickly an action was carried out. In many cases there is a limit to how quickly something can be done; examples are braking with a car or cooling a material - or the notorious Therac-25 accidents (Leveson & Turner, 1992). Note that there is some overlap with the categories of the previous group. For instance, an action that is performed too quickly will most likely be of too short a duration. The choice of one error mode ("speed") or the other ("duration") depends on whether it is the speed of the execution or the time when the action ends that are important for the consequences. Finally, direction denotes the line or course along which an action takes place, such as the movement of a lever or the turning of a knob, or the type of movement.

Table 2: Categories for "action of wrong type".

General effect	Specific effect	Definition / explanation
Force	Too little	Insufficient force.
	Too much	Surplus force, too much effort.
Distance / magnitude	Too far	A movement taken too far.
	Too short	A movement not taken far enough.
Speed	Too fast	Action performed too quickly, with too much speed or finished too early.
	Too slow	Action performed too slowly, with too little speed or finished too late.
Direction	Wrong direction	Movement in the wrong direction, e.g. forwards instead of backwards or left instead of right.
	Wrong movement type	The wrong kind of movement, such as pulling a knob instead of turning it.

Action at wrong object. This sub-group only includes the error mode of object, cf. Table 3. This could, in principle, be expressed in terms of wrong direction and wrong distance or magnitude, since that can be used to account for a position in three-dimensional space. However, it makes sense to use the conceptually simpler description in terms of the wrong object. Action at wrong object is one of the more frequent error modes, such as pressing the wrong button, looking at the wrong indicator, etc.

Table 3: Categories for "action at wrong object".

General effect	Specific effect	Definition / explanation
Wrong object	Neighbour	An object that is in physical proximity to the object that should have been used.
	Similar object	An object that is similar in appearance to the object that should have been used.
	Unrelated object	An object that was used by mistake, even though it had no obvious relation to the object that should have been used.

Action in wrong place. This sub-group only includes the error mode of sequence, cf. Table 4. The "wrong place" thus refers to the relative order of the actions rather than to a physical position, and an alternate term might have been "action out of sequence". Actions that take place at the wrong physical position are covered by the sub-groups of "wrong object" or "action of wrong type". Performing an action at the wrong place in a sequence or procedure is a common erroneous action, and has been described as place-losing and place-mistaking errors or blends (Reason, 1984). Some of the basic features of this has been discussed in Chapter 2 ("omission-commission"). The terminology used here is based on the systematic discussion of phenotypes in Hollnagel (1993b).

Table 4: Categories for "action in wrong place".

General effect	Specific effect	Definition / explanation
Sequence	Omission	An action that was not carried out. This includes in particular the omission of the last action(s) of a series.
	Jump forward	One or more actions in a sequence were skipped.
	Jump backwards	One or more earlier action that have been carried out, is carried out again.
	Repetition	The previous action is repeated.
	Reversal	The order of two neighbouring actions is reversed.
	Wrong action	An extraneous or irrelevant action is carried out.

Finally, an action can of course also go as planned, corresponding to the category of **no erroneous action**. This can either be included in each of the classification groups, or kept as a separate group - depending on how the analysis is performed.

4.1.2 Person Related Genotypes

First among the person related genotypes are the specific cognitive functions. It is a fundamental trait of human cognition that it is **covert**, i.e., that it cannot be observed. While there is wide agreement about the general characteristics of human cognition, both in terms of types of functions and in terms of functional characteristics - and in particular their limitations (Simon, 1972) - there is less agreement about the details. There is no shortage of specific models but it seems that the more detailed a model of cognition is, the less likely it is to be correct - in the sense that it can be substantiated by empirical evidence. The reason for that is our knowledge of human cognition remains limited despite more than a century of psychological research and experimentation. It therefore makes sense to be cautious in the description of the cognitive functions. While it is clearly necessary to refer to a model of cognition, the model should not be unnecessarily complex. The detailed model of cognition used in this classification scheme was described above (Section 2).

For the purpose of the analysis of erroneous actions it is, of course, highly desirable to be able to identify the cognitive functions that can be used to explain observed actions. The motivation for an analysis is the need to determine **why** an event occurred and to find out **what** can be done about it - usually to prevent it from happening again. Consequently, the possible causes should be described at a level where they can be used as a basis for recommending changes. This means that the possible causes should not refer to specific details of a theory of cognition - or a model of cognition - unless the theory or model has a strong link to actual performance. Possible causes that refer to hypothetical "cognitive mechanisms" should be avoided, since hypothetical "cognitive mechanisms" have limited practical applications. The cognitive functions that are used as a basis for the classification should also be arguably correct, i.e., they should conform to the established knowledge about cognition but they should not be too speculative. For this reason, the division into specific cognitive functions is kept as simple as possible.

The cognitive functions that are the basis for thinking and decision making can be categorised in many different ways. One of the simplest is to differentiate between **analysis** and **synthesis**. Analysis refers to the functions that are invoked when a person tries to determine what the situation is, typically including observation, identification, recognition, diagnosis, etc. Synthesis refers to the functions that are invoked when a person tries to decide what to do and how to do it; this typically includes choice, planning, scheduling, etc. In the present classification scheme the category of analysis includes **observation** and **interpretation**. Analysis thus describes all aspects of receiving data and information from the devices, whether it is as a reaction or response to a signal or an event or it is actively looking for information. The category of synthesis includes **planning** and **execution**, where the latter has already been covered by the error modes.

Observation. This group refers to the consequents of failed observations. These consequents need not be observable in the same way as the error modes described above. They are rather consequents in the sense that they are something which by itself either can be or contribute to the explanation of an error mode, i.e., they can be the antecedents of another classification group. The practical distinction between consequents and antecedents, as it is used in the CREAM classification scheme, is explained later in this chapter (Section 4 and following).

Table 5: Categories for "observation".

General consequent	Specific consequent	Definition / explanation
Observation missed	Overlook cue / signal	A signal or an event that should have been the start of an action (sequence) is missed.
	Overlook measurement	A measurement or some information is missed, usually during a sequence of actions.
False observation	False reaction	A response is given to an incorrect stimulus or event, e.g. starting to drive when the light changes to red.
	False recognition	An event or some information is incorrectly recognised or mistaken for something else.
Wrong identification	Mistaken cue	A signal or a cue is misunderstood as something else. The difference from "false reaction" is that it does not immediately lead to an action.
	Partial identification	The identification of an event or some information is incomplete, e.g. as in jumping to a conclusion.
	Incorrect identification	The identification of an event or some information is incorrect. The difference from "false recognition" is that identification is a more deliberate process.

There is a gradual transition from pure perception over recognition to identification and interpretation. The categories denote conjectured functions and represented convenient ways of separating between various stages of elaboration. If a specific information processing model is used, the separation between the categories may be more distinct, but not necessarily more concrete. For the purpose of CREAM, observation is assumed to include also identification (Table 5), whereas interpretation is considered a separate group (Table 6).

Interpretation. This term is used as a common label for understanding, diagnosis, and evaluation. It thus refers to a group of cognitive processes that have to do with the further breakdown or resolution of the observed information. This group may possibly be divided into further detail (sub-groups). Current classification schemes have usually been quite rich in referring to the various facets of analysis. The difficulty with a very detailed classification is, however, to specify the links and dependencies between the sub-groups. In CREAM, interpretation is assumed to include decision making as well as prediction (Table 6); consequently, there is no specific group for decision making.

Table 6: Categories for "interpretation".

General consequent	Specific consequent	Definition / explanation
Faulty diagnosis	Wrong diagnosis	The diagnosis of the situation or system state is incorrect.
	Incomplete diagnosis	The diagnosis of the situation or system state is incomplete.
Wrong reasoning	Induction error	Faulty reasoning involving inferences or generalisations (going from specific to general), leading to invalid results.
	Deduction error	Faulty reasoning involving deduction (going from general to specific), leading to invalid results.
	Wrong priorities	The selection among alternatives (hypotheses, explanations, interpretations) using incorrect criteria, hence leading to invalid results.
Decision error	Decision paralysis	Inability to make a decision in a situation.
	Wrong decision	Making the wrong decision (typically about action alternatives).
	Partial decision	Making a decision that does not completely specify what to do, hence creates a need for further decisions to complete the course of action.
Delayed interpretation	No identification	An identification is not made in time (for appropriate action to be taken).
	Increased time pressure	An identification is not made fast enough, e.g. because the reasoning involved is difficult, leading to a time pressure.
Incorrect prediction	Unexpected state change	A state change occurred which had not been anticipated.
	Unexpected side-effects	The event developed in the main as anticipated, but some side-effects had been overlooked.
	Process speed misjudged	The speed of development (of the system) has been misjudged, so things happen either too slowly or too fast.

Planning. This group includes all functions that have to do with setting out the detailed course of action, i.e., choosing and scheduling. It may be a matter of belief - or preference - whether the actual decision or choice is put together with analysis or with synthesis. In the present classification scheme the decision or choice has been included in the analysis group. The reason is that it may be possible to analyse and interpret a situation without actually doing anything. A process of synthesis is needed before the choice is turned into actual actions.

In CREAM, the synthesis only contains the group of **planning** (cf. Table 7). In the corresponding model of cognition (SMoC), the synthesis also included the implementation or effectuation of the plan (execution of action). In the CREAM classification scheme this is, however, been described in the groups related to error modes.

Table 7: Categories for "planning".

General consequent	Specific consequent	Definition / explanation
Inadequate plan	Incomplete plan	The plan is not complete, i.e., it does not contain all the details needed when it is carried out. This can have serious consequences later in time.
	Wrong plan	The plan is wrong, in the sense that it will not achieve its purpose.
Priority error	Wrong goal selected	The goal has been wrongly selected, and the plan will therefore not be effective (cf. the conventional definition of a mistake).

General person related functions. In addition to the specific cognitive functions, the person related genotypes also include two groups called temporary and transient general functions respectively. General person related functions are not directly linked to a specific cognitive function in the sense of being explicitly described by the associated cognitive model, but rather refer to characteristics of people that have an effect on their performance. It may, however, be a matter of interpretation whether some categories belong to the general person related functions or to the specific cognitive functions. In the case of CREAM this is not a serious issue, since there are clear principles for how changes to the classification scheme can be made. The groups do not include categories that refer to the physiological constitution or anthropometric characteristics of people. If the need arises, such categories may easily be added.

It is useful to make a distinction between **temporary** (transient, sporadic) and **permanent** person related functions. The **temporary person related functions** (Table 8) only have a limited duration and may in particular change or vary during an event or a task.

The **permanent person related functions** are present in all situations, hence exert a constant influence. In some cases a specific function may belong to either group depending on how enduring it is - but it can obviously never belong to both groups at the same time. Memory problems, for instance, can be either temporary or permanent. In the current version of the classification scheme memory problems are considered as belonging to the temporary cause. If they were permanent they might express themselves as a cognitive style, since presumably the operator would learn to cope with them by adopting a suitable strategy. It is quite normal to experience a temporary functional impairment due e.g. to stress, fatigue physical working conditions, etc. (Newell, 1993). In CREAM these cases will, however, not fall under the group of permanent person related functions but rather be treated in terms of their effects. Again, the emphasis in CREAM is on the psychological rather than the physiological constitution of people (Table 9).

The category of cognitive biases is quite important, since a large number of studies have shown how cognitive biases or heuristics have consequences - sometimes serious - for human action (Tversky & Kahneman, 1974; Silverman, 1992). A considerable number of bias types have been described in the literature, but the present group only uses of the main ones.

Table 8: Categories for "temporary person related functions".

General consequent	Specific consequent	Definition / explanation
Memory failure	Forgotten	An item or some information cannot be recalled when needed.
	Incorrect recall	Information is incorrectly recalled (e.g. the wrong name for something).
	Incomplete recall	Information is only recalled partially, i.e., part of the information is missing.
Fear	Random actions	Actions do not seem to follow any plan or principle, but rather look like trial-and-error.
	Action freeze	The person is paralysed, i.e., unable to move or act.
Distraction	Task suspended	The performance of a task is suspended because the person's attention was caught by something else.
	Task not completed	The performance of a task is not completed because of a shift in attention.
	Goal forgotten	The person cannot remember why something is being done. This may cause a repetition of previous steps.
	Loss of orientation	The person cannot remember or think of what to do next or what happened before.
Fatigue	Delayed response	The person's response speed (physically or mentally) is reduced due to fatigue.
Performance Variability	Lack of precision	Reduced precision of actions, e.g. in reaching a target value.
	Increasing misses	An increasing number of actions fails to achieve their purpose.
Inattention	Signal missed	A signal or an event was missed due to inattention. This is similar to "observation missed", the difference being whether it is seen as a random event or something that can be explained by a cognitive function.
Physiological stress	*Many specific effects*	A general condition caused by physiological stress. This may have many specific effects.
Psychological stress	*Many specific effects*	A general condition caused by psychological stress. This may have many specific effects.

4.1.3 Technology Related Genotypes

The technology related genotypes include everything that can be traced directly to technological parts of the system. This includes in particular the various technical malfunctions that may occur, inadequacies of the operational support systems - and in particular of procedures - and general issues of the interface.

A conspicuous feature of the technological system - the process, the control system, and the interface - is that it may fail due to problems with the hardware or the software, together referred to as **equipment failures** (Table 10). The failure of a component or subsystem can be one of the initiating causes for an event, but is rarely a contributing cause for a human erroneous action. Failures of the interface may, on the other hand, directly affect human performance and be seen as the immediate cause of an error mode. A particularly important category is failures of software systems, which today are an integral part of the equipment.

Table 9: Categories of "permanent person related functions".

General consequent	Specific consequent	Definition / explanation
Functional impairment	Deafness Bad eyesight Colour blindness Dyslexia/aphasia) Other disability	These specific effects refer to well-defined functional impairments, mostly of a psycho-physical nature. They are therefore not defined further. Specific physiological disabilities may be added to this group if required by the analysis.
Cognitive style	Simultaneous scanning	Search for data and information is accomplished by looking for several things at the same time.
	Successive scanning	Search for data and information is accomplished by looking at one thing at a time.
	Conservative focusing	Search for data and information starts from an assumption of which the various aspects are examined one by one.
Cognitive bias	Focus gambling	The search for data or information changes in an opportunistic way, rather than systematically.
	Incorrect revision of probabilities	New information does not lead to a proper adjustment of probabilities - either a conservative or a too radical effect.
	Hindsight bias	Interpretation of past events is influenced by knowledge of the outcome.
	Attribution error	Events are (mistakenly) seen as being caused by specific phenomena or factors.
	Illusion of control	Person mistakenly believes that the chosen actions control the developments in the system.
	Confirmation bias	Search for data or information is restricted to that which will confirm current assumptions.
	Hypothesis fixation	Search for information and action alternatives is constrained by a strong hypothesis about what the current problem is.

Table 10: Categories of "equipment failure".

General consequent	Specific consequent	Definition / explanation
Equipment failure	Actuator stick/slip	An actuator or a control either cannot be moved or moves too easily.
	Blocking	Something obstructs or is in the way of an action.
	Breakage	An actuator or a control or another piece of equipment breaks.
	Release	Uncontrolled release of matter or energy that causes other equipment to fail.
	Speed-up / slow down	The speed of the process (e.g. a flow) changes significantly.
	No indications	An equipment failure occurs without a clear signature.
Software fault	Performance slow-down	The performance of the system slows down. This can in particular be critical for command and control.
	Information delays	There are delays in the transmission of information, hence in the efficiency of communication, both within the system and between systems.
	Command queues	Commands or actions are not being carried out because the system is unstable, but are (presumably) stacked.
	Information not available	Information is not available due to software or other problems.

Procedures. Much of human performance in industrial settings is guided by procedures, i.e., by carefully prepared step-by-step descriptions of a set of actions that will accomplish a specific goal, whether it is to bring about a state change or to diagnose a disturbance. Experience has shown that deficiencies in procedures or discrepancies between procedures and the working environment can be an important cause for human erroneous actions. Some of the main categories are listed in Table 11.

Table 11: Categories for "procedures".

General consequent	Specific consequent	Definition / explanation
Inadequate procedure	Ambiguous text	The text of the procedure is ambiguous and open to interpretation. The logic of the procedure may be unclear.
	Incomplete text	The descriptions given by the procedure are incomplete, and assume the user has specific additional knowledge.
	Incorrect text	The descriptions of the procedure are factually incorrect
	Mismatch to actual equipment	The procedure text does not match the physical reality, due to e.g. equipment upgrades.

Man-machine interface. A properly functioning man-machine interface (or human-computer interface) is an important prerequisite for the operators' ability to perform their tasks in an adequate fashion. The interface is usually divided into information presentation and control options. This is one area of human factors that in recent years has received considerable interest, and a significant amount of effort has gone into research on interface design and evaluation. One reason for that is undoubtedly that the immediate (attributed) cause of an erroneous action very often is related to the interface, such as inadequate information presentation. For the purpose of a reliability analysis it is, however, not necessary to go into the minute details of information presentation design. Instead Table 12 lists a number of characteristic ways in which a **temporary interface problem** can show itself, i.e., failures of a shorter duration. If the failure is due to a direct malfunction of the physical equipment it is considered as belonging to the **equipment** group.

Table 12: Categories for "temporary interface problems".

General consequent	Specific consequent	Definition / explanation
Access limitations	Item cannot be reached	An item is permanently out of reach, e.g. too high, too low, or too far away from the operator's working position.
	Item cannot be found	An item is permanently difficult to find. Infrequently used items that are inappropriately labelled fall into this category.
Ambiguous information	Position mismatch	There is a mismatch between the indicated positions of an item and the actual positions, e.g. controls have unusual movements.
	Coding mismatch	There is a mismatch in coding, e.g. in the use of colour or shape. This may lead to difficulties in the use of equipment.
Incomplete information		The information provided by the interface is incomplete, e.g. error messages, directions, warnings, etc.

Failures of the man-machine interface can be either temporary or permanent. As before, these terms should be understood in a sense relative to the duration of the situations being analysed. Something that is the case for all situations under investigation is to be considered permanent. In terms of the interface, **permanent interface problems** (Table 13) are often the result of incomplete or inadequate design. As such they are built into the system and can usually not be corrected on the spot.

Table 13: Categories for "permanent interface problems".

General consequent	Specific consequent	Definition / explanation
Access problems	Item cannot be reached	An item, e.g. a control, cannot be reached, for instance because it is hidden by something or due to a change in the operator's working position.
	Item cannot be found	An item, information or a control, cannot be located when it is needed or it is temporarily unavailable.
Mislabelling	Incorrect information	The labelling or identification of an item is not correct.
	Ambiguous identification	The labelling or identification of an item is open to interpretation.
	Language error	The labelling or identification of an item is incorrectly formulated, or is written in a foreign language.

4.1.4 Organisation Related Genotypes

The third main category of the classification scheme includes all those genotypes that can be attributed to the environment, rather than to the operator or the technological system. They belong more to the socio-technical system than to the technical system. Various terms that have been applied to denote that, but in keeping with the tradition of the MTO-triad, the term **organisation related genotypes** will be applied here. As the details of the groups will show, these categories cover more than the management or administrative parts of the organisation.

The first group contains the factors that have to do with **communication**, i.e., with the exchange of information among the operators or between operators and sources outside the control room (Table 14). Information gathering from the interface is not considered as communication, although it may be so from the point of view of e.g. computerised support systems. Classical examples of communication are found between the air traffic control and the pilots, the main control room and the technical support centre, and in managerial command and control in general.

Table 14: Categories for "communication".

General consequent	Specific consequent	Definition / explanation
Communication failure	Message not received	The message or the transmission of information did not reach the receiver. This could be due to incorrect address or failure of communication channels.
	Message misunderstood	The message was received, but it was misunderstood. The misunderstanding is, however, not deliberate.
Missing information	No information	Information is not being given when it was needed or requested, e.g. missing feedback.
	Incorrect information	The information being given is incorrect or incomplete.
	Misunderstanding	There is a misunderstanding between sender and receiver about the purpose, form or structure of the communication.

Organisation. The second group contains the factors that relate to the organisation as such, i.e., the managerial or administrative structure. As Table 15 shows, this also includes quality assurance procedures and basic social relations. The group of organisational factors can easily become exceedingly large. It is, however, important to keep in mind that the purpose of this classification scheme is to identify possible causes that are proximal genotypes, i.e., which have a direct impact on the operator's performance. This may serve to limit the number of organisation factors.

Table 15: Categories for "organisation".

General consequent	Specific consequent	Definition / explanation
Maintenance failure	Equipment not operational	Equipment (controls, resources) does not function or is not available due to missing or inappropriate management.
	Indicators not working	Indications (lights, signals) do not work properly due to missing maintenance.
Inadequate quality control	Inadequate procedures	Equipment / functions is not adequate due to insufficient quality control
	Inadequate reserves	Lack of resources or supplies (e.g. inventory, back-up equipment, etc.)
Management problem	Unclear roles	People in the organisation are not clear about their roles and duties.
	Dilution of responsibility	There is not clear distribution of responsibility; this is particularly important in abnormal situations.
	Unclear line of command	The line of command is not well defined and control of the situation may be lost.
Design failure	Anthropometric mismatch	The working environment is inadequate, and the cause is clearly a design failure.
	Inadequate MMI	The interface is inadequate, and the cause is clearly a design failure.
		Table continues on next page
Inadequate task allocation	Inadequate managerial rule	The organisation of work is deficient due to the lack of clear rules or principles.
	Inadequate task planning	Task planning / scheduling is deficient.
	Inadequate work procedure	Procedures for how work should be carried out are inadequate.
Social pressure	Group think	The individual's situation understanding is guided or controlled by the group.

Training. It is essential for day-to-day work in e.g. an industrial facility that the operators have the necessary competence (knowledge and experience) to understand the information and control the process. This is so whether the person is driving a car or monitoring a pharmaceutical process. The organisation is usually responsible for ensuring that people are sufficiently well trained or instructed in what they shall do. Lack of sufficient skills or knowledge (Table 16) is often seen as a cause for an erroneous action.

Table 16: Categories for "training".

General consequent	Specific consequent	Definition / explanation
Insufficient skills	Performance failure	Lack of skills (practical experience) means that a task cannot be accomplished.
	Equipment mishandling	Lack of skills (practical experience) means that equipment is incorrectly used.
Insufficient knowledge	Confusion	The person is not quite certain about what to do, due to lack of knowledge
	Loss of situation awareness	The person has lost general situation awareness (understanding) due to lack of knowledge.

The two final groups of categories refer to the conditions in which the work takes place. Here a distinction is made between the ambient conditions and the working conditions. The **ambient conditions** (Table 17) describe mainly the physical working conditions that have a significant impact on performance, although they may not always be fully recognised.

Table 17: Categories for "ambient conditions".

General consequent	Specific consequent	Definition / explanation
Temperature	Too hot	Uncomfortably warm.
	Too cold	Uncomfortably cold.
Sound	Too loud	Noise level is too high.
	Too soft	Signal level is too low.
Humidity	Too dry	Uncomfortably dry.
	Too humid	Uncomfortably humid.
Illumination	Too bright	High luminosity, glare, reflection
	Too dark	Low luminosity, reduced colour and contrast.
Other	Vibration	There may be other "dimensions", depending on the specific type of work.
Adverse ambient conditions	None defined	Highly context dependent, may coincide with some of the Common Performance Conditions

Working conditions. The last group of categories is the working conditions. These describe the psychological working conditions, and do to some extent include the social environment (Table 18). The latter is something that possible should be developed into its own group.

Table 18: Categories for "working conditions".

General consequent	Specific consequent	Definition / explanation
Excessive demand	None defined	Excessive task demands or insufficient time / resources.
Inadequate work place layout	Narrow work space	Available work space is not large enough for the required activities. This is often the case for maintenance work.
	Dangerous space	Work must be carried out in dangerous conditions, e.g. high voltage line work, radiation, unstable mass or energy storage, etc.
	Elevated work space	Work must be carried out where there is a risk of falling down.
Inadequate team	Unclear job description	The roles within the team are not well defined or well understood.
support	Inadequate communication	The distribution of work / responsibilities within the team is not mutually agreed.
	Lack of team cohesiveness	There is little cohesiveness in the team, hence little collaboration.
Irregular working hours	Circadian rhythm effects.	Shift work leading to disturbances of physiological and psychological functions (jet lag, lack of sleep, etc.).

4.1.5 Summary

The classification scheme described in the preceding sections is composed of phenotypes and genotypes. Both phenotypes and genotypes are further divided into more detailed classification groups, each of which has been described in terms of general consequents (or effects) and specific consequents (or effects). The phenotypes are the description of the error modes, which are divided into four classification groups called: (1) **action at wrong time**, (2) **action of wrong type**, (3) **action at wrong object**, and (4) **action in wrong place / sequence**.

The genotypes describe the categories that in the classification scheme serve as antecedents - hence ultimately as attributed causes. These are divided into ten different classification groups, which in turn are assigned to three main categories. One main category is **person related genotypes** that is further divided into: (1) **observation**, (2) **planning**, (3) **interpretation**, (4) **temporary person related causes**, and (5) **permanent person related causes**. The first three of these refer to the underlying model of cognition - where the function of execution is covered by the error modes. The second main category is the **technology related genotypes**, which is further divided into: (1) **components**, (2) **procedures**, (3) **temporary interface problems**, and (4) **permanent interface problems**. Finally, the third main category is the **organisation related genotypes**, which is divided into five more detailed groups: (1) **communication**, (2) **organisation**, (3) **training**, (4) **ambient conditions**, and (4) **working conditions**. The overall structure of the classification scheme was shown in Figure 4.

So far there has been no description of the links between the classification groups. In accordance with the principles behind CREAM, there are no permanent or hierarchical relations between the various classification groups. Instead of the hierarchical organisation, a simple principle for creating links between the classification groups will be described in the following.

5. LINKS BETWEEN CLASSIFICATION GROUPS

The purpose of the classification scheme is to provide an account of how erroneous actions can occur, as part of an event analysis or a performance prediction. It is therefore necessary that the possible relations between the classification groups are clearly described. In doing that two things must be remembered.

Firstly, since the classification scheme is not organised as a strict hierarchy, there are no pre-defined links between the classification groups. Instead, it is necessary explicitly to provide a principle by which the groups can be related to each other. This can be achieved by noting that to each **consequent** described by a classification group must correspond one or more **antecedents**, and that these antecedents must occur in the other classification groups. If not, then the classification scheme is incomplete. This was shown in a simple way by Figure 4: the genotypes were the antecedents of the phenotypes, and the phenotypes (error modes) in turn were the antecedents of the general error consequences. This repeated antecedent-consequent (or cause-effect) relationship must obviously be a feature of any classification scheme that has more than two levels of categories, and can therefore be found both in traditional human factors engineering and in the information processing approaches.

Secondly, when an analysis is made there may be varying amounts of information available. In some cases it will be possible to describe details of the event and be very specific about possible antecedents-consequent links. In other cases it will only be possible to describe the event in broad terms, and hence not possible to be precise in the analysis or in the identification of antecedents-consequent links. In order to account for this, the concepts or terms in the classification groups should provide different degrees of specificity. In practice, this requires at least two degrees, generic and specific. In the preceding this has been used to distinguish between general and specific consequents. In a similar way different degrees of specificity will be used to distinguish between general and specific antecedents.

The relation between the classification groups can be established by providing for each group a list of the antecedents that are likely candidates as explanations for the general consequents of that group, where each of these antecedents in turn either appears as a consequent of another classification group or is a "root" cause. This linking between consequents and antecedents means that the analysis can be carried on as far as one likes - which in practice means as far as there is data to support it and as far as it makes sense. It is, of course, necessary to have a well-defined stop rule, and this will be explained shortly. This principle of mutual linking between consequents and antecedents is illustrated in Figure 7. (For the sake of the example, each group has a single consequent with several associated antecedents.)

Imagine that an event analysis begins by the consequent in group C (in practice this will always be an error mode or an effect). The descriptions in group C provides a number of antecedents for the event. Assume that the most likely antecedent under the circumstances is antecedent.C.2, which also occurs as the consequent of group A. The analysis therefore continues from that point and tries to determine the likely (second-order) antecedent. Assume again that this is antecedent.A.3, which is also found as a consequent of group B. The analysis now continues from group B, trying to find the probable antecedent, and can go on as long as links are defined. If this seems very abstract, it hopefully becomes clearer when the antecedents belonging to the classification groups have been introduced. (Note, by the way, that the direction of the search could have been the opposite, i.e., starting from an antecedent and trying to find the probable consequents of that. This corresponds to making a prediction of how an initiating event will propagate through the system. In this case the stop rule is a bit easier to stipulate, since the chain obviously cannot proceed beyond an error mode. Event analysis and performance prediction will be discussed in more detail in Chapter 7 and Chapter 8.)

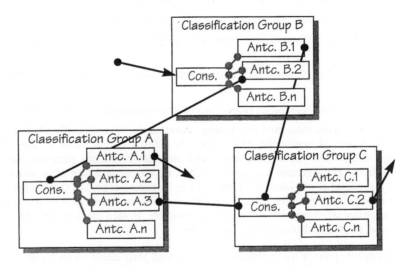

Figure 7: The principle of consequent-antecedent relations in CREAM.

5.1 Consequent-Antecedent Relations In CREAM

The consequent-antecedent relations (or antecedent-consequent relations, the order of the terms is not significant) will be described by going through the classification groups in the same order as before, i.e., error modes (phenotypes), person related genotypes, technology related genotypes, and organisation related genotypes (Figure 4).

5.1.1 Error Modes (Phenotypes)

The error modes are the observable features of actions and are either the starting point of an analysis or the end point of a prediction. From a psychological point of view the error modes are the tip of the iceberg, and the developments that result in a specific error mode can be many and varied. This is reflected in Table 19, which shows that each error mode can be related to a considerable number of general antecedents. Many of these general antecedents are furthermore common for all error modes. This is not really surprising since the general antecedents describe the likely first-order explanations for an erroneous action. The purpose of a systematic analysis method must, however, be to go beyond the immediate explanations and look for what further details can be found given the available information about the event. The general antecedents listed for the error modes are therefore only the starting point for an analysis.

A careful reading of the general antecedents for the error modes will reveal that they all occur as general consequents in the classification groups described above (Section 4). Consider, for instance, the general antecedents described for **timing / duration**. Here we find the following links:

- The general antecedent *"communication failure"* is a general consequent of *"communication"* (Table 14).

- The general antecedent *"faulty diagnosis"* is a general consequent of *"interpretation"* (Table 6).

- The general antecedent *"inadequate plan"* is a general consequent of *"planning"* (Table 7).

- The general antecedent *"inadequate procedure"* is a general consequent of *"procedures"* (Table 11).

- The general antecedent *"inattention"* is a general consequent of *"temporary person related functions"* (Table 8).

- The general antecedent *"observation missed"* is a general consequent of *"observation"* (Table 5).

The set of general antecedents associated with **timing / duration** therefore offers five different candidates from which a more detailed explanation can be built. In addition to that Table 19 also contains two possible specific antecedents. The specific antecedents are used to describe particular conditions that could have contributed to the general consequent. In this case erroneous timing of an action could have been caused by e.g. an earlier omission (of an action) or by a trapping error, i.e., that the person responds to an incorrect or incorrectly perceived cue. The specific antecedents should only be used as part of the explanation if sufficient information about the event is available. In such cases the specific antecedent is also the final cause, i.e., the analysis cannot go any further.

Whereas it is possible to provide a comprehensive set of general antecedents for each general consequent, it is usually less easy to describe the specific antecedents. This is because the specific antecedents quite naturally differ between different domains and task types. The CREAM classification scheme presented here is of a generic nature, i.e., it does not intend to be a classification scheme for a particular domain, although the influence from nuclear power operation probably can be found in several places. The number of general antecedents is therefore larger than the number of specific antecedents. In practical use it is probably necessary to develop more specialised classification schemes for different domains, such as power plants or aviation. In each case it is reasonable to expect that a larger number of specific antecedents can be described, reflecting the practical experiences from the domain. It is an important part of CREAM that the rules for extending the classification scheme are explicitly described. The attempt to make such an extension shall, however, not be made here.

Table 19: General and specific antecedents for "error modes"

General consequent	General antecedent		Specific antecedent
Timing / duration	Communication failure Faulty diagnosis Inadequate plan	Inadequate procedure Inattention Observation missed	Earlier omission Trapping error
Sequence	Access limitations Communication failure Faulty diagnosis Inadequate plan	Inadequate procedure Inattention Memory failure Wrong identification	Trapping error
Force	Communication failure Equipment failure Faulty diagnosis	Inadequate plan Inadequate procedure Observation missed	Ambiguous label Convention conflict Incorrect label
Distance / magnitude	Communication failure Equipment failure Faulty diagnosis	Inadequate plan Inadequate procedure Observation missed	Ambiguous label Convention conflict Incorrect label
Speed	Communication failure Distraction Equipment failure Faulty diagnosis	Inadequate plan Inadequate procedure Observation missed Performance variability	None defined
Direction	Communication failure Faulty diagnosis Inadequate plan	Inadequate procedure Inattention Observation missed	Ambiguous label Convention conflict Incorrect label
Wrong object	Access problems Communication failure Wrong identification Inadequate plan	Inadequate procedure Inattention Performance variability Observation missed	Ambiguous label Incorrect label

5.1.2 Person Related Genotypes

The groups in the category of person related genotypes follow the same order as above (Section 4.1.2). For each group both the general antecedents and the specific antecedents are described. Just as for the **error modes,** the categories listed under general antecedents all occur in other classification groups as general consequents. That is actually a defining characteristic of the whole classification scheme. In no case can a general antecedent in a group be a general consequent of the same group. If that was the case, the analysis - or prediction - could go into an endless loop. Apart from the methodological problems this creates, it would not produce any meaningful results.

The first three tables (Table 20, Table 21, and Table 22) show the general and specific antecedents for the main cognitive functions - observation, planning, and interpretation. This is an area where there has been a considerable amount of research, and where several specific models or theories have been proposed (cf. the descriptions in Chapter 3). It is therefore an area where it is relatively easy to find many proposals for specific antecedents. From the point of view of CREAM, practically all information processing theories and models propose specific antecedents only, since they describe relatively shallow hierarchical category structures. The three tables below have tried to limit the number of specific antecedents to those that have been shown by experience to have a general usefulness. In each case the terms should be understood in their common-sense meaning, rather than as specialised terms (for example, overgeneralisation or violation), and specific literature references have therefore not been provided.

Table 20: General and specific antecedents for "observation"

General consequent	General antecedent	Specific antecedent	
Observation missed	Equipment failure Faulty diagnosis Inadequate plan Functional impairment Inattention	Information overload Multiple signals	Noise Parallax
False observation	Fatigue Distraction	None defined	
Wrong identification	Distraction Missing information Faulty diagnosis Mislabelling	Ambiguous symbol set Ambiguous signals Erroneous information	Habit, expectancy Information overload

Table 21: General and specific antecedents for "interpretation"

General consequent	General antecedent	Specific antecedent	
Faulty diagnosis	Cognitive bias Wrong identification Inadequate procedure	Confusing symptoms Error in mental model Misleading symptoms Mislearning	Multiple disturbances New situation Erroneous analogy
Wrong reasoning	Cognitive bias Cognitive style	Too short planning horizon	False analogy Overgeneralisation Mode error
Decision error	Fear Cognitive bias Distraction Social pressure	Lack of knowledge Mode error Shock	Stimulus overload Workload
Delayed interpretation	Inadequate procedure Equipment failure Fatigue	Indicator failure	Response slow-down
Incorrect prediction	Cognitive bias Ambiguous information Incomplete information	None defined	

Table 22: General and specific antecedents for "planning"

General consequent	General antecedent	Specific antecedent	
Inadequate plan	Distraction Memory failure Wrong reasoning Excessive demand Insufficient knowledge	Error in goal Inadequate training Model error Overlook precondition	Overlook side consequent Violation Too short planning horizon
Priority error	Faulty diagnosis Communication failure	Legitimate higher priority	Conflicting criteria

The next two tables (Table 23 and Table 24) describe the general and specific antecedents for the groups of temporary and permanent person related functions respectively.

Table 23: General and specific antecedents for "temporary person related functions"

General consequent	General antecedent	Specific antecedent	
Memory failure	Excessive demand	Daydreaming Long time since learning Other priority	Temporary incapacitation
Fear	None defined	Earlier error Possible consequences	Uncertainty
Distraction	Equipment failure Communication failure	Boss / colleague Comfort call Commotion	Competing task Telephone
Fatigue	Adverse ambient conditions Irregular working hours	Exhaustion	
Performance variability	Equipment failure Excessive demand Insufficient skills	Change of system character Illness	Lack of training Overenthusiasm
Inattention	Adverse ambient conditions	Temporary incapacitation	
Physiological stress	Adverse ambient conditions Irregular working hours	Boredom	
Psychological stress	Excessive demand Insufficient knowledge	Boredom	

In the case of the permanent person related functions, the general antecedent is in all three cases given as "none". This does not mean that none of the general consequents has antecedents; surely, a specific functional impairment - such as deafness - has a cause, which even may be an environmental one. Yet in the context of the specific application of the method, the condition described by the general consequent is considered permanent, and trying to find the "cause" for it does not enrich the analysis as such. Obviously, if the general antecedent is "none", then there cannot be a specific antecedent either.

Table 24: General and specific antecedents for "permanent person related functions"

General consequent	General antecedent	Specific antecedent
Functional impairment	None defined	None defined
Cognitive style	None defined	None defined
Cognitive bias	None defined	None defined

5.1.3 Technology Related Genotypes

The technology related genotypes include the equipment (Table 25), the procedures (Table 26), and the temporary and permanent interface problems (Table 27, Table 28). In the cases of both **equipment failure** and **temporary interface problems** a relative large number of specific antecedents can be proposed. In these cases the specific antecedents can be understood either as specific instances of the general antecedent or as an additional antecedent. Thus, a maintenance failure can show itself as power failure, fire or flooding, in the sense that neither might have occurred had the system been properly maintained. On the other hand, a tremor - such as an earthquake - can be the initiating event for an equipment failure although it is hardly related to inadequate maintenance.

Table 25: General and specific antecedents for "equipment"

General consequent	General antecedent	Specific antecedent	
Equipment failure	Maintenance failure	Power failure Fire Flooding	Tremor External event Impact / Projectile
Software fault	Inadequate quality control	None defined	

Table 26: General and specific antecedents for "procedures"

General consequent	General antecedent	Specific antecedent
Inadequate procedure	Design failure Inadequate quality control	None defined

Table 27: General and specific antecedents for "temporary interface problems"

General consequent	General antecedent	Specific antecedent	
Access limitations	Equipment failure Design failure	Design Distance Localisation problem Obstruction	Ladder / stair Temporary incapacitation
Ambiguous information	Design failure	Sensor failure	Incorrect coding scheme
Incomplete information	Design failure Inadequate procedure	Indicator failure Display clutter Navigation problems	Inadequate display hardware

Table 28: General and specific antecedents for "permanent interface problems"

General consequent	General antecedent	Specific antecedent
Access problems	Inadequate work place layout	None defined
Mislabelling	Inadequate work place layout Maintenance failure	None defined

5.1.4 Organisation Related Genotypes

The organisation related genotypes include communication (Table 29), organisation (Table 30), training (Table 31), ambient conditions (Table 32), and working conditions (Table 33). For **communication** there are a number of proposals for both general and specific antecedents. These refer both to technological and psychological factors.

Table 29: General and specific antecedents for "communication"

General consequent	General antecedent	Specific antecedent	
Communication failure	Distraction Functional impairment Inattention	Noise Presentation failure	Temporary incapacitation
Missing information	Mislabelling Design failure Inadequate procedure	Hidden information Presentation failure	Incorrect language Noise

For **organisation**, there are no further antecedents proposed. This does not mean that there are no further "causes" to be found in the organisation or that there are no specific antecedents that may be used for event analysis / prediction. On the contrary, more specialised approaches to the description of organisational failures have produced quite long lists - going from MORT (Johnson, 1980) to TRIPOD (Hudson et al., 1994; Reason, 1997). The lack of entries in Table 30 should rather be seen as a reflection of the bias that has characterised first-generation HRA, i.e., the preoccupation with the technological or engineering parts of the system. This is a bias that hopefully will be weakened as alternative approaches begin to flourish. It is most definitely a bias that will have to be remedied in future version of CREAM.

Table 30: General and specific antecedents for "organisation"

General consequent	General antecedent	Specific antecedent
Maintenance failure	None defined	None defined
Inadequate quality control	None defined	None defined
Management problem	None defined	None defined
Design failure	None defined	None defined
Inadequate task allocation	None defined	None defined
Social pressure	None defined	None defined

Table 31: General and specific antecedents for "training"

General consequent	General antecedent	Specific antecedent
Insufficient skills	Management problem	None defined
Insufficient knowledge	None defined	None defined

In the case of the ambient conditions, the general antecedent is in all cases given as "none". This does not mean that none of the general consequents has antecedents, but rather that condition described by the general consequent is considered permanent, and that trying to find a "cause" therefore may not improve the analysis much. Obviously, if the general antecedent is "none", then there cannot be a specific antecedent either.

Table 32: General and specific antecedents for "ambient conditions"

General consequent	General antecedent	Specific antecedent
Temperature	None defined	None defined
Sound	None defined	None defined
Humidity	None defined	None defined
Illumination	None defined	None defined
Other	None defined	None defined
Adverse ambient conditions	None defined	None defined

Table 33: General and specific antecedents for "working conditions"

General consequent	General antecedent	Specific antecedent	
Excessive demand	Inadequate task allocation Adverse ambient conditions	Unexpected tasks	Parallel tasks
Inadequate work place layout	Design failure Communication failure	None defined	
Inadequate team support	None defined	None defined	
Irregular working hours	None defined	Shift work Changing schedule	Time zone change

5.2 The Interdependency Of Consequents And Antecedents

The retrospect analysis of error events, or, alternatively, the prediction of performance in HRA, is based on a distinction between consequents and antecedents using the principles described above. In general, the consequents and antecedents are parts of the background of the observed or predicted performance, hence are not observable themselves. In principle, they are hypothetical constructs whose absolute correctness cannot be proven. The important point is, however, that they provide a consistent basis from which a method can be developed either for event analysis or performance prediction. The usefulness of the method is therefore the main criterion for the appropriateness of description of consequents and antecedents in CREAM. As the above descriptions have shown, the distinction between consequents and antecedents is not absolute. On the contrary, a category or a factor (or whatever it may be called) is usually both a consequent and an antecedent. That is, in fact, an essential part of the structure of the classification scheme and a prerequisite for the method.

In addition to the consequents and antecedents described by the classification scheme there are also what might be called the proper consequences - which in Figure 4 were called **general error consequences**. These are the consequences that can be observed as a result of operators' actions. In a reliability analysis, the emphasis is naturally placed on potential unwanted consequences that result from erroneous actions. Such outcomes are typically described in terms of common error effects (i.e., the erroneous actions are usually not themselves considered as the consequence, but are important as the penultimate stage in the chain of events). In a retrospective event analysis, the consequences usually make up the starting point and the search for the underlying causes involves the determination of the occurrence of a certain "inappropriate" consequent-antecedent link. In the CREAM classification scheme each observable action (error mode) can have one or more antecedents. For example, the antecedent can be an external event or, more likely, an intervening cognitive function. An action is, however, rarely the result of a single cognitive function but rather relates to a complex of functions, each of which also has one or more antecedents. Thus, a misinterpretation may result in an incorrect diagnosis, which may lead to an inappropriate plan, which may lead to an erroneous action. Antecedents and consequents are thus used to account for how cognitive functions depend on each other; in the sense that the consequents of one function are the antecedent of the next - and *vice versa*.

It is useful to make a distinction between general and specific antecedents, and general and specific consequents (see Figure 8; since the arrows are not bi-directional, this shows the relations in the case of an event analysis). The categorisation of both consequents and antecedents as being either general or specific serves to simplify the analysis. If a specific antecedent or consequent is known or if it can be predicted, then it should clearly be used. If, however, there is insufficient information available the general antecedents and consequents should be used as defaults. The specific antecedents and consequents will, of course, be more precise than the general ones, but will also require supplementary knowledge and effort. It may therefore be reasonable to begin a performance prediction with the general antecedents and consequents, and only proceed to the specific ones when it is necessary. This will, among other things, prevent the analysis from becoming more complex than needed. In the case of a retrospective analysis, the starting point is usually a specific event or outcome. It can normally also be assumed that sufficient details will be available to allow the identification of specific antecedents.

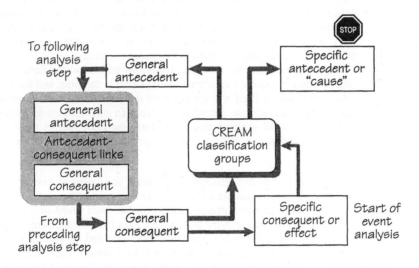

Figure 8: Links between consequents and antecedents for a retrospective analysis.

In the case of an event analysis it is assumed that a relation exists between the consequents of a cognitive function at any stage in the development of an intention / action and the cognitive functions that follow it. For performance prediction the inverse relation is assumed to hold. This is due to the fact that cognitive functions do not have pre-defined directional links in the cognitive model. The specific relations are based on the classification scheme of error modes, consequents, and antecedents as described in separate classification groups. The groups describe, for instance, how the consequent of an erroneous interpretation can be seen as the antecedent of inappropriate planning; and how the inappropriate planning may in turn may be seen as the antecedent of an inappropriate action (as a final consequence). The causal relations between the cognitive functions are very important for how an analysis should be carried out for a real context, as it will be seen below. With this framework in mind it follows that erroneous actions (error modes) can have many different antecedents and that the actual instantiation of the antecedent (as an attributed cause) is important to understand how inappropriate behaviour can come about. The causal relations create the conditions for capturing the dynamic aspects of the MMI that are crucial to carry out a comprehensive qualitative analysis as part of the error and risk management efforts.

As a simple illustration of how this relation works, consider the following example of an erroneous action.

"I intended to pick up a knife to cut the potatoes, but actually picked up the teatowel. I wanted to cut the potatoes I had just peeled for boiling. Normally when I prepare french fries, I dry the potatoes with a teatowel before cutting them. On this occasion I was doing boiled mashed potatoes."
(Reason & Mycielska, 1982, p. 70)

The starting point for the analysis is the picking up of the teatowel instead of the knife. In the group of error modes this belongs to the sub-group of "action at wrong object" (Table 3), and - since the knife was neither similar nor neighbour to the teatowel - more specifically a case of an "unrelated object". The analysis must therefore continue from the general and specific antecedents for "error modes" (Table 19), which lists eight possible general antecedents and two specific antecedents. In the present example, the most likely general antecedent is "inattention" - since the setting was an ordinary kitchen rather than a place of work (presumably), and since the activity was a rather routine one, in the sense that it did not require the person's explicit and concentrated attention. The general antecedent of "inattention" leads further to the person related genotypes and to the group of "temporary person related functions" (Table 23). In this case no plausible general or specific antecedents can be found in the table, and the most likely explanation is therefore that inattention in itself was the cause of the erroneous action. In terms of the analysis it therefore changes status from being an antecedent to being an attributed cause. Although this outcome is not very surprising, the small example nicely illustrates the principles of the consequent-antecedent links in the classification scheme. In Chapter 7 and Chapter 8 more elaborate examples will be given.

As shown in Figure 8, the link is between **general** antecedents and **general** consequents. In a retrospective analysis, the starting point is a specific effect - i.e., the observed consequence. The analysis moves backwards step by step, until a probable general antecedent has been found or until a specific antecedent has been found. In the classification scheme there are no links **from** specific antecedents; once a specific antecedent has been found, the analysis has come to a conclusion and the specific antecedent is considered the *de facto* cause. General and specific consequents may, however, coexist, i.e., it may be possible to identify a specific consequent in addition to a general consequent. Both will, however, point to the same general antecedent.

5.3 Direct And Indirect Consequent-Antecedent Links

In the description of the consequent-antecedent links (Table 19 to Table 33), one or more general antecedents are listed for each general consequent. In the case of an analysis the search principle is that the most probable consequent-antecedent links for the context are used to look further in the classification scheme. However, this does not preclude that other consequent-antecedent links in the same classification group are explored as well. (The discussion here is for analysis, but the same principle must obviously hold for prediction.)

Consider, for instance, a relatively simple group such as the general and specific antecedents for **equipment** (Table 25). This classification group provides two direct links. One is the link between "equipment failure" and "maintenance failure"; the other is the link between "software fault" and "inadequate quality control". These links are called **direct links** because the are intended to be representative of the actual experience. (The reader is reminded that the classification scheme discussed here is a general one, and not particular for any domain.) On the whole, it is very often the case that equipment failures are due to maintenance failures - or to one of the specific antecedents described in Table 25. Similarly, software failures are commonly due to inadequate quality control; in this case there are no specific antecedents listed, because there are no known external events that can create a software fault.

It is, however, also possible that an equipment failure can be due to inadequate quality control, or that a software fault can be due to a maintenance failure (in fact, a large number of software faults are presumably created during maintenance and modifications). These links in the classification group are called **indirect links**, i.e., the possible links between a general consequent and the general antecedents for the **other** general consequents in the group. The indirect links are not uncommon or inconceivable, but they are not the most representative ones either. They should therefore be considered only if the direct links do not seem right for a given context or situation. The distinction between direct and indirect links is introduced to make the analysis easier, in the sense that there normally should be little need to explore all possible combinations of consequents and antecedents. An example of how the indirect links are used will be given in Chapter 7.

5.4 Context Dependence Of Classification Groups

The separation of control and competence recognises that cognitive functions evolve in a **context** consisting of past events as well as anticipated future events. This contrasts with a strictly sequential modelling of cognition, where one action follows the next in a pre-defined pattern. The principles of COCOM, shown in Figure 1, can be used to describe how the execution of a particular action, for instance, can be preceded (or caused) by planning, by interpretation, or by observation, depending on the context and the mode of control. It is these causal connections that must be unravelled by event analysis. There is no *a priori* defined causal chain that links the different cognitive functions. In order to use the classification scheme and the method it is therefore necessary to begin by establishing an understanding of what the likely context is. From this it will be possible to infer the likely mode of cognitive control.

There is a clear difference between using the COCOM as an approach to modelling of cognition, and using it as the basis for analysis of human erroneous actions and HRA. In the use for modelling it is necessary explicitly to account for how control is implemented and which specific mode of control one should assume for a given set of conditions. This is obviously a task that requires substantial effort. In the use for analysis and prediction the impact of using COCOM is less demanding in terms of effort. It simply means that the method does not need to follow a predetermined route through the classification scheme. In the lack of any specific information the method will rely on a standard or default context, corresponding to the general nature of the classification scheme. But it can obviously be made more precise if the conditions of the event can be provided.

The degree of context dependence of the groups and the factors is important for both analysis and prediction. In a retrospective analysis the context of the event is known in advance, and this may be used to select a subset of classification groups and / or antecedents that is particularly relevant. In the case of a prediction no such prior selection is possible, because the nature of the situation is unknown. There are, however, ways in which the context dependence can be ascertained.

The context dependence of the various groups has already been touched upon in the discussion relating to Figure 4. All genotypes are context dependent, although the degree of dependency may vary. In the two cases where a distinction was made between temporary and permanent groups (general person related functions and interface problems), it was reasonable to assume that the groups of temporary factor were more context dependent than the group of permanent factors. For the other groups the variation within groups is probably larger than the variation between groups.

5.5 Possible Manifestations And Probable Causes

As mentioned several times in the preceding, the classification scheme and the classification groups are not intended to be used rigidly. Instead, groups of antecedents and consequents should be adjusted to match the current conditions of the event that is being analysed. On each occasion where the method is used the first step must be to define or characterise the context, and only then go on to describe the likely error modes and their possible antecedents.

In a retrospective analysis, the context is defined as the conditions that existed for the event that is being analysed. The first step should therefore be relatively easy to accomplish. It may, however, be useful to try to formalise the description of the context, i.e., to bring it on a common form. This will ensure both that all necessary information is available, and also that possible comparisons between different analyses are made easier. A suggestion for the common form is the set of Common Performance Conditions (CPCs), which were introduced in Chapter 4. Although the CPCs deliberately have been limited to the basic characteristics of MTO systems, they may still require some work to be filled out. The information that is provided by an event report may, for instance, not be sufficient to describe all the CPCs. An event report concentrates on the actual event but the CPCs also include information about the general conditions and the general quality of the system / organisation. Filling out the CPCs may therefore provide a useful complement to the event report, which in turn may facilitate the event analysis and the identification of the causes.

In the predictive use of the method, such as for an HRA, the expected performance conditions must be described in as much detail as possible. This can usually be done with reference to the set of situations that are addressed, the likely working environment, the people who will be working there, etc. Although many PSAs only provide little explicit information about the working conditions there is usually always something that can be used as a basis for characterising the common performance conditions such as knowledge of the type of plant that is being analysed, the organisations that are involved, and the operating history of similar plants. In any event it is important to acknowledge that human actions do not take place in a vacuum and that a description of the context therefore always must be the first step.

When the Common Performance Conditions have been determined, the next step is to identify the possible error modes. In the case of an analysis the initiating event, of course, describes itself. If the initiating event is the general error consequence (for instance, controlled flight into terrain) it must be related to the possible error modes. If the initiating event is an observed or inferred action, it is still important to limit the number of possible error modes to avoid possible incorrect classifications. In the case of performance prediction the possible error modes refer to the ways in which actions can take place. This will always depend on the nature of the working environment and in particular of the interface technology.

In most cases the analyst can make a distinction between error modes that are **possible**, i.e., which can actually occur, and error modes that are **impossible**, i.e., which cannot occur. An error mode may be impossible because of the working context, the design of the interface, the nature of the task, etc. For example, if the input to the system is provided via touch panels, it is impossible to do something with too much or too little force, barring the extreme case where the touch panel is crushed. The action is either registered (i.e., done) or not registered and does not depend on the pressure exerted. Similarly, the functionality of the control system may make it impossible to do something with the wrong speed, in the wrong direction, etc. In addition to marking certain error modes as impossible, this preparatory step can also be used to define more precisely the conditions for certain error modes. In the case of incorrect timing it is, for instance, important to be able to define the limits for what is too early and what is too late. Depending on the type of application - an aeroplane or a power plant, the limit may be expressed in terms of seconds or minutes. The same goes for the other error modes. Wherever these refer to physical dimensions or to time, the analyst should replace the general descriptors with specific ones.

When it comes to the antecedents or possible causes a similar clear distinction between what is possible and impossible cannot be made. Among the person related genotypes, permanent person related functions do not depend on the context (e.g. the categories of functional impairment in Table 9 and Table 24). The very fact that they are permanent means that they will be the same across different conditions. The temporary person related functions, as well as the specific cognitive functions may, however, be influenced by the context. In the case of specific cognitive functions it is not possible to exclude any of them as antecedents or potential causes. Any cognitive function can serve as an antecedent in the explanation of how an event occurred. There will, however, be some antecedents that are more **probable** than others. The same goes for the technology related and the organisation related genotypes respectively. If, for instance, the specification of the Common Performance Conditions has shown that the adequacy of the man-machine interface is only tolerable, then the antecedents that involve the MMI are more likely. Even though the difference may not be quantifiable, the analysis will be more efficient if the more probable antecedents are examined first.

The specification of possible modes and probable causes is a way of making the analysis more efficient and more precise. Without this focusing of the categories, the analysis will always have to go through the full set of categories. This will not only require effort that is unnecessary, but also fail to use the information that is available in terms of a description of the context.

6. CHAPTER SUMMARY

Chapter 6 began the description of CREAM, following the principles laid out in the previous chapters. The main feature of the method is that it is fully bi-directional, i.e., that the same principles can be applied for retrospective analysis - the search for causes - and performance prediction. The model is based on a fundamental distinction between competence and control that offers a way of describing how performance depends on context. Finally, the classification scheme clearly separates causes (genotypes) and manifestations (phenotypes), and furthermore proposes a non-hierarchical organisation of groups of genotypes. The classification scheme and the model are described in further detail in Chapter 6, while the method is explained in the following chapters.

The model of cognition is an alternative to the traditional information processing models, called COCOM (Contextual Control Model). Rather than describing the human mind as an information processor, the focus is on how actions are chosen. The basic notion is that the degree of control that a person has over his/her actions may vary, and that this to a large extent determines the reliability of performance. Control is described in terms of four distinct modes on a continuum going from no control (erratic performance) to perfect control (highly reliable performance). COCOM is similar to the classification scheme in the sense that it does not define specific "routes" of human information processing, but rather describes how a sequence of actions can develop as the result of the interaction between competence and context. Although COCOM itself contains considerably more detail, these are not required for the purpose of CREAM. On the contrary, it is essential for CREAM that the operator model is kept as simple as possible, so that empirical rather than theoretical needs become the determining factors.

The classification scheme basically consists of a number of a number of groups that describe the phenotypes (error modes, manifestations) and genotypes of erroneous actions, where the latter refer to a fundamental distinction between genotypes of a person related, technology related, and organisation related nature. For each classification group a definition is given of the general and specific consequents that describe the observable or inferred consequences. Following that a definition is given of the links between consequents and antecedents for each classification group. Unlike a traditional hierarchical classification scheme, these links describe a large number of potential pathways through the classification groups. The realisation of a specific pathway depends on the conditions under which the action takes place, hence "automatically" takes the influence of the context into account. This is the case both for backwards propagation (accident analysis) and forwards propagation (performance prediction). In both cases the propagation is constrained by information about the performance conditions and the depth of the analysis is determined by pre-defined stop rules. The details of how the method is applied for analysis and prediction are given in the following chapters.

Chapter 7

The Search For Causes: Retrospective Analysis

When you have eliminated the impossible, whatever remains, however improbable, must be the truth.
Sherlock Holmes

1. ANALYSIS AND STOP RULES

Chapter 6 described two of the essential components of CREAM, the model and the classification scheme. The non-hierarchical nature of the classification scheme made also demonstrated that it required a specific method. A hierarchical classification system can, by default, be used with a simple sequential search principle starting either from the root or one of the leaves. The starting point depends on the nature of the classification system, since few are designed to be used in a bi-directional manner. Once the direction of the search has been defined, the progression through the classification system itself is a simple matter.

In the case of the classification scheme proposed for CREAM, the method is less straightforward. Chapter 6 described the basic principles by which the classification groups can be related to each other using the concepts of antecedents and consequents, and this also provides the foundation for the method - or rather the methods, since CREAM can be used both in a retrospective and a predictive manner. The retrospective method describes how CREAM can be used for accident and event analysis, while the predictive method describes how it can be used for human reliability assessment. Although the two methods have many things in common and are based on the same basic principles, it makes sense to describe them separately. This chapter will describe how CREAM can be used for event analysis in the search for the plausible causes.

The aim of a retrospective analysis is to build up a path of probable cause-effect (antecedent-consequent) relationships by working backwards from the observed effect, using the relations defined by the classification scheme. The search for the probable causes must clearly start from the observed event, specifically from the error modes that characterise the event. The search is guided by the links between the categories that are part of the classification scheme. As described in the preceding chapter, an important aspect of CREAM is the notion of a well-defined stop rule - for analysis as well as for prediction. In the case of a retrospective analysis the stop rule is essential to ensure that the outcome is not determined by the current situation and temporary concerns, but that it reflects the full power and principles of the method.

The ending point for the analysis is, in some sense, arbitrary because it is always possible to take the analysis one step further. As mentioned in Chapter 1, one of the causes identified by NASA for reported shuttle incidents was lightening. This is normally seen as a natural phenomenon, and humans cannot in practice control where and when lightning strikes. It is, however, possible to control whether a shuttle is launched - or indeed landed - in dangerous conditions. The best possible example of that was the decision to launch the Challenger on January 28, 1986, despite the cold weather and the ice on the pad. The cause of the Challenger accident has often been described as a faulty gasket design. While this indeed was the immediate cause of the explosion 73 seconds after launch, the report from the accident investigation committee also pointed out that the decision to launch the Challenger in itself was flawed. Specifically, "the Commission concluded that the Thiokol Management reversed its position and recommended the launch of 51-L, at the urging of Marshall (Space Flight Center) and contrary to the views of its engineers in order to accommodate a major customer" (Rogers, 1986, p. 104). In general, while an analysis may conclude that the immediate cause of an incident was lightening (or a faulty gasket design or something else), this can never be the ultimate cause. Either one could have avoided the condition, or one could have taken precautions so that the system (in this case space shuttle) was not affected by the conditions. The other option always exists.

Another example, also from the space domain, is the failure of the first Ariane 5 launch on June 4, 1996. About 40 seconds after initiation of the flight sequence, at an altitude of about 3.700 m, the launcher veered off its flight path, broke up and exploded. (The initial press release from the European Space Agency euphemistically described this as "the first Ariane-5 flight did not result in validation of Europe's new launcher".) The accident investigation showed that Ariane 5 used the same inertial reference system (IRS) as its predecessor, Ariane 4. The acceleration and initial trajectory of the Ariane 5 launcher was, however, different from that of Ariane 4. This had not been taken into account in the specifications and tests of the IRS, and the system was furthermore tested with simulated rather than real data. During the launch a software failure occurred in the IRS, which led the main computer to make large corrections for an attitude deviation that had not occurred, leading to the break up and destruction of the launcher. In this case the analysis stopped at the decision of not verifying the functioning of the IRS. Although this conclusion is quite reasonable it would also have been possible to continue the analysis and look at the conditions under which the decision was taken. The incorrect design decision did lead directly to the malfunctioning of the IRS and the loss of the launcher, but it cannot be considered the root cause.

Even though an analysis in principle always can be taken one step further, practice may differ considerably from that. An event analysis looks for the acceptable cause(s) rather than the true or ultimate ("root") cause(s). (Philosophically it may also be argued that the notion of a true cause is meaningless, since it implies an unnecessarily simplified concept of causality.) One reason for this is that there always are constraints - in terms of money or time - on how much effort can be spent on the analysis. Another is that the classification systems and the methods used for the analysis determine what the set of acceptable causes is. (This is clearly also the case for CREAM, although the limitations should be easier to see.) A third is that in real life there may be interests at stake that, implicitly or explicitly, bias the analysis or set limits for which directions the analysis can take and how far it can go. Finally, there may simply not be enough information available to take the analysis beyond a given point. Both scientists and professional practitioners obviously have a duty to conduct the analysis in an impartial way and use the tools as thoroughly and assiduously as possible. One useful device for that is a consistent method and a well-defined stop rule.

The basic principle of an event analysis is to start from the description of the initiating event and go backwards step by step until a reasonable cause - or set of causes - has been found. The term "reasonable" is here used deliberately, since even a root cause on closer inspection will reveal itself as being relative rather than absolute (Cojazzi & Pinola, 1994). An important part of the analysis method is the criteria that are used to determine when the analysis has gone far enough. In a hierarchical classification system the stop rule is given implicitly by the fact that when the analysis reaches the level of terminal nodes (leaves), then it is not practically possible to take it any further. It has therefore by definition been completed. Conversely, if the terminal nodes have not been reached, then the analysis must continue. This principle works fine with hierarchical classification schemes, and - coincidentally - corresponds to the basic assumptions of the information processing model.

1.1 Terminal And Non-Terminal Causes

Since the CREAM classification scheme represents a network rather than a hierarchy, it is necessary to provide an explicit stop rule. An important prerequisite for that is the difference between terminal and non-terminal causes. In CREAM, the classification groups are described in terms of general and specific antecedents. Of these, the specific antecedents by definition are **terminal causes**. This means that a specific antecedent is seen as sufficient by itself, and that it therefore need not refer to a preceding antecedent-consequent link. The specific antecedent thereby becomes the attributed cause. An example of that can be found in the table of "General and specific causes for observation", where "ambiguous symbol set" is a specific antecedent to "wrong identification". Thus, when the analysis reaches "ambiguous symbol set" it has reached a natural end point.

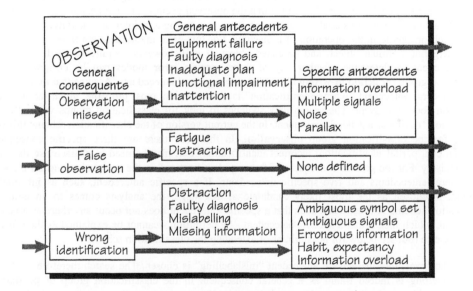

Figure 1: Terminal and non-terminal causes for "observation".

In contrast to the specific antecedents, the general antecedents are considered to be **non-terminal**. This means that it is possible - and indeed necessary - to continue the analysis from a general antecedent. The principle is that a general antecedent in one classification group corresponds to a general consequent in one or more of the other classification groups. (This is actually one of the defining characteristics of the classification scheme and of the distinction between general / specific consequents and antecedents.) If, for instance, the general antecedent of the error mode "action out of sequence" has been determined as "wrong identification", then the analysis can proceed by looking for classification groups where "wrong identification" is among the general consequents. This will in the current version lead to the group for **observation**, where "wrong identification" is found. Since "wrong identification" can have four general or five specific antecedents (cf. Figure 1), the analysis can either stop at this level if one of the specific antecedents, such as "ambiguous symbol set", is chosen or continue to the next level if a general antecedent is chosen.

1.2 Analysis Of A Fictive Event

In general, the analysis comes to a halt when it is not possible to go any further, i.e., when there are no general antecedents for a given consequent. In most cases this also means that there are no specific antecedents. The analysis therefore stops with the general antecedent of the preceding level - which is also the general consequent of the current level. The reason why the analysis does not simply stop at the preceding level, i.e., by changing the general antecedent to a specific antecedent, is that the matching general consequent may include some specific consequents. Noting what these are may be important to provide a full description of the event. If there are specific antecedents, but no general antecedents, then one of the specific antecedents may be chosen - but only if sufficient information is available to warrant that choice.

The analysis principles and the stop rule can be illustrated by the following fictive example, where the main steps are illustrated by Figure 2. Assume that the observed event is that the person chose the **wrong object**. (In a concrete case, this could be the wrong spare part during a maintenance activity, or the wrong control knob or handle in the operation of a machine. For this error mode there are three specific error modes: neighbour, similar object, and unrelated object. It will, however, require detailed information about the event to determine whether any of these specific error modes can be designated.) The classification group for **action at wrong object** lists eight general antecedents. Assume that in the current case the general antecedent "wrong identification" is chosen as the most probable, given the available information about the event. The search now switches to look for classification groups where "wrong identification" occurs as a general consequent. In the current version of the classification scheme this only happens for the classification group **observation**. In this case there are three specific consequents (mistaken cue, partial identification, incorrect identification) that may be used to characterise the event more precisely. For the general consequent of "wrong identification" in the group **observation** there are four general antecedents and five specific antecedents. If a specific antecedent, such as information overload, is chosen as the most likely candidate at this stage, the analysis comes to an end. The classification scheme is constructed such that a specific antecedent does not occur anywhere as a general consequent, which means that a specific antecedent - by definition - cannot be explained as the effect of something else (but see below for possibilities of extending the classification scheme).

Assume, however, that the general antecedent "mislabelling" is chosen. The search therefore continues, and mislabelling is indeed found as a general consequent in the classification group of **permanent interface problems**. In this case there are three specific consequents that can be used if detailed information is available, as well as two possible general antecedents but no specific antecedents. In the example it is assumed that the available information points to "maintenance failure" as the most likely general antecedent. "Maintenance failure" is found as a general consequent in the group of **organisation** and at present there are neither any general nor any specific antecedents (but see Section 2.5 for a further discussion). The analysis therefore stops with "maintenance failure" as the final cause. This might be further specialised by one of the specific consequents (incorrect information, ambiguous identification, language error), but in this case it is assumed that none of them are found useful. The analysis has thus come to an end with the last general antecedent, now the current general consequent, as the outcome.

For each step of this fictive example, it was assumed that there was enough information available to choose one candidate general antecedent over the others. This is not an unreasonable assumption to make. The analysis is about an actual event, i.e., about something that has happened, and that by itself will limit the possible causes or causal pathways. (This will be demonstrated clearly in the real life examples in the following.) It may be that sometimes two or more general antecedents are equally likely, even given knowledge about what actually happened. In such cases both - or all - paths must followed, and when they all have reached their stopping point, the candidate explanations should be seen together. It may be that one of these stands out as the more plausible, or that several remain so that a unique explanation cannot be found. The analysis principle does not guarantee that a single cause will be found. The strength of the method is rather that only a few paths have to be checked so that the analysis is not a simple combinatorial exercise, and that the choice is made in a systematic manner.

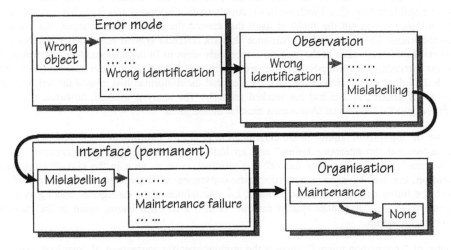

Figure 2: Analysis of a fictive event

1.3 Analysis Of A Real Event

In the case of a real event that has already been analysed there is the distinct advantage of knowing what the officially correct answer is. Even so it is useful to apply the method, firstly because it may illustrate in greater detail how the event occurred, and secondly because it can serve as a support for the method - at least if the same causes are found. Descriptions of actual events have the indisputable advantage of being correct in a rather absolute sense of the word.

There is, of course, no shortage of events that can serve as examples; a few of these - such as Three Mile Island, Challenger, and Herald of Free Enterprise - are notorious and have been well analysed, often several times, while the majority are little known and usually not analysed in any great detail. The example chosen here is a relatively recent accident that had a rather simple, but quite unusual explanation. The accident happened on June 5, 1995, when a NYC subway train on the Williamsburgh Bridge crashed into the rear end of another train that had stopped at a station. The collision speed was about 14-18 mph. According to the first reports from the accident, the motorman ran through a red light and was still applying power at the time of the crash. The accident had serious consequences as the motorman was killed and 54 passengers were injured.

Modern trains in many parts of the world are equipped with an emergency braking system that goes into action whenever a train runs through a red light. The NYC subway train was equipped with such a system, but it did not appear to work - at least it did not manage to stop the train on time. The first explanation was therefore, quite reasonably, (1) that the motorman had missed the red light, i.e., a "human error", and (2) that the emergency braking system had not worked properly. (To that can be added the almost universal statement by the officials in charge of the subway, that they had never seen that failure mode before!)

In terms of an analysis using CREAM, it is necessary first to identify the error mode. In this case it is obviously "distance", specifically "distance, too far"; one train hit the other, meaning that one train had gone too far and thereby collided with the other. An alternative is "speed, too fast", but that is not quite as appropriate since it was not the speed in itself that caused the specific consequence, viz. the collision between the trains. The error mode of "distance" has six different general antecedents and three specific antecedents. In this case none of the specific antecedents seem to fit into the accident description. Of the six general antecedents the two possible ones are "equipment failure" and "observation missed". This fits well with the initial explanation of the accident, i.e., that the motorman had missed the red light and that the emergency braking system had not worked properly. The analysis method, however, does not stop here, but requires that additional antecedents must be investigated.

In the case of "equipment failure" in the classification group for **equipment**, there is a direct link to the general antecedent "maintenance failure" and an indirect link to the general antecedent "inadequate quality control". The "maintenance failure" is a terminal cause as seen in the classification group for **organisation**, but does have the associated specific consequent of "equipment not working", which goes well with the initial explanation of the accident, i.e., the failure of the emergency braking systems. The other possible cause, "inadequate quality control" is also a terminal cause, again in the classification group for **organisation**.

In the case of "observation missed", there is the obvious specific consequent of "overlook cue / signal". Associated to "observation missed" is a direct link to five general antecedents and four specific antecedents. In addition, there is an indirect link to an additional five general antecedents. The available information about the accident is insufficient to suggest one of the specific antecedents. Of the five direct general antecedents there are two possible candidates, namely "equipment failure" and "inattention" respectively. One other theoretically possible candidate is "functional impairment". Since, however, this is a permanent person related cause, any functional impairment that could have explained the motorman passing a red light would presumably have been detected earlier. It is therefore not used as a candidate cause in this case.

Even though "observation missed" points to "equipment failure" as a possible cause, it is not the same kind of failure that was pointed to by "distance too far". In relation to "observation missed", the failed equipment would have been the signal; in the case of "distance too far", the failed equipment would have been the braking system. The cause, and therefore also the further causes, are nevertheless of the same type. It is conceivable that the observation was missed because the stop light did not function properly, which again could be due to a number of causes. The other possible cause for "observation missed" is "inattention". As the general consequent in the **temporary person related causes** it has the specific consequent of "signal missed". The direct link is to the general consequent called "adverse ambient conditions", which in turn suggest several likely causes, such as "too bright illumination". A summary of the main steps in this analysis is shown in Figure 3, which shows the links between the general consequents and antecedents.

In the actual accident it was determined that the motorman did pass the stop light, but also that the emergency braking system had worked. The assumption of a maintenance failure for the braking system can therefore be rejected. The CREAM classification scheme points to "inadequate quality control" as another possible cause. In the investigation of the accident it turned out that the train had passed the stop light with a speed of 32 mph, which is not excessively high. At this speed the stopping distance of the train was around 360 ft., but since the distance to the train ahead was only 299 ft. the collision was unavoidable. A further investigation revealed that the signal spacing had been specified in 1918! At that time trains were shorter and travelled at a lower speed. Due to a combination of developments in train technology and traffic needs, the stopping distance that was sufficient in 1918 was quite inadequate in 1995. Unfortunately, no one had paid any attention to that. The cause was therefore, in a sense, inadequate quality control in the organisation. (From another perspective, this is also an excellent example of the influence of the so-called latent failure conditions, e.g. Reason, 1991 & 1997; Hollnagel, 1995b.)

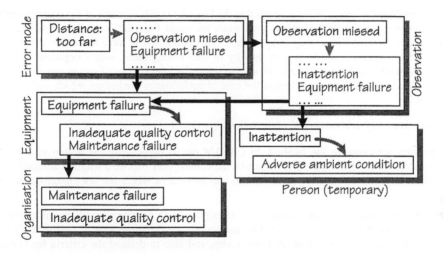

Figure 3: Analysis of NYC subway train crash.

The outcome of this analysis points to two causes that together were sufficient for the accident to occur. The analysis could look further into either cause; in terms of human reliability it would be of interest to take a closer look at the human related cause, i.e., the passing of the stop signal. This is something that happens with unfortunate regularity in train operation (Lucas, 1992) and the aetiology of this is rather well known. In the present case there was not enough information available to warrant a further analysis without resorting to speculation.

This small example has hopefully shown how the method and classification scheme of CREAM can be applied, and how a systematic use of it can be of help in finding the likely causes of an accident. The analysis obviously did not find the complete set of causes, but neither should this be expected. As pointed out previously, the classification scheme described here is of a rather general nature, hence may lack categories that are useful for a particular domain. In practice, an accident analysis should use a classification scheme that has been adjusted to take the specific characteristics of the domain into account. In the present example the causes found by the analysis were the correct ones, accepting that they were of a general nature. The analyses of both the fictive and the real event have shown how the end-point of an analysis is determined by the method and the classification scheme, rather than by the insights or experience of the individual analyst. The specification of an explicit stop rule in particular serves to minimise the effects of individual preferences and therefore also to reduce variations in performing the analysis.

2. OVERALL METHOD

The description given in the following is aimed to be sufficient for a manual use of the analysis method, although it will clearly be more efficient if the analysis can be supported by a computerised version of the classification scheme. Such as computerised tool has been developed in a prototype form, and has served as a useful vehicle for refining the concepts and testing the internal consistency of the classification scheme.

It is fundamental for the retrospective analysis that the first step is a description of the context; in this case it must be the context that existed at the time when the event occurred, i.e., an actual rather than an assumed context. In the NYC subway train example, the context would have to include the time of day, traffic conditions, weather, signal system, quality of line management, safety organisation, etc. It is necessary to know the context in order to be able to select the relevant classification groups and to focus the analysis. This is even more important in the case of performance prediction, which will be discussed in Chapter 8.

The retrospective analysis is composed of the following steps.

1. **Determine or describe the context.** In CREAM this is done using the notion of the Common Performance Conditions (CPC). To describe the context may require a detailed analysis of aspects of the application that are not usually taken into account or which are not contained in the event report.

2. **Describe the possible error modes**. This description is to be given for all possible actions, i.e., without considering a specific action. The description uses the knowledge of the application and the context to produce a limited set of error modes, and also to defined the criteria for certain error modes (e.g. when is an action too late).

3. **Describe the probable causes.** Given the knowledge about the context it is normally possible to identify categories of causes that are more probable than others. In the case where categories refer to cognitive functions, it is never possible completely to rule out any of them. For any given context there will, however, be some that are more likely than others. Thus, the work context may enforce compliance with rules, encourage deviations, support learning of skills, have a bad interface, hence promote misunderstandings or execution errors, etc.

4. **Perform a more detailed analysis of main task steps**. This stage will try to trace the possible consequent-antecedent links for the selected error modes.

The principal stages of this method are illustrated in Figure 4 below. In the following each of the three first stages of the retrospective method will be described, followed by the details of the main analysis step.

2.1 Context Description

A description of the context of an event is always a simplification, even if it is a narrated account. When the context is described in terms of a limited, and usually rather small, number of factors or aspects the simplification is substantial and great care must be taken to avoid loosing potentially important information.

Chapter 4 provided a discussion of the notion of performance shaping factors (PSF) and introduced the idea of the Common Performance Conditions (CPC). The CPCs were proposed as a way of capturing the essential aspects of the situation and the conditions for work which through long experience are known to have consequences for how work is carried out and in particular for how erroneous actions occur. The label Common Performance Conditions was not chosen because the individual CPCs are different from the classical PSFs, but because there is a difference between how the PSFs and the CPCs are used in the analysis.

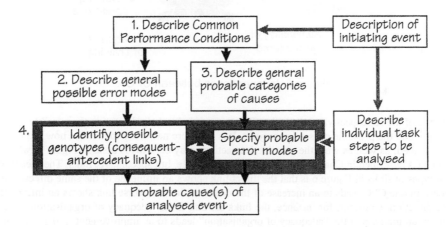

Figure 4: Overall method for retrospective analysis.

The first stage in the retrospective analysis is to describe the Common Performance Conditions. The purpose of the CPCs is to provide an adequate description of the context using a limited number of factors or aspects. The name given to this set of factors as a whole points out that they describe the general determinants of performance, hence the **common modes** for actions in a given context. The number of CPCs is kept relatively small in order to make the analysis manageable. The CPCs are intended not to be overlapping, hence they should not easily be confused with each other. On the other hand they are clearly not **independent** of each other. Chapter 4, Table 4, provided the definition of nine different CPCs together with the qualitative descriptors that are recommended for each.

As mentioned above, the CPCs cannot conceptually be independent of each other. First-generation HRA approaches tacitly seem to adopt the assumption of independence, as illustrated by Figure 5. In accordance with this the total effect of the PSFs / CPCs can be derived simply by a summation of the effects of the separate PSFs / CPCs using appropriate weights. In practice this is unfortunately not a valid assumption. It should not be necessary to put forward any complicated arguments for that; any simple example from the everyday experience of how performance depends on the context should suffice. This does not exclude that situations may occur where one or a few CPCs - such as, for instance, available time or number of goals - have the dominating influence and that these therefore may be used as approximate indicators of the influence of the context. Yet in order to develop a more comprehensive description of human performance, and in particular the reliability of human performance, it is necessary to propose an account of how the CPCs are coupled.

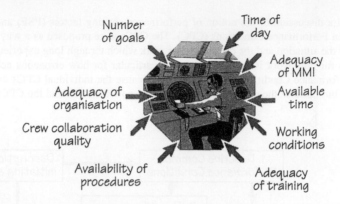

Figure 5: Independence between CPCs.

For the sake of the retrospective analysis the CPCs are needed to help determine the possible error modes / phenotypes and the probable causes / genotypes. It is therefore necessary to indicate how the interactions can be described, as well as what the possible interactions may be. This is done in Figure 6, which shows the nine CPCs described in Chapter 4 as well as some possible, but speculative, links. The meaning of the different types of links in Figure 6 is that the direct link shows an influence of the same direction e.g. that an increase in one CPC leads to an increase of the other, while the inverse link shows an influence of the opposite direction. Consider, for instance, the links starting from "adequacy of organisation". According to Figure 6, an increase in the "adequacy of organisation" leads to an improvement in the "availability of procedures"; conversely, a decrease in the "adequacy of organisation" would lead to a decrease in the "availability of procedures". As an example of the opposite, an improvement in the "availability of procedures" leads to a decrease in the "number of goals", because the operators will be better prepared for the situations that may occur. A decrease in the "number of goals" leads to an increase in the "available time" which in turn leads to improved reliability of performance, i.e., fewer erroneous actions and fewer accidents / incidents.

The links shown in Figure 6 illustrate a set of potential couplings among the eight CPCs. The set is, of course, chosen so that it does not conflict too violently with common sense, but there is no specific empirical evidence for it. In Chapter 9, the illustration will be extended and revised, since the need for a reasonable model of the coupling between CPCs is more important for performance prediction than for event analysis.

2.2 Possible Error Modes

The second stage in the retrospective analysis is to describe the general possible error modes. The complete list of error modes was given in Chapter 6. Based on that a list of possible error modes can be constructed, using information about the domain and specific application. Since the error modes describe how actions can go wrong, information about the interface is particularly important. More than just an assessment of whether the MMI is adequate, this requires information about the modes of interaction. For instance, is control implemented via knobs and dials mounted on panels or walls, or is control achieved via a graphical interface using a pointing device such as a mouse or a tracker ball? Is the duration of an intervention determined by how long time a button is pressed, or does the button press activate a delay circuit? In the latter case the error mode of duration can most likely be ruled out, while in the former it cannot.

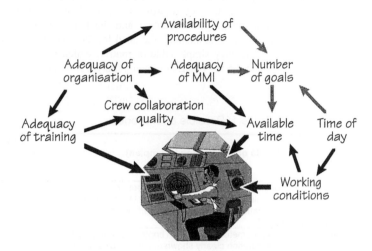

Figure 6: Illustration of possible interactions between CPCs.

The possibility of each error mode can be described by examining them in turn using some well-established principles of human factors and ergonomics. The characterisation can be done in a number of ways, but for the purpose of the retrospective analysis it may be sufficient to determine whether an error mode is impossible, whether it is possible, or whether it is very likely. Clearly, the error modes that are impossible can be excluded from the analysis, thus making it simpler. Similarly, the analysis should investigate the error modes that are very likely before looking at the ones that are just possible. Table 1 shows an example of how the possibility of each error mode can be determined by checking some essential aspects. The final assessment of the possibility must obviously be a matter of expertise, and requires good knowledge of the application as well as of human factors.

The points of clarification shown in Table 1 illustrate the basic principles of the assessment. As a general rule, the questions address such issues as whether the aspect or action described by the error mode is a **control parameter**, whether there is a clear indication of **when** it shall be used and **how much** it shall be used, and whether the physical interface **facilitates** or enables the use of that particular control parameter. If an aspect, such as duration, is an important control parameter and if the interface does not make it easy to use, then it must clearly be considered as a very likely error mode. Conversely, if an aspect is used very little, or if it is not required at all, then it need not be considered as part of the actual error modes. Finally, the description of the possible error modes must be given for all possible actions and not be limited to a specific action, even though the starting point for the analysis may be a specific event.

2.3 Probable Error Causes

In the third stage of the retrospective analysis the probable genotypes or categories of causes are described. This serves to identify in advance the genotypes that are more likely than others to be relevant as part of an explanation. As before, the purpose of this is to simplify the analysis by enabling the analyst to focus or concentrate on the causes that are likely to be part of the explanation.

The determination of the probable causes is obviously related to the description of the context, as it was developed above. The result of assessing the CPCs can therefore be used as the basis for determining the probable genotypes. Ideally, this should be done by referring to an explicit model that describes the dependencies between candidate causes in a specific context and the manifest characteristics of that context. Such models are, however, not available although the need for them is slowly being realised. In lieu of that a starting point can be a relatively simple table or matrix, which indicates the relationship between, on the one hand, the CPCs and, on the other, the main genotypes. A possible version of that is shown in Table 2.

Table 1: Possible error modes.

Error mode	Points of clarification (examples)	Possibility 0 := impossible, 1 := possible, 2 := very likely
Timing	Does the control of the process require timing of actions? Are there clear indicators / signals for the timing of actions? Does the system include lead time indications? Does a signal clearly identify the corresponding action?	
Duration	Is duration a control parameter? Is duration controlled manually or automatically? Is the duration clearly shown or indicated? Does the indication show elapsed time or remaining time, or both?	
Force	Is level of force / effort a control parameter? Is the required / applied level of force clearly indicated? Is there a minimum/maximum limit of force for a control? Can force be controlled without changing position?	
Distance / magnitude	Is distance or magnitude a control parameter? Is the required / applied distance or magnitude clearly indicated? Can distance / magnitude be controlled without changing position?	
Speed	Is speed a control parameter? Is speed controlled manually or automatically (set-point / rate)? Is the required / applied speed clearly indicated?	
Direction	Is direction a control parameter? Is there a direct relation between (movement) direction of controls and direction of system response? Is the required / applied direction clearly indicated?	
Wrong object	Are different objects clearly separated or coded (colour / shape)? Are objects clearly and uniquely identified? Can objects easily be reached and seen when use is required?	
Sequence	Is the sequence of actions / next action clearly indicated? Is the direction of the sequence reversible? Can out-of-sequence actions easily be recovered?	

The tick marks in Table 2 indicate where it is reasonable to assume a relation exists between the Common Performance Conditions and the main genotypes. A double tick mark represents a strong relation, while a single tick mark represents a weaker relation. Consider, for instance, the case of "MMI & operational support". If this CPC is characterised as only "tolerable" or even as "inadequate", then the technology related genotypes will be the most important source of causes for observed erroneous actions. However, both person related genotypes and organisation related genotypes may also play a role, although possibly a smaller one. In common terms this means that if the MMI is found to be inadequate for the job in a given situation, then explanations for observed erroneous actions are more likely to involve causes relating to the technological groups than to the others, although no main group *a priori* can be considered to be unlikely. As another example, consider the case of "organisation". If the adequacy of the organisation is found to be low for a given situation, then it is most likely that acceptable causes for observed erroneous actions are to be found in the group of organisation related genotypes.

Table 2 is not the final description of the relations between CPCs and the main groups of genotypes. The primary purpose of Table 2 is to provide an illustration of how these relations can be described in a straightforward manner. When the CREAM analysis method is going to be used in practice for a given domain, it will be necessary to develop a specific version of the matrix in Table 2 for that domain. The effort required to do that is easily recovered because the ensuing analyses will be simpler to make. The description of the context enables a prior sensitisation of the various causes, but it cannot lead to an exclusion of any of them. There is thus an important difference between **possible** error modes **probable** error causes. If some error modes are possible it means that others are impossible, hence that they need not be taken into consideration. Yet if some error causes are probable it only means that other error causes are improbable, but not that they are impossible. It is never possible to rule out any of the genotypes, even though they are unlikely to be among the immediate or frequent causes. In the case of the specific cognitive functions, in particular, it is **never** possible to rule out any of them. For any given context there will, however, be some that are more likely than others. Thus the context may enforce compliance with rules, encourage deviations, support learning of skills, and include a bad interface, hence promote misunderstandings or execution errors, etc. It is this difference in possibility or likelihood that is addressed by the third stage of the analysis.

Table 2: Relationship between CPCs and main genotype groups.

Common Performance Conditions	Main genotypes		
	Person related genotype	Technology related genotype	Organisation related genotype
Adequacy of organisation			✓ ✓
Working conditions	✓	✓ ✓	✓
Adequacy of MMI and operational support	✓	✓ ✓	✓
Availability of procedures / plans		✓ ✓	✓ ✓
Number of simultaneous goals	✓ ✓	✓ ✓	
Available time	✓ ✓		✓
Time of day	✓		
Adequacy of training and preparation		✓ ✓	✓ ✓
Crew collaboration quality	✓		✓ ✓

2.4 Detailed Analysis Of Main Task Steps

The fourth stage of the analysis method consists of the more detailed analysis of main task steps that traces the consequent-antecedent links for a selected error mode. For each error mode one should expect that the analysis will produce a set of candidate causes rather than a single (root) cause, as shown by the previous examples.

The detailed analysis begins by looking for the most likely error mode associated with the event in question. In the NYC subway train example, the event was that one train had crashed into another. The error mode was therefore clearly one of distance, since this precisely characterised the accident in context neutral terms. The event was not a human action in itself, and the phenotype or error mode does not necessarily have to describe an erroneous action. As the analysis of the example showed, the erroneous action on the part of the train driver was most likely that the train was not stopped at an early enough time, which corresponded to an error mode "timing" or possibly of "sequence", specifically "omission".

In general, the description of the error modes must be taken from one of the four classification groups: (1) action at wrong time, (2) action of wrong type, (3) action at wrong object, and (4) action in wrong place. If there is sufficient information available, the general error mode (that is, the general consequent) may be supplemented by a description of the specific error mode (or specific consequent); but the specific consequent can never replace the general consequent. Once the general / specific error modes for the initiating event have been described, the analysis can proceed to find the likely causes. This is achieved by a recursive search through the classification scheme that begins by selecting one of the antecedents linked to the error mode, either a general antecedent or a specific antecedent.

- If the outcome of this step is the identification of a **specific antecedent**, then the analysis has been completed. It is quite conceivable that the analysis only needs to go one step to find an acceptable cause. If, for instance, the general consequent was "action out of sequence" and the specific antecedent was found to be "trapping error", then the analysis has been completed in a single step.

- If the outcome is the identification of a **general antecedent**, then the analysis must continue. The next step is to check the classification groups to find are any general consequents that match the general antecedent. (It is guaranteed that there will always be a match, since this is one of the construction criteria for the classification groups and the classification scheme as a whole.) In other words, the general antecedent on one level of the analysis must match a general consequent on the next level, as discussed in Chapter 6. When a relevant general consequent has been found, the analysis continues from there.

The analysis can be extended by supplementing the general consequent with the selection of a specific consequent, provided that the necessary information is available. As noted above, a specific consequent can, however, not replace the general consequent. When the general and/or specific consequents have been selected, the corresponding general antecedent(s) or specific antecedent(s) are identified. As before, if the outcome is the identification of a specific antecedent, then the analysis has come to an end. Similarly, if there are no general antecedents - in which case there most probably are no specific antecedents either - then the analysis must stop. In all other cases the outcome will be a general antecedent, and which then is matched to the general consequents of the classification groups, as described before. In this way the analysis continues by applying the same principle recursively, until a stop criterion is reached. The principles of the detailed analysis are shown in Figure 7.

The above description provides a simple account of the analysis, but it is necessary to extend it in several ways to make the method practical. The first extension refers to the notion of direct and indirect links, as they were defined in Chapter 6. The principle of the analysis is that the search for an antecedent within a classification group looks first at the direct links between consequents and antecedents. If, however, there are no satisfactory relations between consequents and antecedents along the direct links, then the indirect links should be investigated. Similarly, if the indirect links suggest a consequent-antecedent relation that is as reasonable as the one(s) suggested by the direct links, then the indirect link should also be explored. The NYC subway train crash described above provided an example of the application of this rule. In the case of "equipment failure" there was a direct link to "maintenance failure" as well as an indirect link to "inadequate quality control". As it turned out, the "inadequate quality control" was part of the correct explanation of this accident.

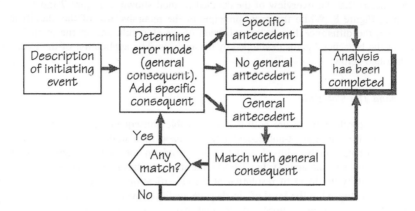

Figure 7: Detailed method for retrospective analysis.

The second revision has to do with tracing the path from consequents to antecedents. As the description of the method made clear, any step in the analysis may show that there is more than one plausible general antecedent for the general consequent of that group. This means that the analysis must explore both paths, and that it therefore is necessary to keep track of the branching points. The NYC subway train example again demonstrated how this worked on a small scale. In this case it was not difficult manually to keep track of the branching points and to retrace the path. In principle, this need not be a difficult issue at all even when the number of branching points grows large, since it is part and parcel of the search techniques developed by Artificial Intelligence. Thus, if and when CREAM is implemented as a software system, the administration of the searches can rely on well-proven techniques. In the manual application of the analysis method the principle applied can either be a depth-first or breadth-first, depending on the preferences of the analysts. The important point is, however, that the analysis is complete, i.e., that **all possible** paths are explored. The use of the CPCs and the initial selection of possible error modes and probable causes serves to limit the analysis so that it does not have to go through **every** path as well.

In the preceding description of the analysis method, the following stop rules were applied:

1.   If a general consequent points to a specific antecedent as the most likely candidate cause, then the analysis is stopped.

2.   If a general consequent does not have a general antecedent and if it does have a specific antecedent, then the analysis is stopped.

The second rule must, however, be extended since it may be that none of the general antecedents are acceptable or good enough. This can be due to unavailability of sufficiently detailed information or because the chosen version of the CREAM classification scheme simply is not fully appropriate for the application or domain. In such cases it is not sensible to continue the analysis. While the method does not force the analysis to continue **beyond** the point where it makes sense, it does ensure that it is taken to that point, i.e., that the analysis is not prematurely stopped. The second of the above stop rules can therefore be amended to read as follows.

2'.   If a general consequent does not have a general antecedent and if it does have a specific antecedent, or if none of the antecedents in the classification group seem reasonable in the given context, then the analysis shall be stopped.

Altogether this means that the overview of the detailed method shown in Figure 7 must be extended to look as shown in Figure 8. While the basic principle of the recursive use of the classification groups remains the same, the difference lies in the extension of the stop rule and in the explicit reference to keeping track of the branches / paths developed by the analysis. The application of the extended method will be demonstrated below.

2.5 Going Beyond The Stop Rule

The purpose of a stop rule is to ensure both that all probable consequent-antecedent links are explored, and that the search for consequent-antecedent links is made in a uniform and consistent way. It is important that the depth and breadth of an analysis depend on the method used rather than on the preferences or experience of the analyst. In CREAM, the stop rule means that the depth of the analysis in essence is determined by the level of resolution of the classification groups. In the NYC subway train example, the analysis stopped at the level of "inadequate quality control" because the classification group **organisation** did not contain any antecedents for that, neither via direct links nor indirect links. It is, however, entirely possible to continue the analysis and, in a sense, go beyond the stop rule. It requires that the classification scheme is suitably extended, but the method remains unchanged.

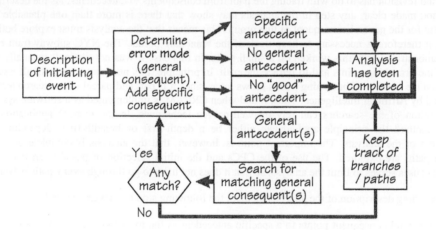

Figure 8: Revised method for retrospective analysis.

This extension of the classification scheme can be achieved in two different ways. One possibility is that a specific antecedent is changed so that it becomes a general antecedent. Consider, for instance, the classification group **observation**. Here "information overload" is provided as a specific antecedent of "wrong identification". If therefore the analysis points to "information overload" as the most likely antecedent for "wrong identification", the stop rule means that the analysis will not continue along this specific path. Assume, however, that "information overload" is defined as a general rather than a specific antecedent. In that case the search will continue by looking for a match between "information overload" as a general antecedent and "information overload" as a general consequent in one of the other classification groups. In the present version of the CREAM classification system this search will fail, since "information overload" does not occur anywhere else. In general, a specific antecedent cannot be changed into a general antecedent without considering the possible consequences for other classification groups. Such a change can therefore not be accomplished without revising the whole classification system. This is nevertheless not a very difficult thing to do, since the rules for accomplishing that can be clearly defined - although they have not been included here. (In practice, the classification scheme can be defined as a database and the necessary consistency checks can be done automatically.)

The other extension is to provide a new general antecedent, either in addition to existing general antecedents or in the cases where no general antecedents are given at present. Returning to the case of "inadequate quality control" in the **organisation** group (cf. above), the classification scheme currently has no general antecedents assigned. This is because the development of CREAM has been focused on the issue of human reliability rather than, say, organisational reliability. It is, however, not difficult to suggest possible general antecedents for "inadequate quality control"; some obvious candidates are "insufficient development time", "insufficient funding", "inadequate safety culture", "shallow decision making", etc. (A good example of that is the Ariane-5 failure, which pointed to an incorrect design decision as the "root" cause. The question might well be asked what the working conditions were when the decision was taken in relation to e.g. concerns about time and cost. In accident analysis it is usually "the poor wot get the blame".) Adding additional general antecedents to the **organisation** group may, in fact, lead not only to an extension of the general consequents in one or more of the existing classification groups, but also to the creation of wholly new classification groups. There is no reason why this should be avoided, provided there is a clear need to do so. In fact, the explicit principle according to which the classification groups of CREAM have been defined makes such an extension relatively easy to do. The only thing that must be observed is that the links between general consequents and general antecedents are properly maintained, so that the classification scheme as a whole remains consistent - no conflicts in the use of specific terms and no loose ends. The basic recursive analysis principle can be retained and the stop rules remain valid. This is necessary not only to be able to use the same analysis method, but also - and more importantly so - to maintain the common basis for analysis and prediction.

3. EXAMPLE OF RETROSPECTIVE ANALYSIS

The retrospective use of CREAM will be illustrated by analysing a part of the steam generator tube rupture incident at the Ginna nuclear power plant. This event has been documented in detail in a report from the Institute of Nuclear Power Operations (INPO, 1982), as well as in NUREG-0909 (1982). The purpose of looking at the incident at Ginna is not to reanalyse the incident as such, but to illustrate how the retrospective analysis methods can be used for a realistic case.

As basis for the analysis, selected parts of the description of the incident are reproduced below. Since the analysis made here only is concerned with the first of several events, the descriptions have been abridged accordingly.

3.1 Tube Rupture

About 09:25 on January 25, 1982, a single steam generator tube ruptured in the "B" steam generator at the R. E. Ginna Nuclear Plant. The plant, which had been operating normally at full power conditions, suddenly lost pressure in the primary coolant system as coolant rushed through the ruptured tube into the "B" steam generator (SG). Alarms indicated a possible steam generator tube rupture, and the operators began reducing power. A reactor trip on low pressure, followed by initiation of safety injection flow and containment isolation, occurred within three minutes. The reactor coolant pumps were stopped as required by procedures. During the first five minutes of the event, reactor coolant system (RCS) pressure decreased from about 2200 p.s.i. to 1200 p.s.i.

In a pressurised water reactor, such as Ginna, water coolant is pumped under pressure through the reactor core, where it is heated to about 325° C. This is referred to as the primary loop. The superheated water continues through a steam generator (a specialised heat exchanger), where a secondary loop of water is heated and converted to steam. This steam drives one or more turbine generators, is condensed, and pumped back to the steam generator. (Turbine generators and condensers are not shown in Figure 9.) The secondary loop is normally isolated from the reactor core and is not radioactive. The Ginna plant had two steam generators and therefore two secondary cooling loops, called A and B.

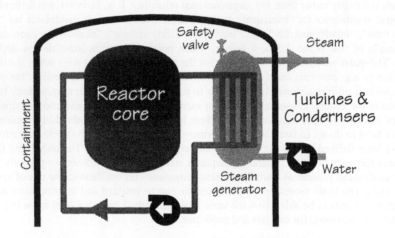

Figure 9: Simplified diagram of a pressurised water reactor.

When a tube ruptures in the steam generator, radioactive water from the primary loop may get into the secondary loop, and from there into the atmosphere. If water from the primary loop continues to flow into the steam generator after it has been isolated from the environment, the pressure will increase and the safety valve may open.

In the case of a tube rupture, the instructions require the operators to: (1) identify the location of the rupture, i.e., in which steam generator(s) it occurred; (2) isolate the identified steam generator(s) which are inside the containment from the secondary loop, i.e., the turbines and the condensers that are outside the containment; and (3) reduce the pressure of the primary loop so that it is lower than the pressure in the faulty steam generator in order to eliminate the flow through the rupture. Once these conditions have been established, a number of other goals must be achieved, but they are not necessary for the example.

Seven minutes after the rupture, at 09:32, the motor-driven auxiliary feedwater pump to the "B" steam generator was stopped and the steam supply valve from the "B" steam generator to the turbine-driven auxiliary feedwater pump (TDAFWP) was closed. Either at 09:32 or a short time later, the auxiliary feedwater flow from the TDAFWP to the "B" steam generator was isolated. Despite this the water level of the "B" steam generator continued to rise. Fifteen minutes after the rupture, at 09:40, the operators were satisfied that they had identified the steam generator with the ruptured tube, and therefore closed the main steam isolation valve (MSIV) for the "B" steam generator. This action isolated the steam generator and halted further spread of primary radioactive nuclides downstream of the "B" MSIV. (Unfortunately, it also led to the inadvertent opening of the Main Steam Line Safety Valve.) Auxiliary feedwater continued to the "A" steam generator until level was established between 75-80 % on the narrow range instrumentation and then stopped at 09:48. Since the analysis here looks at the isolation of the steam generator only, the continued developments are not included in the description.

3.2 Isolation Of Ruptured Steam Generator - How Soon?

The first operational problem encountered by the Ginna staff was how soon to isolate the suspected SG. Operating procedures required early isolation of the ruptured SG as soon it was positively identified to minimise the spread of primary coolant contamination to secondary loop systems and possible releases of radioactivity to the environment. However, isolation of the wrong SG would require a delay to open the MSIV bypass valve in order to repressurize the downstream piping and re-establish the SG as functional. Since Ginna was a plant with only two cooling loops (i.e., two steam generators), a failure of the bypass valve to open would remove the capability for normal cool-down via the condenser dump valves and would require the power-operated atmospheric steam dump valve (ASDV) of the unaffected SG to be used instead. A consequence of using the SG power-operated ASDV for cool-down was an extended cool-down time since its flow capacity was considerably less than that of the condenser steam dump valves. Also, use of the ASDV would increase the possibility of radiological releases in the event that the auxiliary feedwater supply should become contaminated or both steam generators should rupture. Therefore, cool-down via the condenser was preferred during tube rupture events.

Although the Control Operator was convinced, based on preliminary indications, that the rupture had occurred in the "B" SG, the Shift Supervisor wanted more confirming information before isolating the steam generator. The Ginna procedures required positive identification prior to isolating a steam generator. After the reactor trip, with auxiliary feedwater flow to both SGs, the operators noted that the "B" SG water level was increasing more rapidly than the "A" SG water level.

After the auxiliary feedwater pump supplying the "B" SG was stopped at 09:32, the water level continued to rise at approximately 4 per cent per minute. It was this continuing increase in the level that convinced the Shift Supervisor that isolation was prudent. After observing the "B" SG water level increase steadily for eight minutes with feedwater secured, the Shift Supervisor ordered the "B" MSIV closed. A short time later the Health Physics Technician entered the control room and reported that radiation readings on the "B" SG blow-down line were 9 mrem/hour, compared to less than 1 mrem/hour on the "A" SG blow-down line. A high radiation reading is a positive indication of a tube rupture. The SG blow-down lines were isolated by the containment isolation signal that occurred 3-4 minutes after the tube rupture, and the readings by the Health Physics Technician were made several minutes after isolation of the blow-down lines. An auxiliary operator and a health physics technician were dispatched to obtain radiation readings on the "B" SG steam piping. (The use of technicians to monitor radiation conditions for diagnosis of the ruptured SG was in part due to the unavailability of portions of the radiation monitoring system designed to monitor radiation from the steam line.) They reported a reading of 30 mrem/hour upstream of the "B" MSIV and a reading of 2 mrem/hour on the steam header 3 feet downstream of the "B" MSIV.

A number of operational problems and dilemmas complicated the recovery process and demonstrated the need for improvements in operating procedures, training, and design for tube rupture events. The crew had to cope with a novel event. The high cost of making an incorrect action (an error of commission), the inadequate design of the procedures, the various distractions by the arrival of extra personnel to the control room, and the fact that the shift had been on duty for an hour and a half only, placed additional stresses on their performance. The occurrence of operational problems for which the operating procedures and the operator's experience were incomplete or vague led the operators to take conservative actions to protect the reactor core. These actions, while appropriate to maintain core cooling, were not necessarily appropriate to minimise releases from the steam generator safety valves. With the advantage of hindsight and post-event analysis, it can be seen that the reactor core was in no real danger, and actions could have been taken to avoid the steam generator safety valve opening. However, given even perceived threats to the reactor core and the limitations of the operating procedures and training for the problems encountered, the operators made the prudent choices.

It should be emphasised that the Ginna operating staff performed well under stressful and sometimes novel circumstances. The operators had correctly isolated the ruptured SG in 15 minutes. Although the operators achieved their goal, the investigation after the incident pointed out that the isolation of the rupture steam generator could have occurred faster, and thereby prevented the opening of the steam generator safety valve.

3.3 Event Analysis

The analysis of the event will follow the principles described above, cf. Figure 4 and Figure 8.

3.3.1 Describe Common Performance Conditions

The context is described using the Common Performance Conditions, as presented in Chapter 4. Based on the description of the situation given by the official reports from the investigation, and supplemented by general knowledge about the typical conditions in nuclear power plant control rooms, the following characterisation results (Table 3). For each CPC the descriptor shown in **bold** is the one that is considered appropriate for the situation.

The assessment of the CPCs shows that the situation, on the whole, was not a very advantageous for the operators. They had little experience with this type of incident, and there was inadequate support from the organisation, both in terms of providing training, in securing an adequate working environment, and in making effective tools (as instructions) available. Referring to the description of the CPCs and performance reliability in Chapter 4, it means that four out of the nine CPCs would be expected to lead to a reduced performance reliability, while none indicated an improved performance reliability. (This would have been the case if "crew collaboration quality" had been assessed as "very efficient".)

3.3.2 Describe The Possible Error Modes

The possible error modes must be described for the scenario as a whole, rather than for specific events. In this case there was only little detailed information about the design of the control room, but assuming that it was not significantly different from a typical nuclear power plant control room, it is possible to make an assessment using the criteria described in Table 1. The result of that is shown in Table 4.

On the basis of the available information, general as well as specific, it is only possible to rule out two error modes: force and distance / magnitude. All the other error modes must be considered as possible. There was, however, nothing in the conditions for this scenario that pointed to one error mode as being more likely than the others.

Table 3: CPCs for the Ginna incident.

CPC name	Evaluation
Adequacy of organisation	Apparently there were many distractions; this is not efficient for an emergency situation.
Descriptors	*Very efficient / Efficient / **Inefficient** / Deficient*
Working conditions	There were too many people in the control room, and too many disturbances of the operators' tasks.
Descriptors	*Advantageous / Compatible / **Incompatible** /*
Adequacy of MMI and operational support	Insufficient details are provided by the investigation reports; MMI may be considered adequate for normal conditions but possibly less so for an emergency
Descriptors	*Supportive / Adequate / **Tolerable** / Inappropriate*
Availability of procedures / plans	Inadequate design of procedures, which contained conflicting goals and priorities.
Descriptors	*Appropriate / Acceptable / **Inappropriate***
Number of simultaneous goals	This is assumed to be the case for an emergency situation in general. The need to concentrate limits the scope.
Descriptors	*Fewer than capacity / Matching current capacity / **More than capacity***
Available time	The incident occurred early in the shift period. Furthermore, the instruction calls for a quick response, which means time was short.
Descriptors	*Adequate / **Temporarily inadequate** / Continuously inadequate*
Time of day (circadian rhythm):	The incident occurred at mid-morning. Information about the shift cycle is not available.
Descriptors	***Day-time (adjusted)** / Night-time (unadjusted)*
Adequacy of training and preparation	The event happened in 1982, where extensive simulator training was still in its beginning. The experience with the SGTR incident was therefore limited.
Descriptors	*Adequate, high experience / **Adequate, limited experience** / Inadequate*
Crew collaboration quality	The official investigation pointed out that the crew had "performed well under stressful and sometimes novel circumstances".
	*Very efficient / **Efficient** / Inefficient / Deficient*

Table 4: Possible error modes for the Ginna incident

Error mode	Points of clarification (examples)	Possibility *0 := impossible,* *1 := possible,* *2 := very likely*
Timing	In this event scenario, both isolation and depressurisation require correctly timed actions.	*Possible*
Duration	The duration of separate steps in e.g. the depressurisation or cooling may be important for how well the incident is handled.	*Possible*
Force	Level of force is not a control parameter in a typical NPP control room.	*Impossible*
Distance / magnitude	Distance or magnitude is usually not a control parameter in NPP control rooms.	*Impossible*
Speed	The speed of some actions (depressurisation and, later, cooling) was important for this event scenario.	*Possible*
Direction	Direction may be a control parameter in changing setpoints. This may, however, also be a case of "wrong object", depending on how the controls were designed.	*Possible*
Wrong object	A nuclear power plant control room (in 1982) would offer several possibilities for choosing a wrong object	*Possible*
Sequence	The sequence of actions in the recovery procedure is important, but was, on the other hand, laid out in the procedure.	*Possible*

3.3.3 Describe The Probable Causes

Considering the description of the context, as summarised in Table 3 above, it is possible to distinguish between more and less probable categories of causes. The main aspects were the novelty of the situation, the ambiguity of the procedures, the concerns of the operators for not making the wrong decision, and the general working conditions. The probable error causes can either be marked on the separate tables, or be summarised as shown below in Table 5. The difference from Table 2 is that the assigned values of the CPC descriptors have been added (shown in **boldface**), and that CPCs assignments that will not have a significant effect on performance reliability are shaded (cf. Chapter 4).

Table 5: Probable error causes for the Ginna incident.

Common Performance Conditions	Main genotypes		
	Person related genotype	Technology related genotype	Organisation related genotype
Adequacy of organisation: **Inefficient**			✓ ✓
Working conditions: **Incompatible**	✓	✓ ✓	✓
Adequacy of MMI and operational support: **Tolerable**	✓	✓ ✓	✓
Availability of procedures / plans: **Inappropriate**		✓ ✓	✓ ✓
Number of simultaneous goals: **More than capacity**	✓ ✓	✓ ✓	
Available time: **Temporarily inadequate**	✓ ✓		✓
Time of day: **Daytime**	✓		
Adequacy of training and preparation: **Adequate, limited experience**		✓ ✓	✓ ✓
Crew collaboration quality. **Efficient.**	✓		✓ ✓

3.3.4 Detailed Analysis Of Main Task Steps

The more detailed analysis is accomplished through a number of steps, as described below (cf. Figure 8).

◆ **Describe Initiating Event**: The initiating event for the analysis was the delay in closing the Main Steam Isolation Valve in loop "B".

◆ **Identify Error Mode (General consequent)**: According to the description of the initiating event, the most likely error mode was timing. The specific consequent was a delay. No other error modes were applicable for the initiating event.

◆ **Find The Associated Antecedents**: According to the table for "General and specific antecedents for error modes" in Chapter 6, there were six possible general antecedents and two specific antecedents. Of these, two of the general antecedents were chosen for further analysis: "inadequate procedure" and "inadequate plan".

The delay was due to the ambiguity of the procedure, which demanded both fast action and a high level of certainty. That in turn led to problems in identifying the steam generator to be closed. There were no problems in diagnosis of the event as SGTR, nor in communication or observation. As Table 5 has shown, there was furthermore no reason to assume that one group of genotypes was more likely than the others.

Since the outcome in either case is a general rather than a specific antecedent, the analysis is continued. First, the possibility of an "inadequate procedure" is investigated.

♦ **Match Antecedent With Other Classification Groups:** A search for a match to "inadequate procedure" led to the "procedures" group.

Going to the table of "Categories for procedures", the specific consequent of "ambiguous text" seemed relevant. Going on to the table of "General and specific antecedents for procedures" there are two possible general antecedents, "design failure" and "inadequate quality control". According to the table of "Categories for organisation", "inadequate quality control" is clearly relevant and even has "inadequate procedure" as a specific consequent; "design failure" does not convey the proper meaning and is therefore not considered further. Both general antecedents, however, point to the table of "General and specific antecedents for organisation" and in neither case are there further general or specific antecedents. This means that the analysis of "inadequate procedure" has come to an end. The findings of this analysis can be shown as in Figure 10. In each box, the boldface term denotes the **error mode** or **classification group**. In the case of a classification group, the category written in normal text denotes the general consequent, while the category written within parentheses denotes the specific consequent.

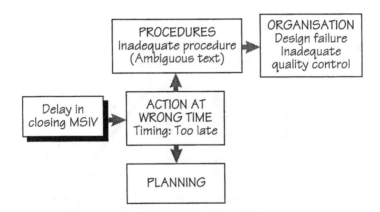

Figure 10: Analysis of Ginna example - first candidate causes.

♦ **Repeat Search For Causes:** After having completed the analysis of "inadequate procedure", the guidelines are applied to the second general antecedent "inadequate plan".

This leads in the first hand to the table of "Categories for planning". Two possible relevant specific consequents are "wrong plan" and "wrong goal selected", the latter being a result of an indirect link. Continuing to the table of "General and specific antecedents for planning" and following the direct link, the most likely general antecedent is "wrong reasoning". In the table of "Categories for interpretation", a relevant specific consequent is "wrong priorities". Going on to the table of "General and specific antecedents for interpretation", none of the direct causes appear likely. Instead, an indirect link from "faulty diagnosis" to "inadequate procedures" seems a possibility, and this does in fact lead back to the path found by the first iteration of the analysis. This means that the analysis of "inadequate plan" ends in much the same way as the analysis of "inadequate procedure", as shown in Figure 11:

(As a further iteration, the indirect link denoted by "wrong goal selected" can also be followed. This is left to the reader as an exercise. It will soon lead to "conflicting criteria" as a specific antecedent.)

3.3.5 Summary Of Analysis

It is interesting in this case that the analysis of the two possible antecedents to "action too late" ends by pointing to the same underlying cause, namely "inadequate procedure" that, in turn, is caused by "inadequate quality control" in the organisation. One causal chain is longer than the other because it contains an additional step. The usual principle followed in scientific research is to choose the simpler of two explanations; this is generally known as Ockham's razor. However, in the case of finding an explanation for a human erroneous action, the purpose is to find the most complete or reasonable explanation rather than the simplest. This means that one cannot apply Ockham's razor in a mechanical fashion - but neither can the inverse principle be invoked. The determination of which explanation is the most complete or most reasonable must be based on the experience of the analyst. The analysis guidelines described here serve to facilitate the analysis and to ensure greater consistency. They cannot, and should not, be used as a complete algorithmic procedure that can be used blindly. Retrospective event analysis will always require a modicum of expertise, yet the analysis should not rely on subjective expertise alone.

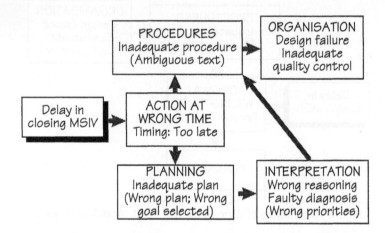

Figure 11: Analysis of Ginna example - second cause.

In the present case, the detailed description of the event actually makes it more likely that the second - and longer - explanation is the correct one. It does, however, lead to the same recommendation for remedial action, namely to improve the procedures both in terms of their content and in how they shall be interpreted and used.

4. CHAPTER SUMMARY

Chapter 7 provides a detailed description of how CREAM can be applied for retrospective use, i.e., for the analysis of accidents and incidents that have occurred. The retrospective analysis describes how the classification scheme can be used to find the likely causes for a given accident or event. One important feature is the distinction between general and specific antecedents, where only the former give rise to a continued search. Another is the definition of explicit stop rules. Since the classification scheme cannot be complete for every possible application, the chapter also describes the principles according to which the classification groups can be extended, thereby effectively increasing the depth of the analysis. Such extensions should, however, only be done when the need is clearly documented.

The retrospective analysis has four basic steps. Firstly, the common performance conditions (CPC) are described, using a small set of general descriptors. These differ from the traditional performance shaping factors by explicitly addressing all aspects of the MTO-triad, and by acknowledging that they are not independent of each other. To illustrate that a possible coupling between the CPCs is described. The second and the third steps describe the possible error modes and probable error causes, respectively. The possible error modes serve to identify the error modes that are actually possible for a given work situation. The probable error causes point to the sets of causes that are most likely given the performance conditions. In both cases the purpose is to use the available information about the event to focus the search for causes. The fourth step is the concrete identification of consequent-antecedent links for a specific event, using the basic principles of backward propagation through the classification groups. The advantage of this way of analysis is that it is neutral with regard to where the "root" cause will be found, i.e., it is not assumed *a priori* that e.g. a "human error" is more likely than a failure of the organisation. The retrospective method is illustrated by two different examples.

Chapter 8
Qualitative Performance Prediction

Predictions of the future are never anything but projections of present automatic processes and procedures, that is, of occurrences that are likely to come to pass if men do not act and if nothing unexpected happens; every action, for better or worse, and every accident necessarily destroys the whole pattern in whose frame the prediction moves and where it finds its evidence.

Hannah Arendt (1906 - 1975)

1. PRINCIPLES OF PERFORMANCE PREDICTION

The above quotation refers to political issues and affairs of the state, but is in many ways also quite pertinent for performance prediction and human reliability analysis. In HRA, and particularly in HRA within PSA, the main purpose is to predict which sequences of events are likely and what the outcomes will be, provided that nothing happens that is not part of the descriptions. This is another way of saying that in order for predictions to be correct, models and reality must correspond. Fortunately, the whole foundation is not destroyed every time an accident happens, although some accidents do give rise to extensive revisions of the conceptual foundations.

The art of performance prediction is to provide a description of what is likely to happen **if** a specific initiating event occurs, **and if** the model of the world bears an acceptable correspondence to the real world. This kind of performance prediction is not confined to HRA, but can be found in many fields where planning and prediction play a role, such as economics, weather forecast, fiscal legislation, environmental protection, global warming, etc. It is furthermore an important part of system design in general. In the situations addressed in this book, the interest is focused on human actions, hence on the underlying model of the person (the operator, the user, etc.), rather than on the model of a process, an engineering system or an organisation. HRA must, however, be less interested in modelling individual performance and more interested in modelling the interaction between humans and technology. Consequently, the details of the human information processing system are less important than how a person can maintain control over a situation.

1.1 Scenario Selection

When an HRA is performed, it is taken for given that there is a scenario or an event sequence that requires analysis. It may nevertheless be useful to consider for a moment how the selection of this scenario takes place, and how it is determined that the scenario requires analysis. The outcome of the prediction reflects the characteristics and constraints of the scenario, and therefore critically depends on the correctness and

---

completeness of the scenario. In retrospective analyses the scenario selection is rather easy to justify because there is an event or an accident that has happened. This not only identifies **what** should be analysed, but also **why** an analysis is necessary, and Chapter 7 has discussed the details of event analysis.

In the case of performance prediction the issues are more complex. When HRA is performed as part of a PSA, the scenario is normally based on the PSA event tree or may even be a simple subset of it. The PSA event tree in turn may have been derived in several ways. Typically, it involves drawing up a complete list of all potential system failures that can reasonably be expected given factors such as the prior experience of the analyst with the system in question, or the specific requirements imposed by the industry's regulatory body. From this list one particular scenario at a time will be selected as the target for the HRA. More formal techniques may also be used, such as a fault tree analysis or a failure mode and effect analysis (FMEA or FMECA). To the extent that the PSA event tree is taken as a given the scenario selection is not an issue. There is furthermore little need to enquire either whether it is relevant, since that is guaranteed by the origin in the PSA. It may nevertheless be worthwhile to consider whether the scenario constitutes a **reasonable** and psychologically realistic basis for the analysis. From the cognitive systems engineering point of view it can easily be argued that an alternative starting point, such as a systematic task analysis, may produce scenarios in addition to those that come from the PSA and in particular suggest alternative sequences of the events. These can either be new sequences, or variations of already known sequences. For HRA as a discipline in its own right, it therefore makes sense to combine the two approaches, since either alone may lead to oversights.

1.2 The Role Of Context

The context refers to the circumstances or conditions under which an event occurs. In order to understand both how an event **has** developed and how an event **may** develop, it is necessary to know the context. Although the description of the context can never be complete, it is quite possible to provide a description that covers all the important aspects - corresponding to the notion of requisite variety from cybernetics (Ashby, 1956). In the case of event analyses or accident analyses, the first step is invariably to describe the situation or the circumstances. The analysis is an attempt to find the most probable causal chain, going backwards from the observed event. This is usually possible because information is available about the conditions in which the events took place. In the investigation of a plane crash, for instance, considerable effort is put into collecting all the evidence and describing the context, almost on a second-by-second basis (Bryant, 1996). Only when this has been done will it be possible to identify the most likely cause (Cacciabue et al., 1992). Exactly the same is seen in the reports on accident investigation such as in the nuclear power plant domain (NUREG-1154, 1985) or the Challenger accident (Rogers et al., 1986).

In the case of performance prediction, the first step must also be to provide a context description. It stands to reason that a discussion of what is likely to happen in a given situation must be based on a description of what the circumstances are expected to be. If no details are provided, practically anything can happen. Human actions are neither produced by an independent deterministic "mental mechanism" nor are they completely stochastic. Human actions are - usually - intentional and directed towards a specific goal. The goals that people assume and how they try to reach them will all depend on the context - or more precisely on the perceived context, since there may often be differences between the actual conditions and what people assume to be the case. In particular, since there normally are several different ways in which a goal can be reached, humans characteristically try to optimise their efforts with regard to some aspect of the situation, be it time, energy, risk, etc. The resulting performance can therefore not be understood unless the performance conditions and performance criteria are known.

Figure 1 shows the principle of retrospective and predictive analyses relative to the context. The potential propagation paths are illustrated by the mesh and the context is represented by the "cloud". Propagation paths that are part of the context are shown in black, while paths that do not are shown in grey. An **event** or **accident analysis** is concerned with events (accidents) that have occurred and tries to find the most probable causes (root causes). The purpose is to understand better something that has already happened. Because the context is known, it is possible to follow the development of the event step-by-step in the direction **from** the focal event (the observed consequence) **to** the probable cause (Cojazzi & Pinola, 1994). If the information is incomplete, inferences about missing details can be made with reasonable certainty. Note, however, that Figure 1 does not show the links between the focal event and the probable cause(s) as a simple event tree, but rather as a network or mesh. The event tree rather can, of course, be seen as the instance of a limited set of paths through the causal network.

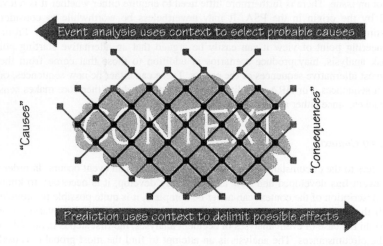

Figure 1: The central role of context in performance analysis and prediction.

A **performance prediction** tries to describe how a scenario may possibly develop, and how this development depends on the conditions. If the propagation of the events is made entirely within the representation of the scenario, it is basically a question of using the rules of mathematical logic to calculate the truth value of the conclusion or the effect. In many cases the representation of the scenario only provides the basic structure of the events but not the detailed conditions that may influence how an event develops. In order to make the prediction, the scenario description must therefore be supplemented by information about the conditions or factors that may influence the propagation of events. One of these is the variability of human performance, which in itself depends on the general performance conditions - including the previous developments (cf. Figure 2). The purpose of the prediction is to find out what **may** possibly happen under given conditions, for example if a component fails, if there is insufficient time to act, or if a person misunderstands a procedure. Even when both the scenario and the performance conditions have been described in sufficient detail, a mechanical combination of taxonomic categories will soon generate so many possibilities that the total quickly becomes unmanageable. The focus can be improved only if the context can be defined, because the context can be used to limit the number of combinations that need to be investigated. Performance prediction must therefore describe the likely context **before** it goes on to consider the actions that may occur.

(Note that Figure 2 contains two loops, indicated by the shaded backgrounds. The first loop indicates that the scenario description may depend on the likely performance, while the other suggests that human performance reliability also depends on the likely performance. The purpose of the loops, which in principle are similar to the ones in the ATHEANA method (cf. Chapter 5), is to point out that the description of the context and the performance must be developed iteratively.)

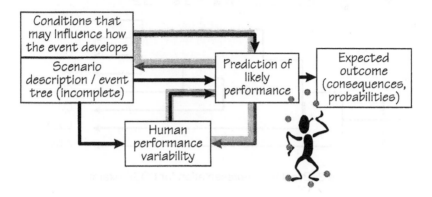

Figure 2: Basic dependencies in performance prediction.

In principle, the task is "simply" to find a path between antecedents and consequences, as in the stylised representation of Figure 1. Any classification scheme will define a large - and sometimes exceedingly large - number of possible paths. Only a subset of these will realistically be probable for a given context or set of conditions. If the conditions are changed, the path may look completely different. When the event actually occurs, only one of the probable paths will actually be taken - although to the chagrin of analysts it may sometimes be a path that has not been anticipated! The basic prerequisite for performance prediction is therefore that a probable context has been described. The essence of a performance prediction, such as HRA, should therefore be to estimate the probable performance conditions rather than to predict specific events!

1.3 Performance Prediction In First-Generation HRA

The traditional approach to performance prediction in HRA is based on the event tree representation of the scenario. As an example, consider the event tree shown in Figure 3. (Experts in nuclear PSA are kindly asked to disregard that the tree is semantically and syntactically incomplete, and accept it simply as an example. An equivalent form is the THERP tree, which usually is drawn vertically with the initiating event at the top.) In performing the HRA it is assumed that the event tree representation contains all possible predictions for a given event sequence, i.e., all the possible combinations of events and conditions that should be considered.

This particular type of representation makes two rather strong assumptions, which in turn serve to make the performance prediction manageable. The first assumption is that the sequences of events can be determined in advance; the other is that any event included in the sequence can be adequately described by means of the binary classification of "succeed or fail".

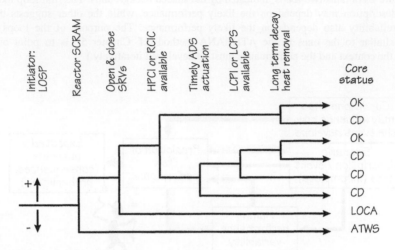

Initiator: LOSP | Reactor SCRAM | Open & close SRVs | HPCI or RCIC available | Timely ADS actuation | LCPI or LCPS available | Long term decay heat removal | Core status

OK
CD
OK
CD
CD
CD
LOCA
ATWS

Figure 3: Event tree representation for LOSP example.

1.3.1 Pre-Defined Sequence Of Events

Considering the sequence shown in Figure 3, the same information can be represented as the fault tree in Figure 4. Whereas Figure 3 appears to describe one sequence of events, as shown by the top row, Figure 4 makes clear that it actually contains a number of different sequences leading to four possible outcomes, some of which have been concatenated in Figure 4.

While it is obvious that the set of 5.040 possible permutations of the seven events of Figure 3 will contain many sequences that are unreasonable and physically impossible according to the constraints of the underlying process, it is not given that only one sequence deserves further analysis. In the case of a PSA the sequence to be analysed is, of course, selected with considerable care and reflects accumulated engineering judgement. It is also true that many of the physical events described by the event tree can only occur in a certain sequence due to, e.g. the nature of the physical process or built-in limitations in the control systems. A similar argument cannot, however, be made for events that involve or depend on human action. On the contrary, in each case where human actions are included it is possible that they are carried out at the wrong time, in the wrong sequence, etc., as discussed in Chapter 6. It is therefore necessary critically to review the scenario to determine whether alternative event sequences should be considered. If so it may have consequences for the event tree, i.e., it may no longer be possible to maintain a single event tree.

The PSA event tree assumes that the events depicted by it cannot reasonably occur in a different order. The events can therefore be described as a simple sequence, and the seeming complexity of an event tree, or an operator action tree, arises because the representation includes the branch-points and the possible alternative developments. The prediction addresses **how** likely a specific outcome is, rather than **what** may actually happen in the sense of the events in the sequence (Heslinga & Arnold, 1993). Within the context of PSA, HRA is concerned with assigning probabilities to the set of specified events, but not with defining the events as such. First-generation HRA thus considers only **one** sequence of events. This means that it does not really address the problem of performance prediction in the sense of trying to predict how the situation may develop after the initiating event has occurred, how the events may propagate through the system, and what the different consequences may be. Yet even assigning the probabilities requires detailed knowledge of the context. In first-generation HRA this important condition is usually neglected. While the qualitative effects of the context may be captured in the initial task analysis, the quantitative effects are brought to bear only after the probabilities have been assigned, by means of adjustments with the PSFs / PIFs.

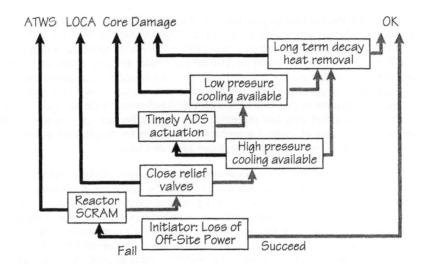

Figure 4: Fault tree for LOSP example.

1.4 Success And Failure

The second assumption inherent in the PSA event tree is that any event included in the sequence can be adequately described by distinguishing between the simple categories of success and failure. This distinction represents a convenient simplification of how the outcome of functions may be categorised, but the simplification often hides a complex chain of reasoning. Consider, for instance, the fourth event of Figure 3, "High pressure cooling available". In reality, this need not be an all-or-none phenomenon but cooling may be partly available in terms of effect or duration. In practice, most system functions can exist to a varying degree and it is normally an oversimplification - but in this case a convenient one - to describe them as either being present or not. The same goes, with even more emphasis, for psychological states and events, such as being tired, making a correct diagnosis, etc. Even apparently simple things, such as whether a response was made in time, may be difficult to determine since it requires an unambiguous criterion for the meaning of too early and to late. Reality is grey rather than black and white.

The graphical notation of an event tree / fault tree imposes a binary classification, as do most hierarchies. Therefore, in the choice of event representation something has to give. It is necessary either to simplify the event descriptions so that they match the representation formalism, or to find a better way of representing events. In PSA-*cum*-HRA the first solution has been the dominant one, even though this clearly results in an over-simplified representation. In order to have an adequate basis for performance prediction it is, however, necessary to adopt the second solution. This means that the scenario, and in particular the possible human actions, should be represented in a more realistic fashion. It is not difficult to find suggestions for that, e.g. in the representations used by the various task analysis methods (Kirwan & Ainsworth, 1992). As Chapter 9 will show, it is quite possible to use a task representation of this kind as the basis for a HRA, without sacrificing the possibility of linking the results to the PSA.

1.5 The Separation Between Analysis And Prediction

In first-generation HRA, the accepted purpose is to calculate the probability that operator actions fail and various classification schemes and models have been developed to support that. For historical reasons, first-generation HRA adapted the approach used in reliability analysis for technical systems. Thus, Miller & Swain (1987) noted that "the procedures of THERP are similar to those employed in conventional reliability analysis, *except that human task activities are substituted for equipment outputs*" (italics added). The development of operator models has been almost incidental to the HRA approach, and it is therefore not surprising that the models only contain enough detail to satisfy the demands from PSA/HRA applications. Models associated with first-generation HRA approaches are appropriate for performance prediction in the narrow sense of supporting probability estimates for simple error manifestations, but cannot easily be applied to the broader type of prediction advocated here.

In the information processing approaches the emphasis is on the analysis of events and the explanation of the psychological causes for erroneous actions. The specific paradigm for explanation, i.e., the information processing system, was taken as a starting point, and the main effort was put into reconciling this with detailed introspective reports. The results, as described in Chapter 3, were a number of rather detailed theoretical accounts, although for many of them the validity has not been established. While the ability to explain events was well developed, the information processing approaches on the whole showed little concern for performance prediction. Even in the case of the more detailed accounts, such as the step-ladder model, the descriptions referred to how decision making **should** take place, but could not easily be used to predict exactly how it **would** happen. Furthermore, the majority of information processing models have done little to support directly event analysis, the most notable exception being Pedersen's (1985) guide. The reason for this seems to be that the models try to explain the causes of actions from the point of view of the information processing "mechanism", i.e., as a process. Although this can serve as the basis for an analysis method, it requires a reformulation which few of the approaches have bothered to make.

In the cognitive engineering systems approach, and particularly in the case of the phenotype-genotype approach that is the basis for CREAM, the emphasis is on a principled way of analysing and predicting human erroneous actions. The cognitive systems engineering approach is consistent with the MCM framework described in Chapter 4, and therefore provides a basis for supporting both retrospective analysis and performance prediction. In particular, CREAM is based on non-directional links between classification groups or tables. This means that the same system can be used for retrospective and predictive purposes alike.

The preceding chapter described in detail how CREAM could be used to support event analysis, and in particular gave an account of the method that should be used. With respect to performance prediction, the importance of describing the context before looking at the details of how an event may develop has been repeatedly emphasised. In this respect the role of the Common Performance Conditions (CPCs) is crucial. In the retrospective method the CPCs were used to delimit the possible effects and the probable causes. In the method for performance prediction, the principle must be exactly the same, i.e., the CPCs must be used as a means of constraining the propagation of events, by effectively eliminating some of the links between causes and effects.

2. PREDICTIVE USE OF THE CLASSIFICATION SCHEME

As argued in above, the essence of performance prediction is to describe the event sequences that may likely occur, rather than to calculate or assign probabilities to individual events that have been provided by a pre-defined description. At the very least, the two things should be done in that order. The distinction can be emphasised by naming them **qualitative performance prediction** and **quantitative performance prediction** respectively (Figure 5). The purpose of qualitative performance prediction is to find out which events are likely to occur, in particular what are the possible outcomes. The purpose of quantitative performance prediction is to find out how probable it is that a specific event will occur and expressing this as a number between 0 and 1.

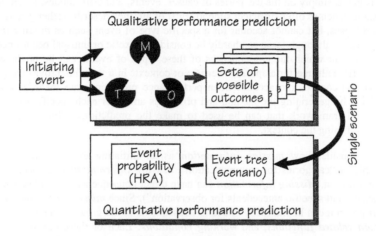

Figure 5: Qualitative and quantitative performance prediction

The qualitative performance prediction produces the set of outcomes that is likely to be the result of various event developments. The **validity** of this set depends on the models or the assumptions on which the predictions are based, in particular the details of the descriptions that make up the MTO triad. If the assumptions are reasonable, the set of outcomes will by itself provide a good indication of the reliability of the system, and show whether unwanted outcomes can occur at all. The qualitative analysis may also, in the first instance, be sufficient. It may only be necessary to proceed to a quantification of specific probabilities if significant unwanted consequences are part of the set.

2.1 Combinatorial Performance Prediction

The basic problem of performance prediction is that any classification that is used in a simple, mechanical fashion will produce far more alternatives than are needed. Consider, for instance, the case of missing information. In general, if there is a need to consider the consequences of missing information in the operation of a system, it is necessary to refer to a description of how the system functions or of how a specific event may have consequences for other events. In other words, it is necessary to know what role missing information plays in the network of antecedents and consequences that is characteristic for the

---

domain. If we consider the typical models of human performance or of human information processing that have been applied in HRA (cf. Chapter 3) they only offer limited assistance to consider this case. In fact, they offer little assistance for any case that uses categories or terms that do not directly relate to the components of the model. For instance, the model proposed by Rouse & Rouse (1983) may be used to predict the consequences of a failure of a specific step in the sequence. The associated classification scheme contains categories for each of the steps or phases in the model ("observation of system state", "choice of hypothesis", "testing of hypothesis", "choice of goal", "choice of procedure", and "execution of procedure"). Yet a situation such as missing information, denoting that not all the information needed by the operators is available, cannot easily be treated by the model in its present form. This illustrates the general principle that the scope of the prediction is constrained by the details of the model. (That is obviously also the case for CREAM, but in this case the principles for extensions have been explicitly included as part of the approach.)

If a prediction is based simply on the categories of causes, events, and error modes, it can be argued that none of the existing schemes offer sufficient support. First of all, they describe rather general phenomena, such as slips and lapses, but cannot account for a specific type of event such as missing information or failed communication - although these can hardly be considered exotic or unusual occurrences. Secondly, there is no easy way in which the consequences of these types of events can be predicted, except by developing a specific fault tree for a specific scenario or context. In a way this proves the point that it is necessary to take the context into account when performance predictions are made, although it seems awkward to do that by generating the method and principles anew for each specific case. Rather, there should be a general framework that can be used to make the predictions, which then further can be focused or improved using the context.

As an example, consider how *missing information* as an initiating event is treated by CREAM. According to table of "categories for communication", *missing information* is a specific instance of a failure of communication. *Missing information* can have direct consequents in the *observation* group (in the table of "general and specific antecedents for observation"). Since *missing information* is part of the *communication* group it can be indirectly linked to consequents in the groups of *error modes*, *planning*, *temporary person related functions*, and *working conditions*. Each of these can be linked to further consequents through one or several iterations. This kind of combinatorial prediction may quickly generate too many alternatives to be useful. Furthermore, a simple combination will not show whether one alternative is more probable or reasonable than another. Yet for any specific condition, some alternative developments will clearly be more likely than others - since clearly not all alternatives will be equally likely. The efficiency of performance prediction depends on how effectively the important consequences can be separated from the unimportant ones. It is argued that this can only be done if the context is sufficiently well known.

To illustrate the perils of combinatorial performance prediction, consider the above example a little more closely. First of all, the *missing information* itself can have three general antecedents, namely *mislabelling*, *design failure*, and *inadequate procedure*. (In addition, there can also be a number of specific antecedents as described in the table of "general and specific antecedents for communication".) For each of the three general antecedents further antecedents can be sought for, but in this discussion it shall be assumed that *missing information* is the initiating event, hence that the prediction starts there. Moving forward in the classification system, using *missing information* as the starting point (while remaining on the level of general consequents), the first iteration can lead to five different consequents in the groups mentioned above. Of these only *error modes* (*execution*) does not propagate further, since the event sequence ends by the manifest error mode. The other general consequents are all "internal", i.e., they appear as general antecedents of other groups in the classification scheme. If each of these "internal" consequents is considered in a second iteration, the outcome for all of them is either an execution error or an "internal" consequent. Most of these lead back to a previous "internal" consequent. This is due to the recursive structure of the classification scheme - which in turn reflects the cyclical nature of cognition, as described by both the SMoC and COCOM. One of the consequents (*interpretation*) has not occurred previously and therefore requires a third iteration, but after that the already identified antecedent-consequent links can be repeated. The complete example is shown in Table 1. Note, however, that the categories used in Table 1 refer to the general antecedents / consequents. If the specific consequents had been added, the table would have been considerably larger.

Table 1: Example of combinatorial performance prediction.

Initiating event	1st iteration	2nd iteration	3rd iteration
Fault in communication (communication failure)	Fault in execution (error mode).		
	Fault in observation *False observation /*	Fault in execution (error mode).	
	Observation missed / Wrong identification	Fault in interpretation *Decision error / Delayed interpretation / Faulty diagnosis /*	Fault in execution (error mode).
		Incorrect prediction /	Fault in observation
		Wrong reasoning	Fault in planning
	Fault in planning *Inadequate plan /*	Fault in execution (error mode).	
	Priority error	Fault in observation	
	Fault in working conditions *Excessive demand / Inadequate work place layout / Inadequate team support /*	Temporary, person related consequents	
	Irregular working hours	Fault in planning	
	Temporary, person related functions *Memory failure / Fear Distraction / Fatigue /*	Fault in execution (error mode).	
	Performance variability / Inattention	Fault in communication	
	/ Psychological stress	Fault in interpretation	Same as above
	/ Physiological stress	Fault in observation	
		Fault in planning	

Entries shown in *italics* describe the detailed general consequents of a classification group.

2.2 Context Dependent Performance Prediction

The example in Table 1 shows that a combinatorial approach to performance prediction can be a futile exercise. The alternative is to develop the prediction relative to a description of the likely context or the likely working conditions, which in turn must be based on a valid description of the tasks. The context description must, however, be in a form that matches the performance description. In first-generation HRA, performance was described by means of the nodes in the PSA event tree, and the context was characterised in terms of the factors that could influence performance. The approach was furthermore predicated on the PSA requirement to express the results in a quantitative fashion. The performance shaping or performance influencing factors have usually been a conglomerate of factors that empirically have been recognised as important but with few attempts either to structure them systematically - the best known exception being STAHR (Phillips et al., 1983) - or to consider dependencies and overlaps. The use of the performance shaping factors also seems to be confined to the traditional human factors approaches. Although the information processing approaches implicitly recognise the importance of performance conditions, there have been few attempts to integrate them into the models. This is probably because the concepts used by the information processing approaches offer little possibility of including the effects of conditions that cannot be expressed as information input. Instead, there has been a further specialisation of the field of organisational risk and reliability (Reason, 1992).

For analytical purposes it is clearly necessary to use some kind of simplification or abstraction. It also seems quite reasonable to describe the context with reference to a limited number of factors or dimensions, as long as the properties of these dimensions are explicitly defined. The description of the retrospective analysis method made use of the set of Common Performance Conditions (CPCs, introduced in Chapter 4). The retrospective analysis demonstrated how the CPCs could be related to the classification scheme, and how this could be used to focus the analysis - and, incidentally, also to support the interpretation of the conclusions.

In relation to performance prediction it is therefore necessary to consider whether the CPCs can be used in a similar way and, if possible, to provide a detailed description of how this should be done. Considering the discussions in the preceding, and in particular the distinction between qualitative and quantitative performance prediction, an overall approach can be described following the principles shown in Figure 6.

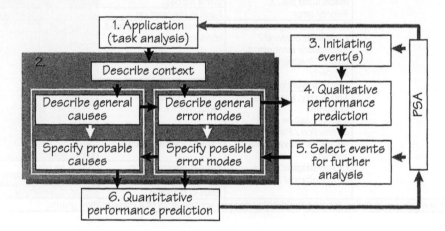

Figure 6: General method for performance prediction.

1. **Application analysis**. It is first necessary to analyse the application and the situation. This may in particular involve a task analysis, where the tasks to be considered can be derived from e.g. the PSA. The analysis must, however, include considerations of the organisation and the technical system, rather than being confined to the operator and the control tasks. Information sources may vary depending on whether the analysis is made for an existing system or for one that is being designed. If the system in question does not yet exist, it is important to use information from the design specifications, from similar systems and from operating experiences in general.

2. **Context description**. The context is described by using the CPCs (cf. Chapter 4 and Chapter 6). The principle for the context description is exactly the same as for the retrospective analysis, the

difference being that the level of detailed information may vary. In some cases it may be necessary to make assumptions about aspects of the design or the process that are not precisely known, as well as about aspects of the organisation.

The general context description can be used: (1) to prime the classification groups, just as for the retrospective analysis, by specifying more precisely the probable external and internal antecedents, and (2) to specify the possible error modes. The specification of the internal antecedents is, in particular, important and considerable care should be taken to ensure that a balanced result is achieved. It is a question of striking a proper equilibrium between, on the one hand, constraining the analysis to avoid unnecessary propagation paths and, on the other, to ensure that potentially important paths are not neglected or eliminated. This is, of course, the dilemma that is faced in any kind of HRA, but in this case it is more pertinent because the event tree is to be produced by the method rather than simply taken over from the PSA.

3. **Specification of initiating events.** The initiating events for the human actions / performance can be specified from several points of view. An obvious candidate is the PSA, since the PSA event trees will define the minimum set of initiating events that must be considered. Another is the outcome of the application and task analysis. A task analysis will, in particular, go into more detail than the PSA event tree, and may thereby suggest events or conditions that should be analysed further. The outcome of this step is the set of initiating events for which a performance prediction should be made.

    Initiating events are usually restricted to what has been called "actions at the sharp end". While it is fully acknowledged that organisational and management failures - and their latent consequences - may play a significant role in the aetiology of accidents, "actions at the blunt end" are rarely initiating events by themselves.

4. **Qualitative performance prediction**. The qualitative performance prediction uses the classification scheme, as modified by the context, to describe how an initiating event can be expected to develop. Initially this may be done manually using a paper-and-pencil representation of the events and the classification scheme. Eventually this is a step that should be supported by a software tool, since it might otherwise become too laborious to be accomplished in practice.

    In a manual version, the performance prediction can be done using the matrix shown in Table 2. In this matrix the rows show the possible consequents, while the columns show the possible antecedents. In both cases only the main classification groups are given, since the matrix otherwise would be too large for practical use. The matrix obviously shows the complete set of categories; the priming or filtering will have to be done in each case, for instance by marking the rows / columns that can be excluded in the current context. The analysis starts by finding the classification group in the column headings that correspond to the initiating event (e.g. for *missing information* it would be "Communication"). The next step is be to find all the rows that have been marked for this column. Except for the rows for the various error modes, each row will point to a consequent which in turn may be found among the possible antecedents. In this way the prediction can continue in a straightforward fashion until there are no further new paths, cf. the example in Table 1. As the preceding illustration showed, this is easy to do but not necessarily very useful in terms of practical results, unless knowledge of the context is applied to constrain the propagation.

5. **Selection of task steps for analysis**. The qualitative performance prediction, properly moderated by the context, may in itself provide useful results, for instance by showing whether there will be many or few unwanted outcomes. If a quantitative performance prediction is going to be made, it is necessary to select the cases that require further study. This can be done from the set of outcomes of the qualitative performance prediction, or from the PSA input.

Table 2: Main forward links between classification groups.

General consequent (Classification group)	Ambient condition	Communication	Equipment	Interface, permanent	Interface, temporary	Interpretation	Observation	Organisation	Person, permanent	Person, temporary	Planning	Procedures	Training	Working conditions
Timing / duration		↙				↙	↙			↙	↙	↙		
Sequence		↙		↙	↙	↙				↙	↙	↙		
Force		↙	↙			↙	↙				↙	↙		
Distance / magnitude		↙	↙			↙	↙				↙	↙		
Speed		↙	↙			↙	↙			↙	↙	↙		
Direction		↙				↙	↙			↙	↙	↙		
Wrong object		↙			↙	↙	↙			↙	↙	↙		
Ambient condition	■													
Communication		■	↙					↙	↙	↙		↙		
Equipment			■					↙						
Interface, permanent				■				↙						
Interface, temporary		↙			■			↙				↙		
Interpretation		↙		↙		■	↙	↙	↙	↙		↙		
Observation	↙	↙	↙		↙		■	↙	↙	↙				
Organisation								■						
Person, permanent									■					
Person, temporary	↙	↙	↙							■			↙	↙
Planning		↙				↙			↙		■			↙
Procedures								↙				■		
Training								↙					■	
Working conditions	↙	↙						↙						■

6. **Quantitative performance prediction.** The last step is the quantitative performance prediction. The issue of quantification is, of course, the philosophers' stone of HRA (and in a very practical sense too, since it will have the power to turn base material into gold!). The lesson to be learned from the previous discussion is that one should not attempt the quantification without having first established a solid qualitative basis or description. If the performance prediction identifies potentially critical tasks or actions, and if the failure modes can be identified, then it is perhaps not necessary to quantify beyond a conservative estimate. In other words, the search for specific HEPs for specific actions may be somewhat unnecessary. To the extent that a quantification is required, the qualitative analysis may at least be useful in identifying possible dependencies between actions. This will be described in more detail in Chapter 9. The description of the context in terms of the CPCs may also serve as a basis for defining ways of preventing or reducing specific types of erroneous actions through barriers or recovery.

These six steps provide a high-level description of how the CREAM classification scheme can be used for performance prediction. The description demonstrates that the principles of the classification scheme are equally well suited to retrospective and predictive applications, and thus confirms the arguments presented previously. The detailed categories will most certainly have to be modified and extended, since the present version was developed with the retrospective application as a primary concern. However, the main principles of the classification scheme - the non-directedness and the dynamic development of the links between classification groups - should provide the necessary basis for further refinement.

3. PRINCIPLES OF QUALITATIVE PERFORMANCE PREDICTION

As discussed above it is important that the process of performance prediction avoids the perils of a "mechanical" combination of categories. The only way this can be achieved is by making the performance prediction depend on the context, i.e., to let the probable context determine the path between the classification groups. As described for the retrospective analysis, it is an essential feature of CREAM that the classification groups do not have a fixed, hierarchical order. The retrospective analysis used the principles of CREAM to create a specific path - or link - between the classification groups depending on the context. The performance prediction must, in principle, do the same.

3.1 Forward Propagation From Antecedents To Consequents

In order to achieve this it is necessary first to outline the paths or links that are possible, given the contents of the classification groups. This can be done by noting the cases where a consequent of one group matches an antecedent of another. For instance, "equipment failure" appears as an antecedent in the "interpretation" group. This means that an equipment failure may lead to a delayed interpretation, which in turn may lead to a planning failure, and so on. The basis for the performance prediction is therefore to establish the possible forward links between the classification groups, and then select from these using the context description given by the Common Performance Conditions.

In order to assist the analyst, a table can be constructed which show the forward links. As an example, Table 2 shows the main forward links in terms of the group names rather than in terms of specific antecedents and consequents. The principle of Table 2 is that the columns describe the antecedents with the categories listed in the top row while the rows describe the consequents with the categories listed in the left column. For instance, Table 2 shows that a failure in **planning** (= an antecedent) can either be linked to specific **error modes**, or be an antecedent for **observations** (which then is considered as a consequent). If, in turn, **observation** is considered as an antecedent, Table 2 shows that it can be linked to **interpretation**. Each of these can be taken a step further until a complete event tree has been constructed. The forward propagation clearly comes to halt only when an error mode has been reached.

In order to perform the performance prediction it is, however, necessary to have a table or matrix that shows the complete forward links between antecedents and consequents. Such a table can be constructed by using the links described by the classification scheme, cf. the tables in Chapter 6. Clearly, if the classification groups change, the table of forward links must also change. A printed version of the table of forward links will naturally be cumbersome to use, but illustrates well the principles of the performance prediction. In practice, a simple computerised tool could make the process much easier.

3.2 Example: The Consequents Of Missing Information

The use of the complete table can be illustrated by considering the possible consequences of a communication failure. By using the main links shown in Table 2, it can be seen that a communication failure - which is part of the *communication* group - can propagate forwards as shown in Figure 7. Since the classification groups are not ordered hierarchically, loops will occur which means that the propagation

---

can go through a very large number of steps. This corresponds to a chain of cognitive functions that form a cascade, for instance when the consequences of a misinterpretation show themselves in later tasks.

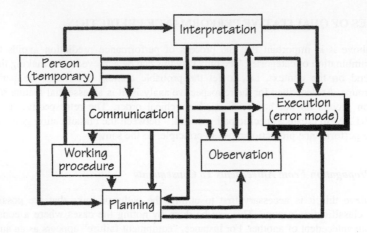

Figure 7: Forward propagation of fault in communication.

For the purpose of making a prediction, the unconstrained cascading is naturally of limited value. In this form it offers little advantage over the "mechanical" combination of categories which - in the worst case - leads to a prediction of all outcomes. This unconstrained cascading is clearly not a desirable result. The situation can, however, be improved by taking the context into account. Knowing - or assuming - what the performance conditions will be, makes it possible to consider only those consequents that are consistent with the situation. This will dramatically restrict the forward propagation of antecedents and consequents, hence serves to focus the prediction.

Table 3: Assignment of CPCs to illustrate prediction.

CPC name	Evaluation
Adequacy of organisation	*Very efficient / Efficient / **Inefficient** / Deficient*
Working conditions	*Advantageous / Compatible / **Incompatible** /*
Adequacy of MMI and operational support	*Supportive / **Adequate** / Tolerable / Inappropriate*
Availability of procedures / plans	*Appropriate / Acceptable / **Inappropriate***
Number of simultaneous goals	*Fewer than capacity / **Matching current capacity** / More than capacity*
Available time	*Adequate / **Temporarily inadequate** / Continuously inadequate*
Time of day (circadian rhythm):	See text.
Descriptors	***Day-time (adjusted)** / Night-time (unadjusted)*
Adequacy of training and preparation	*Adequate, high experience / **Adequate, limited experience** / Inadequate*
Crew collaboration quality	*Very efficient / **Efficient** / Inefficient / Deficient*

As an example, consider a hypothetical assessment of the expected Common Performance Conditions shown in Table 3. The CPCs characterise a situation where the procedures are difficult to follow, where the organisational support is inadequate, where working conditions are incompatible, and where the situation otherwise is a run-of-the-mill type, i.e., with no outstanding (positive) features. (The use of the CPCs for performance prediction will be described in greater detail in Chapter 9.)

The pattern of CPCs shown in Table 3 suggests that temporary person related functions will be affected, due to the incompatible working conditions and inadequate organisational support. Planning is also likely to be affected as is the execution of actions, due to the inappropriate procedures. If these hypothetical consequences are applied to the antecedent-consequent links described in Table 1, the outcome may be as shown in Table 4. (The consequences are entirely hypothetical, because the example is deliberately simplified and because only the main principles of the prediction are being used here. More will be said about that in Chapter 9.)

Table 4: Reduced forward antecedent-consequent links.

Step 0	Step 1	Step 2	Step 3
Fault in communication (communication failure)	Fault in execution (error mode).		
	Fault in observation	Fault in execution.	
		Fault in interpretation	Fault in execution.
			Fault in observation
			Fault in planning
	Fault in planning	Fault in execution (error mode).	
		Fault in observation	
	Fault in working conditions	Temporary, person related consequents	
		Fault in planning	
	Temporary, person related functions	Fault in execution	
		Fault in communication	
		Fault in interpretation	
		Fault in observation	
		Fault in planning	

If, for the sake of illustration, the content of Table 4 is combined with the propagation paths shown in Figure 7, the result may look as shown in Figure 8. Here the dark grey boxes indicate the functions that are mostly affected by the assumed CPCs. The most probable propagation paths are similarly shown by the thick, black arrows. The main consequence of the hypothetical situation is that a failure in communication may lead to an incorrect action either directly or through the adverse effects of the working conditions interacting with temporary person related faults and faults in planning. It is evident from a comparison of Figure 7 with Figure 8 that the number of paths that must be considered has been considerably reduced.

3.3 Discussion

This small example has illustrated how the effects of the context, described in terms of CPCs, can be used to curtail the forward propagation of the antecedent-consequent links and thereby avoid the drawbacks of a combinatorial performance prediction. Ultimately, any performance prediction will end with some of the error modes. The interesting part is, however, how the error modes are reached, and how an initial failure (say, of communication) can have effects for other antecedent-consequent links.

The reader should remember, however, that this example is used only to show the **principles** of performance prediction. Furthermore, that this method is not intended to be used in a mechanical fashion, i.e., without understanding fully the situation that is being analysed. In formal terms, the correctness of the predictions depends on the correctness of the classification groups, as well as the appropriateness of the CPCs and the specific values they have been assigned. In both cases practical experience plays an important role. Thus, for a given application and scenario, the classification groups may have to be modified to reflect the distinct features of the system. Similarly, the evaluation of the CPCs requires a good deal of experience and understanding. While computerised tools may go some way towards facilitating performance prediction, the process can never be automated as a whole.

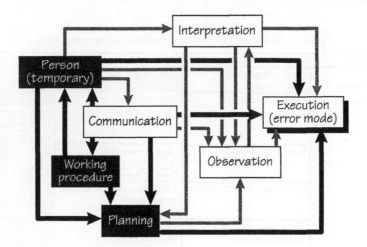

Figure 8: Constrained forward propagation of fault in communication.

As argued in Chapter 6, a main advantage of the CREAM classification system is that the same principles can be used for retrospective and predictive analyses. The use of the classification system for retrospective analysis - event analysis - will gradually lead to a refinement of both the categories and of the potential links between the groups. This will probably have to be done separately for each major domain or application, although a more general set of classification groups may also emerge. This coupling between event analysis and prediction, mediated by the classification groups, is of utmost importance for the predictions since it provides the best possible assurance that they reflect realistic assumptions about antecedent-consequent relationships. This book can only present the current state of development and indicate how the classification system can be used. There is clearly much work to be done, particularly in the predictive applications, and in applying the principles of CREAM to a proper quantitative human reliability analysis.

4. CHAPTER SUMMARY

The fundamental issue in HRA is the prediction of the likely performance under given conditions, and the last two chapters of the book provide a detailed account of how this can be accomplished. Chapter 8 has discussed the basic principles for performance prediction. In a typical HRA, the outcome of the prediction is, in a sense, given because it must follow the set of paths described by the PSA event tree. This means that a typical HRA is reduced to finding the probabilities of already specified events. A second generation HRA must clearly go beyond that by than providing a qualitative analysis that can generate a set of possible outcomes.

A problem with most classification schemes is that performance prediction takes place as a mechanical combination of the various categories. Since this eventually leads to a prediction of every possible occurrence (leaf event), the practical value of such an approach is limited. In order to be useful, the prediction must be constrained to provide only the outcomes that are likely given the circumstances or performance conditions.

The general method for performance prediction that is part of CREAM has six main steps. The first step is a detailed task analysis - cognitive or otherwise - of the application. The second step is a description of the context or common performance conditions, similar to what is done in the retrospective analysis. This step also includes the delineation of the possible error modes and the probable causes. The third step is the specification of the initiating event, i.e., the starting point for the particular set of event paths that requires analysis. The fourth step is the qualitative performance prediction, which corresponds to the generation of the most likely event tree - or set of event trees - using the forward propagation principles of the classification scheme. This is effectively a kind of screening which in itself may provide valuable information, for instance whether the initiating event can create situations that are potentially dangerous. If that is the case it may be necessary to continue with the fifth and the sixth steps, which identify the task steps or events that require further analysis and which provide the quantification of the performance prediction, respectively.

Chapter 9
The Quantification Of Predictions

In the practice of HRA it is generally agreed that the qualitative analysis is the most important part, and that the benefits of quantification often may be relatively small. In terms of providing a sufficient basis for evaluation of system performance and possible suggestions for design changes, a qualitative analysis may in many cases be all that is needed. However, when HRA is carried out in the context of PSA, the need to express the results in quantitative terms is seemingly unavoidable. It therefore behoves the HRA practitioner to provide a practical method by means of which the results of the analysis can be expressed in a meaningful form that can be used by the PSA. Quantification has always been the Achilles' heel of HRA, and while CREAM is no panacea it nevertheless offers a clear and systematic approach to quantification. The quantification problem is, in fact, considerably simplified because CREAM is focused on the level of the situation or working conditions rather than on the level of individual actions.

CREAM approaches the quantification in two steps, by providing a **basic** and an **extended** method. The basic method corresponds to an initial screening of the human interactions. The screening addresses either the task as a whole or major segments of the task. The extended method uses the outcome of the basic method to look at actions or parts of the task where there is a need for further precision and detail. The relationship between the basic and the extended method is illustrated in Figure 1.

1. CREAM - BASIC METHOD

The purpose of the basic method is to produce an overall assessment of the performance reliability that may be expected for a task. The assessment is expressed in terms of a general action failure probability, i.e., an estimation of the probability of performing an action incorrectly for the task as a whole. This provides a first screening of the task, either for the complete task or for the main task segments. The screening can be used to determine whether there is a need to continue with an extended analysis that looks at task segments or specific actions in further detail, or whether the situation appears to be one where the probability of action failures is acceptably low. The basic method consists of the following three steps:

- **Describe the task or task segments to be analysed**. The first step of an HRA **must** be a task analysis or another type of systematic task description. Unless the task is known, it is impossible to appreciate the consequences of individual task steps and actions.

- **Assess the Common Performance Conditions**. The CPCs are used to characterise the overall nature of the task, and the characterisation is expressed by means of a combined CPC score.

- **Determine the probable control mode**. The probable control mode is a central concept of the underlying Cognitive Control Model (COCOM). The probable control mode is determined from the combined CPC score. It is assumed that a control mode corresponds to a region or interval of action failure probability.

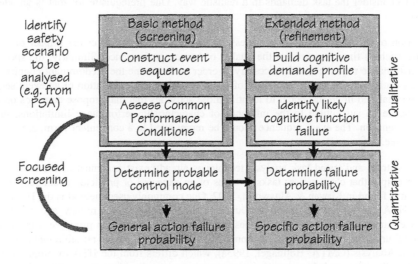

Figure 1: CREAM - basic and extended methods.

There are two differences between the basic CREAM method and other screening methods. Firstly, the characteristics of the situation that may influence performance are the starting point for the reliability assessment, together with the task analysis. The reliability - or failure probability - of individual actions is only considered after the situation as a whole has been adequately characterised. Secondly, the screening refers to an explicit, although simple, model of cognition. This model provides the fundamental rationale for the results.

1.1 Construct The Event Sequence

The first step in the application of the method requires the identification of the scenarios or events for which a reliability analysis is needed. As described in Chapter 8, this will typically involve drawing up a comprehensive list of potential system failures that are serious enough to warrant further study. Such a list will include failures that reasonably can be expected, given the prior experience of the analyst with the type of system, the general operational experience, or the specific requirements imposed by the industry's regulatory body. This is normally done as part of the overall PSA, or as part of a more specific risk analysis. (For particular domains where Human-computer Interaction is the main issue this may not always be done, although it is surely and sorely needed.) From this list one particular scenario must be selected at a time as the focus for the analysis.

Following the identification of a scenario to be analysed, the first step of the basic CREAM method is a task analysis in which the objective is to produce a description of the task, with sufficient details to support the following steps. The scenario or situation to be analysed may come from e.g. a PSA, the system design test cases, or from the specifications of a safety case. For the purpose of HRA in general it is important that the task representation being provided by the external source is not uncritically accepted and subjected to the HRA. In the case of a PSA, for instance, the description of the task is subsumed to the objectives of the PSA. Typically, only the main steps of the task are described - often referred to as the Human Interaction (or HI). This means that the level of detail may be quite limited, and furthermore that the organisation of the task description is that of the PSA event tree. Yet for the purpose of a HRA it is important to consider the task demands in a realistic way. One prerequisite for that is an adequate task description.

A description of the event sequence is necessary for the analyst to predict the impact of performance conditions, such as time pressure, on the quality of human performance. Similarly, the specification of human performance requirements is necessary in order to determine the predominant cognitive demands imposed by each part of the task. The result of this analysis is the specification of an event sequence that then can be analysed in greater detail to identify those activities that may impose demands on the human operator, hence be susceptible to the influence of characteristic performance limitations, in particular relating to cognition. The event sequence should be represented in a convenient form, e.g. as a time-line or an event tree.

A number of tried and tested task analysis techniques already exist and most of the available methods provide an output that characterises tasks in terms of an expected or predicted event sequence; a recent compilation is found in Kirwan & Ainsworth, 1992. One commonly used method is the Hierarchical Task Analysis (HTA) developed by Annett & Duncan (1967). (For a recent review and comment on this method, see Annett, 1996.) A HTA describes the main task steps that, in principle, can be analysed in further detail until the most elementary actions have been found. Another possible method is the Goals-Means Task Analysis (GMTA; Hollnagel, 1993a), which differs from the HTA by emphasising the goal structure, i.e., the relations between goals and means that are applicable to the situation. The set of goals-means relations can be used as a basis for defining the steps or actions that are necessary (and sufficient) to achieve the overall goal (e.g. Lind & Larsen, 1995).

1.2 Assess Common Performance Conditions

The CPCs have been introduced in Chapter 4. The CPCs provide a comprehensive and well-structured basis for characterising the conditions under which the performance is expected to take place. As mentioned in the preceding discussion of the CPCs, it is a basic assumption that they depend on each other. One consequence of that is that a combined CPC score cannot be produced simply as a sum of the individual CPCs, not even if they are weighted. Instead, the derivation of the combined CPC score must take into account the way in which the CPCs are coupled or dependent. This effectively corresponds to applying an underlying description or model of how the CPCs interact in the way they affect human performance. Although it should not really be necessary to argue that a description or model of this kind is necessary such practice does, unfortunately, seem to be the exception rather than the rule.

Chapter 4 gave an outline of a basic model for the dependency between the CPCs. In order to apply that in a practical manner to derive a combined CPC score, it is necessary to explain how the dependencies can be concretely treated. It is important to emphasise that the model serves the purpose of enabling a derivation of the combined CPC score. The model is clearly not the only one possible, nor is it the final version of one. However, by proposing such a model it becomes possible to see clearly how the underlying dependencies affect performance reliability, and therefore also to assess whether or not they are reasonable.

A possible relation between the CPCs and performance reliability was already described in Chapter 4. In the light of the previous discussion this description was limited because it represented the relations for each CPC individually. This can also be seen as expressing the direct or immediate relation between a CPC and performance reliability. As the description of the dependencies between the CPCs has shown there may also be an indirect or mediated relation in the cases where one CPC depends on one or more of the others. In terms of the indirect effects on performance reliability it is reasonable to assume that these only have consequences for the cases where the direct effect was assessed to be "not significant". As an example, if the "working conditions" were assessed as compatible, the effect on performance reliability was given as "not significant", i.e., neither reducing nor improving performance reliability. Conversely, performance reliability was assumed to be reduced when "working conditions" were incompatible, and improved when "working conditions" were advantageous.

If we consider the case of "working conditions", this CPC depends on five other CPCs according to the proposed description. These are "adequacy of organisation", "adequacy of MMI and operational support", "available time", "time of day", and "adequacy of training and experience", as described in Chapter 4. (For practical reasons, the table from Chapter 4 showing the relationship between CPCs and performance reliability is reproduced as Table 3 at the end of this section.) In terms of considering how these CPCs may exert a possible indirect effect, it is reasonable to assume that a majority of these CPCs must be synergistic in order to produce an effect on the "working conditions" that can change the primary effect. As a starting point, I will assume that at least four out of these five CPCs will have to be synergistic, in the sense that four of them must point in the same direction (reduce or improve) in order to have an effect on the working condition. This can be expressed by a rule that looks as shown in Figure 2.

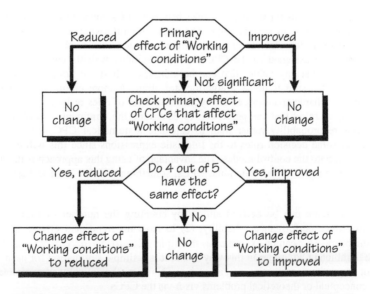

Figure 2: Rule for assessing dependency of "working conditions".

When the same principle is applied to all the CPCs, it turns out that there are only four that depend on more than one other CPC. These four are therefore the only ones of interest, since it is assumed that the indirect effect of a single CPC is insufficient to overrule the direct effect of another. The four CPCs in question and their dependencies to other CPCs are shown in Table 1 below, and are also indicated with *italics* in Table 3.

Table 1: Rules for adjusting the CPCs.

CPC	Depends on the following CPCs				
Working conditions	Adequacy of organisation	Adequacy of MMI and operational support	Available time	Time of day	Adequacy of training and experience
Number of goals	Working conditions	Adequacy of MMI and operational support	Availability of procedures/plans		
Available time	Working conditions	Adequacy of MMI and operational support	Availability of procedures / plans	Number of simultaneous goals	Time of day
Crew collab. quality	Adequacy of organisation	Adequacy of training and experience			

In the case of both "working conditions" and "available time" the criterion for changing the primary assignment is that four out of five of the related CPCs are synergistic, i.e., point in the same direction (indicating reduced or improved reliability). In the case of "number of goals" the criterion is that two out of three points are synergistic. In the case of "crew collaboration quality" the criterion is that both the related CPCs point in the same direction. These simple rules provide a way of taking the interaction between the CPCs into account, without making unnecessarily complex modelling assumptions and without requiring excessively difficult computations.

After having made adjustments for the dependencies between CPCs, the next step is to derive a combined CPC score. This can be done in several ways. The most straightforward, but also the one that is most vulnerable to criticism, is to calculate or compute the weighted sum of the CPCs. This requires that numerical values can be assigned to the CPCs and that suitable weights can be found. What speaks strongly against this is that any values that can be assigned at best are nominal. This means that the calculation of a weighted sum is not a straightforward affair. A more complex approach would be to combine the linguistic characterisations either by means of a set of rules or by considering them as fuzzy descriptors. Keeping in mind that the purpose is to determine the dominant control mode and find the general action failure probability rather than to calculate a numerical CPC score as such, it seems reasonable to apply some decision rules to the linguistic expressions since this will effectively map the linguistic description onto the control modes. The advantage of using this approach is that there is no need to define and assign weights to the individual CPCs, something which - even at best - is fraught with uncertainty.

The combined CPC score can be derived simply by counting the number of times where a CPC is expected: (1) to reduce performance reliability, (2) to have no significant effect, and (3) to improve performance reliability. This can be expressed as the triplet $[\Sigma_{reduced}, \Sigma_{not\ significant}, \Sigma_{improved}]$. The advantage of deriving the combined CPC score in this way is that it is a simple summation for each category, hence that it does not require a combination of categories. This is mathematically very simple, and does not introduce any conceptual or theoretical problems vis-à-vis the CPCs.

Altogether, the steps in assessing the CPCs can be described as follows:

♦ Determine the expected level of each CPC by using the descriptors given in Table 3.

♦ Determine the expected effects on performance reliability, using the outcomes listed in Table 3.

- Determine whether "working conditions", "number of goals", "available time" and "crew collaboration quality" should be adjusted for indirect influences, using the principles described in the rule above. Make the adjustment if necessary.

- Make a total or combined score of the expected effects and express it as the triplet [$\Sigma_{reduced}$, $\Sigma_{not\ significant}$, $\Sigma_{improved}$].

The final step of the basic CREAM method is to go from the combined CPC score to a general action failure probability.

1.3 Determine The Probable Control Mode

The final step of the basic method is to determine the probable or likely control mode. The basis for doing that is the assessment of the CPCs and the determination of the combined effect on human performance reliability.

For a given situation, the description of the CPCs will result in a specific value of the combined CPC score, expressed as the triplet [$\Sigma_{reduced}$, $\Sigma_{not\ significant}$, $\Sigma_{improved}$]. As Table 3 shows, not all values are possible; for instance, the "number of simultaneous goals" and "time of day" cannot result in an improvement on performance reliability. There are therefore only 52 different values of the combined CPC score. Of these 52 values the triplet [9, 0, 0] describes the least desirable situation, in the sense that all CPCs point to a reduced performance reliability. Similarly the triplet [0, 2, 7] describes the most desirable situation because the maximum number of CPCs point to an improved performance reliability. (This is, of course, under the assumption that all CPCs are equally important for performance reliability. If this assumption cannot be maintained, the approach suggested here must be modified.)

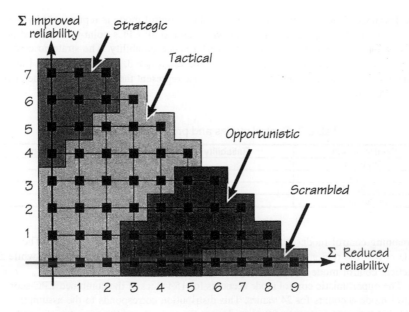

Figure 3: Relations between CPC score and control modes.

It is reasonable to assume that the least desirable situation corresponds to the scrambled control mode, and that the most desirable situation corresponds to the tactical or strategic control modes. (The control modes were introduced and defined in Chapter 6.) In situations where every CPC is inadequate or inferior there will be many demands and little support or preparation to meet them. In this type of situation the operators are likely to lose control and performance reliability is expected to be low. Conversely, in situations where every CPC has a high rating, there will be few surprises in carrying out the task, and the task demands will not seriously challenge the operators' capacity or ability to meet them. The whole working environment, from training and preparation to interface and organisation will furthermore be advantageous; performance reliability is therefore expected to be high, and the operators should be able to plan and act in a strategic manner.

Whereas the end points can relatively easily be accounted for, it is less obvious how the regions in between shall be seen. In between the end points the combined CPC score will show every possible value of the triplet [$\Sigma_{reduced}$, $\Sigma_{not\ significant}$, $\Sigma_{improved}$]. We can begin by assuming that the value of $\Sigma_{not\ significant}$ will not make a serious difference. In other words, it is the values of $\Sigma_{reduced}$ and $\Sigma_{improved}$ that are important. Although this does not reduce the number of values of the combined CPC score, it reduces the components to two. This suggests that the possible values of the CPC score can be plotted in a Cartesian co-ordinate system. If this is done, the control modes can be defined as regions or areas in the system. (If necessary, the CPC scores could, of course, also have been plotted in a three-dimensional co-ordinate system.)

The resulting plot is shown in Figure 3. The two axes show the possible values of $\Sigma_{reduced}$ and $\Sigma_{improved}$, respectively, according to the contents of Table 3. The 52 different values of the combined CPC score are shown as black dots, and the four regions that correspond to the four control modes are shown by the grey lines.

Expressed in language, Figure 3 shows that the scrambled control mode is represented by the four cases where $\Sigma_{improved} = 0$ and $\Sigma_{reduced} > 5$, i.e. the cases where almost all CPCs point to a reduced performance reliability and where none point to an improved performance reliability. The strategic control mode is represented by nine cases; in four of these $\Sigma_{reduced} = 0$ and $\Sigma_{improved} > 3$, in three $\Sigma_{reduced} = 1$ and $\Sigma_{improved} > 4$, while in the last two $\Sigma_{reduced} = 2$ and $\Sigma_{improved} > 5$. These represent the end regions of the distribution of the combined CPC score.

Table 2: Control modes and probability intervals.

Control mode	Reliability interval (probability of action failure)
Strategic	0.5 E-5 < p < 1.0 E-2
Tactical	1.0 E-3 < p < 1.0 E-1
Opportunistic	1.0 E-2 < p < 0.5 E-0
Scrambled	1.0 E-1 < p < 1.0 E-0

The two remaining control modes, the opportunistic and the tactical, are less regular. The opportunistic control mode covers the region where - generally - $\Sigma_{reduced}$ is moderately high to high, while $\Sigma_{improved}$ is low. The tactical control mode covers the region where $\Sigma_{reduced}$ is low, but where $\Sigma_{improved}$ can be either low or high. The opportunistic control mode accounts for 15 values of the combined CPC score, while the tactical control mode accounts for 24 values. This distribution corresponds to the assumption (described in Chapter 6) that the most frequent control modes are the tactical and the opportunistic and also that the strategic control mode is more frequent than the scrambled one. These expectations are based on the semantics of the model and should, of course, be adjusted by the accumulated practical experience from a domain.

The final step in the basic CREAM method is to find a general action failure probability that corresponds to how the situation has been characterised by the CPCs. Since the mapping of the combined CPC score onto the control modes has narrowed the possible performance regions - hence also the possible reliability intervals - to only four, the objective is to propose four reliability intervals that correspond to the four control modes. One proposal for that is shown in Table 2. The basis for the intervals shown here are commonly accepted estimates in the available HRA literature. The lowest value corresponds to 0.5E-5, while the highest value obviously corresponds to the certainty for failure. The intervals are overlapping; if need be, the middle of the interval can be used as a point estimate. However, at this stage of the analysis, the aim is to find a good indicator for the probability of a general action failure, rather than a precise value for the probability of failing in a specific action.

Table 3: CPCs and performance reliability.

CPC name	Level / descriptors	Expected effect on performance reliability
Adequacy of organisation	Very efficient	Improved
	Efficient	Not significant
	Inefficient	Reduced
	Deficient	Reduced
Working conditions	Advantageous	Improved
	Compatible	Not significant
	Incompatible	Reduced
Adequacy of MMI and operational support	Supportive	Improved
	Adequate	Not significant
	Tolerable	Not significant
	Inappropriate	Reduced
Availability of procedures / plans	Appropriate	Improved
	Acceptable	Not significant
	Inappropriate	Reduced
Number of simultaneous goals	Fewer than capacity	Not significant
	Matching current capacity	Not significant
	More than capacity	Reduced
Available time	Adequate	Improved
	Temporarily inadequate	Not significant
	Continuously inadequate	Reduced
Time of day (circadian rhythm)	Day-time (adjusted)	Not significant
	Night-time (unadjusted)	Reduced
Adequacy of training and experience	Adequate, high experience.	Improved
	Adequate, limited experience.	Not significant
	Inadequate.	Reduced
Crew collaboration quality	Very efficient	Improved
	Efficient	Not significant
	Inefficient	Not significant
	Deficient	Reduced

This last step of determining the probable control mode can be refined by doing it for major task segments rather than for the task as a whole. The major task segments may be determined either from experience or from a goals-means decomposition. This process of refinement may produce a differentiation between the task segments and lead to more precise estimates of the control mode and the corresponding probability intervals. It is clear, however, that only some of the CPCs can be expected to change through this refinement. Typically, the assessment CPCs such as the organisation or the time of day will remain the same, regardless of which task segment is being considered. Most, if not all, of the other assessments may change - in particular training, procedures, and the adequacy of the MMI. These CPCs are generally the ones that are most likely to differ between task segments since their effect is very situation dependent.

1.4 The Control Mode For The Ginna Example

The determination of the control mode can be illustrated by means of the example used in Chapter 8. If the ratings of the CPCs for the Ginna example are transformed into a CPC score, the result is the triplet (4, 5, 0). According to Figure 3 this corresponds to the opportunistic control mode, which is consistent with what actually happened. If such an analysis had been done prior to any accident of this nature, it would have indicated that there was an appreciable probability that something could go wrong and that at the very least the situation should be examined in further detail.

2. CREAM BASIC METHOD: AN EXAMPLE

In order to illustrate the basic method in more detail, it will be applied to an example. The task is a typical application from the process control domain. Since it does not refer to a specific process installation, there are many details that are missing. Whenever additional information is needed by the analysis, reasonable assumptions will be made. If the basic CREAM method was applied to an actual task, such information would naturally be available from the context.

2.1 Construct Event Sequence

The task used in this example is that of "restarting a furnace following a system trip". In lay terms, this describes the steps that must be taken to restore the function of a furnace.

Figure 4 shows the results of an HTA that has been created for the furnace warm-up task. (Note: The shaded boxes represent tasks and sub-tasks that are elementary for the given application, hence not decomposed further.) As can be seen from Figure 4 the overall operation (task 0.) involves four basic tasks. First, the system and ancillary services need to be prepared for start-up (task 0.1). This involves checking that the plant is generally ready (task 0.1.1), ensuring that gas and oil are available (task 0.1.2), and checking whether the oxygen measuring equipment is working (task 0.1.3). If these systems all check out then the operator can start the air blower (task 0.2) and the oil pump (task 0.3). In order to complete the final task of heating the furnace to 800 degrees Centigrade (task 0.4), the operators must increase the set-point of the temperature controller according to the instructions provided on a chart located on the control panel (task 0.4.1). They must then monitor and regulate oxygen levels as required (task 0.4.2), and monitor the furnace heat until the necessary temperature has been attained (task 0.4.3). Finally, the furnace can be switched to automatic (task 0.4.4) whereupon the start-up procedure is completed.

In addition to defining the various task elements, Figure 4 also shows that three work plans govern the actual sequence of operations that are required to perform the warm up operation. According to the results of the HTA, level 1 activities (0.1 to 0.4) need to be performed in strict sequence. In the case of the level 2 actions that are required to prepare the system for operation (0.1.1 to 0.1.3) each sub-task can be performed in any order. Finally, the first three level 3 sub-tasks relating to furnace heating (steps 0.4.1 to 0.4.3) are performed iteratively as required until the furnace temperature reaches 800 degrees, after which the system can be switched to automatic.

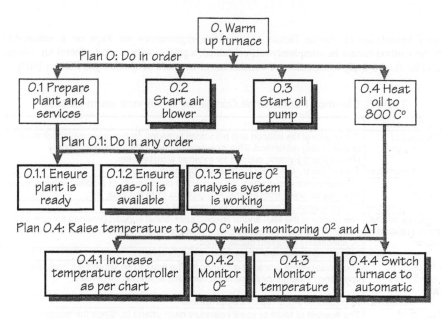

Figure 4: Hierarchical Task Analysis for the furnace warm-up task

Based on the outcome of the task analysis, it is possible to produce a list of the activities that the operator must perform to accomplish the task. This list effectively constitutes a procedure for the task, although a written set of operating instructions may not actually exist in this form.

Table 4: Steps in furnace warm-up task

#	Goal	Task step or activity
0.1.1	Prepare plant and services	Ensure plant is in a ready state
0.1.2		Ensure that gas-oil supply is available
0.1.3		Ensure that O_2 analysis system is working
0.2	Start air blower	Start air blower
0.3	Start oil pump	Start oil pump
0.4.1	Heat to 800 C	Increase temperature controller as per chart
0.4.2		Monitor O2
0.4.3		Monitor temperature
0.4.4		When temperature = 800 C, then switch furnace to automatic control

2.2 Assess Common Performance Conditions

The second step of the basic CREAM method involves an examination and assessment of the work conditions under which the task is performed. Since the example does not refer to a specific process installation, it is necessary to make a number of assumptions about the performance conditions in order to characterise the CPCs. The outcome of that is shown in Table 5.

The actual filling out of Table 5 should not represent a problem, since each CPC is clearly described. It is important for the CREAM method, as indeed it should be for every HRA approach, that the requirements to specialist knowledge of human factors or cognitive ergonomics are kept on a reasonable level. Although the method cannot be completely self-contained, a person with a good general knowledge of the field should be able to apply it regardless of whether his or her field of speciality is PSA or HRA.

Table 5: Common Performance Conditions for furnace warm-up task.

CPC name	Evaluation
Adequacy of organisation	The quality of the support and resources provided by the organisation for the task or work being performed. This includes communication systems, Safety Management System, support for external activities, etc.
Descriptors	Very efficient / Efficient / **Inefficient** / Deficient
Working conditions	The conditions under which the work takes place, such as ambient lighting, glare on screens, noise from alarms, interruptions from the task, etc.
Descriptors	Advantageous / **Compatible** / Incompatible /
Adequacy of MMI and operational support	The quality of the MMI and/or specific operational support provided for operators. The MMI includes control panels, workstations, and operational support provided by specifically designed decision aids.
Descriptors	Supportive / Adequate / **Tolerable** / Inappropriate
Availability of procedures / plans	The availability of prepared guidance for the work to be carried out, including operating / emergency procedures, routines, & familiar responses.
Descriptors	Appropriate / Acceptable / **Inappropriate**
Number of simultaneous goals	The number of tasks or goals operators must attend to. Since the number of goals is variable, this CPC applies to what is typical / characteristic for a situation.
Descriptors	Fewer than capacity / **Matching current capacity** / More than capacity
Available time	The time available to complete the work; or the general level of time pressure for the task and the situation type. How well the task is synchronised to the process dynamics.
Descriptors	**Adequate** / Temporarily inadequate / Continuously inadequate
Time of day (circadian rhythm):	The time at which the task is carried out, in particular whether the person is adjusted to the current time.
Descriptors	**Day-time (adjusted)** / Night-time (unadjusted)
Adequacy of training and preparation	The level of readiness for the work as provided (by the organisation) through training and prior instruction. Includes familiarisation to new technology, refreshing old skills, etc. as well as the level of operational experience.
Descriptors	Adequate, high experience / Adequate, limited experience / **Inadequate**
Crew collaboration quality	The quality of the collaboration between crew members, including the overlap between the official and unofficial structure, the level of trust, and the general social climate among crew members.
	Very efficient / **Efficient** / Inefficient / Deficient

The specific basis for making the assignments of the CPCs is the preceding task analysis. This underlines that the task analysis should not simply refer to an event sequence taken over from PSA, but that it should rather look at the task as a whole in a realistic fashion, using as many available sources of information as possible. By virtue of having performed the task analysis, the analyst will be in a good position to assess the various CPCs such as the control panel information available to monitor, the complexity of the plans or means needed to achieve a task, the number of simultaneous operations, and so forth. This is helpful also for the later determination of the types of erroneous actions that may possibly occur under particular work conditions. The CPC analysis is also very important for the assessment of error probabilities as it will be seen at a later stage.

In the present example, it has been assumed that the conditions as a whole are of an average character, except for the availability of procedures/plans and the adequacy of training and experience. Here it is assumed that the warm-up task does not have any procedural support, nor that it is one that has been specifically trained. In other words, it is considered a relatively unimportant task (at least from the safety point of view) and whatever training is given occurs as part of the job (apprentice, learning from more experienced colleagues, etc.) This also means that the organisation must be characterised as inefficient, at least for the case of illustration.

In order to find the combined CPC score, the given assignments to the CPCs must first be entered into Table 3. When this has been done it must be checked whether the CPCs need to be adjusted, using the rules described in Table 1; in the example and with the assignments given here this is not necessary. (I leave it as an exercise for the reader to find out why.) As a result, the combined CPC score in this case is [3, 5, 1]. This means that three CPCs point to a reduced performance reliability, that four CPCs indicate there is no significant influence, and finally that one CPC points to improved performance reliability.

2.3 Determine The Probable Control Mode

The third and last step of the basic CREAM method is to determine the probable control mode, and thereby also the general action failure probability. Using the diagram in Figure 3, the result is that the operator is expected to be in an opportunistic control mode. This corresponds well with the assumption that the operator has little training or experience in the task, and that there is inadequate operational support. Under those conditions it is not unreasonable to assume that the operator may resort to almost a "try-this-and-see-what-happens" type of performance. This may particularly be the case in the most complicated of the tasks, the controlled increase of the temperature (tasks 0.4.2 and 0.4.3).

All that remains is to determine the probability interval for the expected control mode. Table 2 shows that in this case the general action failure probability is in the range [1.0 E-2, 0.5 E-0]. Since this can hardly be seen as an acceptable range, it means that there is a clear need to continue the analysis. This will therefore be done after the extended CREAM method has been described.

3. CREAM - EXTENDED METHOD

The purpose of the extended method is to produce specific action failure probabilities. The actions may either be those that have been defined by the PSA event tree, or actions that have been noticed during the screening process using the basic method. The extended and the basic methods are both based on the fundamental principle that the failure probabilities are identified on the background of a characterisation of the task as a whole; actions occur in a context and not as isolated or idealised cognitive or information processing functions. The extended method consists of the following three steps:

*   **Build or develop a profile of the cognitive demands of the task**. In this step the description of the event sequence is refined by identifying the cognitive activities that characterise each task step or action. The cognitive activities are then used to build a cognitive profile for the main task segments, based on the functions described by the underlying cognitive model.

*   **Identify the likely cognitive function failures**. Once the cognitive activities and the corresponding cognitive functions have been identified, it is possible to propose the cognitive function failures that could occur, using the main categories of the CREAM classification scheme.

*   **Determine the specific action failure probability**. The probability of a failure corresponding to a specific cognitive function failure can be determined by adjusting a set of basic - or *a priori* - action failure probabilities with the impact of the performance conditions, expressed in terms of the CPCs.

The extended method should not - and indeed, cannot - be used independently of the basic method, since it makes use of information that is generated by the basic method. The two initial steps of the extended method continue the qualitative analysis started by the basic method. The final step provides the quantification. As argued above it may in many cases be sufficient to perform the qualitative analysis, since that by itself provides an excellent basis for describing the overall performance reliability.

3.1 Build A Cognitive Demands Profile

The first step in the extended method is to build a cognitive demands profile. The purpose of a cognitive demands profile is to show the specific demands to cognition that are associated with a task segment or a task step. This serves to indicate whether the task as a whole is likely to depend on a specific set of cognitive functions. If so, the conditions where these cognitive functions are required should be further analysed to determine whether it is likely that they will be performed correctly. The first part is, however, to characterise the task steps in terms of the cognitive activities they involve. This is basically an addition to the event sequence description that categorises each task step using a list of characteristic cognitive activities.

Table 6: List of critical cognitive activities.

Cognitive Activity	General Definition
Co-ordinate	Bring system states and/or control configurations into the specific relation required to carry out a task or task step. Allocate or select resources in preparation for a task/job, calibrate equipment, etc.
Communicate	Pass on or receive person-to-person information needed for system operation by either verbal, electronic or mechanical means. Communication is an essential part of management.
Compare	Examine the qualities of two or more entities (measurements) with the aim of discovering similarities or differences. The comparison may require calculation.
Diagnose	Recognise or determine the nature or cause of a condition by means of reasoning about signs or symptoms or by the performance of appropriate tests. "Diagnose" is more thorough than "identify".
Evaluate	Appraise or assess an actual or hypothetical situation, based on available information without requiring special operations. Related terms are "inspect" and "check".
Execute	Perform a previously specified action or plan. Execution comprises actions such as open/close, start/stop, fill/drain, etc.
Identify	Establish the identity of a plant state or sub-system (component) state. This may involve specific operations to retrieve information and investigate details. "Identify" is more thorough than "evaluate".
Maintain	Sustain a specific operational state. (This is different from *maintenance* that is generally an off-line activity.)
Monitor	Keep track of system states over time, or follow the development of a set of parameters.
Observe	Look for or read specific measurement values or system indications.
Plan	Formulate or organise a set of actions by which a goal will be successfully achieved. Plans may be short-term or long-term.
Record	Write down or log system events, measurements, etc..
Regulate	Alter speed or direction of a control (system) in order to attain a goal. Adjust or position components or subsystems to reach a target state.
Scan	Quick or speedy review of displays or other information source(s) to obtain a general impression of the state of a system / sub-system.
Verify	Confirm the correctness of a system condition or measurement, either by inspection or test. This also includes checking the feedback from prior operations.

The current version of the list of characteristic cognitive activities is shown in Table 6. It has been developed from several sources, in particular the similar type of list described by Rouse (1981) and the Human Action Classification Scheme (Barriere et al., 1994). Since such lists have been in use for many years it is reasonable to assume that they are exhaustive in a pragmatic sense, but since they are not derived from an analytical principle it is impossible to prove that they are complete, consistent, or even correct. The list contains a number of characteristic cognitive activities that are relevant for work in process control applications and also provides a pragmatic definition for each. Experience has shown that the definitions in most cases allow a cognitive activity to be assigned uniquely to a task step. There may, however, be cases where the assignment requires some degree of judgement. In cases where the analyst is uncertain about which assignment to make it is recommended that the reasons for the final choice are documented as part of the analysis, in order to provide an adequate audit trail.

The notion of a cognitive profile is based on the idea that the cognitive activities listed Table 6 put different requirements to the cognitive functions that are part and parcel of human activity. Some activities are mostly made up of operations directly on the physical interface (i.e., manipulation of control equipment), for instance, "regulate" or "execute". Conversely, other cognitive activities mainly involve manipulation of knowledge of the world (i.e., a symbolic representation of the system and the environment) and depend little on direct operations, for instance, "identification" or "evaluation". They are, however, all cognitive activities in the sense that they do involve an irreducible level of mental activity. Only tropisms and reflexes, conditioned or unconditioned, can be considered as non-cognitive activities.

The purpose of the cognitive profile is to represent the demand characteristics of the task and sub-tasks, and to indicate the kind of failures that should be expected. There are clearly differences between tasks that predominantly involve manipulation of control equipment according to a pre-defined procedure, and tasks where the operator does not immediately know what to do. Obviously, if the answer to a problem is easily available the solution does not require much effort. Similarly, if the answer is difficult to find, the solution may need considerable effort. The actual cognitive profile is based on a table of the cognitive functions associated with each of the cognitive activities. This table, shown in Table 7 below, is based on a Simple Model of Cognition (SMoC) which has been applied in a number of cases (Hollnagel & Cacciabue, 1991; Cojazzi et al., 1993). The functions are, of course, the same that are found in COCOM, described in Chapter 6.

The model underlying Table 7 assumes that there are four basic cognitive functions that have to do with observation, interpretation, planning, and execution. Each typical cognitive activity can then be described in terms of which combination of the four cognitive functions it requires. As an example, co-ordination involves planning as well as execution: the planning is used to specify what is to be done, and the execution is used to carry it out or perform it. Conversely, a task step that requires the operator to monitor aspects of the performance of the system will primarily impose a demand on observation and interpretation as cognitive functions that is indicated by a mark in the corresponding cells. Similarly, communication refers to execution only, i.e., performing the act of communicating. Note that it is not possible to make unique assignments of the cognitive functions to the cognitive activities, because the cognitive functions cannot be combined in an arbitrary way. Thus *diagnose* and *evaluate* both refer to the cognitive functions of interpretation and planning. The reason why they are separate cognitive activities is that they refer to different characteristic tasks on the level of performance.

It is entirely possible to use a different model to define the cognitive functions that constitute the substratum for the cognitive activities. There is, however, some advantage in using a model that is as simple as possible, since it is less likely to be proven wrong. An alternative model might also include additional cognitive functions. In that case Table 7 would have to be redefined, using the functions of the replacement model. This would also have consequences for some of the following steps of the CREAM method, but it would not change the basic principles of it.

Table 7: A generic cognitive-activity-by-cognitive-demand matrix

Activity type	COCOM function			
	Observation	Interpretation	Planning	Execution
Co-ordinate			◆	◆
Communicate				◆
Compare		◆		
Diagnose		◆	◆	
Evaluate		◆	◆	
Execute				◆
Identify		◆		
Maintain			◆	◆
Monitor	◆	◆		
Observe	◆			
Plan			◆	
Record		◆		◆
Regulate	◆			◆
Scan	◆			
Verify	◆	◆		

Once each of the cognitive activities has described in terms of the associated cognitive functions, it is straightforward to provide a summary for the task. The simplest solution is to count the number of occurrences of each of the cognitive functions for the task as a whole. In many cases it may be more informative to calculate the totals for major segments of the task in order to identify possible differences. The totals can easily be shown in a graphical form, e.g. as vertical bars, which then provides a visual representation of the cognitive profile.

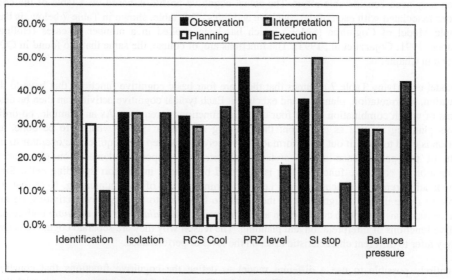

Figure 5: Example of a cognitive demand profile.

As an example of what a cognitive demands profile could look like, Figure 5 shows the result of analysing an emergency operation procedure for a nuclear power plant (further details can be found in Hollnagel et al., 1996). Although it is of little importance here, the procedure in question is for the handling of a rupture in a steam generator. More importantly, the cognitive task analysis identified 66 detailed task steps, which were classified in accordance with the principles described above. These 66 task steps referred to six major task segments, which in Figure 5 are called Identification, Isolation, RCS cool(ing), PRZ level (restoration), SI stop, and Balancing of pressure (again, the exact meanings of these terms are not important for the illustration of the method). The result of the analysis is a cognitive demands profile which shows the differences among the six task segments.

As it can easily be seen from Figure 5, the first task segment (identification) is dominated by interpretation and planning, but involves no observation and very little execution. (In the concrete task segment, the dominating cognitive activity is evaluation.) In contrast to that, the last task segment (balance pressure) is dominated by execution, together with observation and interpretation. This suggests that there are some important differences in the nature of these task segments, and therefore also that performance reliability may be different depending on the quality of the performance conditions. It is precisely the purpose of the cognitive demand profile to indicate the parts of the task where there may be a specific need to carefully analyse performance conditions and assess the expected performance reliability.

The cognitive demand profile for the furnace warm-up task is described later in this chapter.

3.2 Identify Likely Cognitive Function Failures

The second step of the extended CREAM method is to identify the likely cognitive function failures. Based on the phenotype-genotype classification of erroneous actions, it is possible to produce a complete list of cognitive function failures (Chapter 6). In principle, the complete list should be used as the basis for finding the likely cognitive function failures. However, in order to make the method practical in use it is proposed to consider only a subset of the list. It may be necessary to develop a specific subset for a specific domain to ensure that important cognitive function failures and manifestations are included in the analysis. For the purpose of describing the method it is, however, sufficient to look at a small set of generic cognitive function failures. As shown in Table 8, the cognitive function failures are defined relative to the four cognitive functions in the associated model. This means that a different set of cognitive function failures must be defined if another model is used.

The purpose of identifying the likely cognitive function failures is not to consider all the possible ways in which each step - or a specific step - of the task can fail, but rather to look at what the predominant type of failure is expected to be for the task as a whole. The cognitive function failures assigned to the task steps are selected from Table 8. The assignment is based on the description of the scenario and likely performance conditions produced by the corresponding step of the basic CREAM method. In order to make a sound assignment it is nevertheless required to have some familiarity with, and understanding of, the characteristic cognitive function failures.

Consider, for instance, the case where a task step has been characterised as corresponding to an evaluation (as a cognitive activity). The **evaluation** is described in terms of two cognitive functions, namely **interpretation** and **planning** (cf. Table 7). In assigning the likely failure mode for the evaluation, it is therefore necessary to consider the three possible failures of interpretation and the two possible failures of planning before choosing the one that is most likely under the given conditions. This can obviously not be done without knowing some details about the of nature of the task and the performance conditions. The former have been provided by the initial task analysis, the latter by step two of the basic CREAM method. This information may make it possible to determine for the specific case, firstly whether a failure of **interpretation** is more likely than a failure of **planning**; and secondly which specific type of failure one can reasonably expect.

Table 8: Generic cognitive function failures.

Cognitive function		Potential cognitive function failure
Observation errors	O1	Observation of wrong object. A response is given to the wrong stimulus or event.
	O2	Wrong identification made, due to e.g. a mistaken cue or partial identification.
	O3	Observation not made (i.e., omission), overlooking a signal or a measurement.
Interpretation errors	I1	Faulty diagnosis, either a wrong diagnosis or an incomplete diagnosis.
	I2	Decision error, either not making a decision or making a wrong or incomplete decision.
	I3	Delayed interpretation, i.e., not made in time.
Planning Errors	P1	Priority error, as in selecting the wrong goal (intention)
	P2	Inadequate plan formulated, when the plan is either incomplete or directly wrong.
Execution Errors	E1	Execution of wrong type performed, with regard to force, distance, speed or direction.
	E2	Action performed at wrong time, either too early or too late.
	E3	Action on wrong object (neighbour, similar or unrelated)
	E4	Action performed out of sequence, such as repetitions, jumps, and reversals.
	E5	Action missed, not performed (i.e., omission), including the omission of the last actions in a series ("undershoot").

Overall, knowledge of the task at hand and of the characteristic working conditions is more important than specialised knowledge of human factors or cognitive psychology and cognitive ergonomics. As I have argued for the previous steps it is impossible, and probably also undesirable, that the method is completely self-contained, in the sense that it can be implemented as a software system and carried out by a computer. The method is intended to be a help in performing a human reliability analysis, but a rigid adherence to the method cannot and should not replace the knowledge, understanding and insight that a human being has. (The same, of course, goes for the method for accident analysis.) The major purpose of the method is therefore to ensure that the predictions are made in a systematic fashion, that the individual steps are well documented, and that the reliability of the method itself is ensured.

In analogy with the cognitive demands profile it is also possible to construct a distribution of cognitive function failures. The steps needed to do so are simple. For each step of the task the analyst must assess which cognitive function failure is most likely, given the knowledge of the task and the Common Performance Conditions. The details of that will be illustrated later in the chapter. As an illustration, the distribution of cognitive function failures for the emergency operating procedure that was the basis for Figure 5, is shown in Figure 6.

A comparison between the cognitive demand profile and the distribution of cognitive function failures shows an important difference, which is due to the fact that only one cognitive failure can be assigned to each action or task step. In the case of the second task segment, isolation, the cognitive demands profile showed an equal level of *observation* and *planning* - in addition to *execution*. In the distribution of cognitive function failures, however, failures of *observation* rather than *planning* were seen as likely. The reason for this was the choice made in selecting the likely cognitive function failures. In the case of the fourth task segment, PRZ level, the cognitive demands profile showed a larger proportion of *observation* than of *interpretation*. Yet the distribution of cognitive function failures in Figure 6 shows that failures of *interpretation* were deemed to be more likely. As the example makes clear, there is no simple transformation between the cognitive demands profile and the distribution of cognitive function failures. The distribution of the cognitive function failures depends both on the cognitive demands profile and an assessment of the performance conditions.

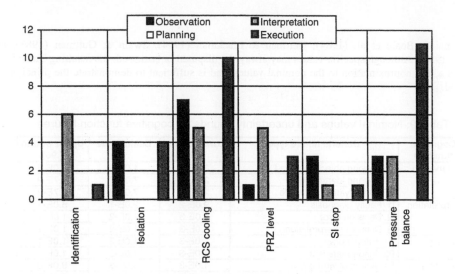

Figure 6: Distribution of cognitive function failures.

Figure 6 shows that there is a clear difference between the various segments of the procedure. From this it is not unreasonable to assume that some parts of the procedure will be susceptible to the failures associated with observation and execution, while others will be more likely to exhibit failures of e.g. identification. This indicates that the procedure is vulnerable in ways that depend on the characteristics of the segment being considered. Consequently, different types of preventive or remedial actions may be required for different segments of the procedure. In each case the likely failure types may need to be further investigated e.g., by using more detailed time estimates and/or information from engineering simulations.

3.3 Determine Failure Probability

Once the likely cognitive function failures have been assigned for each task element in the task description, it is possible to assess the probability of failure for each cognitive failure type. This can be called the **Cognitive Failure Probability (CFP)** in analogy with the traditional Human Error probability (HEP). The resulting CFP values can subsequently be inserted into a fault tree or an Operator Action Event Tree (OAET) in order for them to be integrated with the results of the engineering analysis. The quantification stage therefore comprises the following steps, that are described in detail in the following sections:

- Determine the nominal Cognitive Failure Probability (CFP) for each of the likely cognitive function failures.

- Assessing of the effects of Common Performance Conditions on the nominal CFP values.

- Incorporating the adjusted CFP values into event trees.

From a review of exiting approaches to PSA/HRA it is possible to compile a data base with nominal CFPs and uncertainty bounds for most of the generic cognitive function failures that were described in Table 8. Table 9 has made extensive use of the established data sources for proceduralised behaviours such as *observation* and *execution*. While these CFPs are relatively well established, CFPs for *interpretation* and *planning* behaviour are mostly based on expert judgements. The values have been taken from a variety of

---

sources, mainly Beare et al. (1984), Gertman & Blackman (1994), Swain & Guttman (1983), and Williams (1989). Although specific values may always be disputed or criticised, the resulting Table 9 can be used as a first approximation to the nominal values, and is sufficient to demonstrate the principles of the method.

Table 9: Nominal values and uncertainty bounds for cognitive function failures

Cognitive function	Generic failure type	Lower bound (.5)	Basic value	Upper bound (.95)
Observation	O1. Wrong object observed	3.0E-4	1.0E-3	3.0E-3
	O2. Wrong identification	2.0E-2	7.0E-2	1.7E-2
	O3. Observation not made	2.0E-2	7.0E-2	1.7E-2
Interpretation	I1. Faulty diagnosis	9.0E-2	2.0E-1	6.0E-1
	I2. Decision error	1.0E-3	1.0E-2	1.0E-1
	I3. Delayed interpretation	1.0E-3	1.0E-2	1.0E-1
Planning	P1. Priority error	1.0E-3	1.0E-2	1.0E-1
	P2. Inadequate plan	1.0E-3	1.0E-2	1.0E-1
Execution	E1. Action of wrong type	1.0E-3	3.0E-3	9.0E-3
	E2. Action at wrong time	1.0E-3	3.0E-3	9.0E-3
	E3. Action on wrong object	5.0E-5	5.0E-4	5.0E-3
	E4. Action out of sequence	1.0E-3	3.0E-3	9.0E-3
	E5. Missed action	2.5E-2	3.0E-2	4.0E-2

3.4 Accounting For The Effects Of Common Performance Conditions On CFPs

The basic notion in CREAM is that there is a coupling or dependency between the CPCs and the cognitive functions, due to the fact that human cognition and performance take place in, hence are determined by, a context. It is therefore necessary to describe, in general terms, the nature of this coupling. This can be done by combining the description of the CPCs with the cognitive functions as they are defined by COCOM, thereby adjusting the nominal CFP value to account for the effect of specific performance conditions. As a basis for that it is necessary to examine and describe the relationships between CPCs and cognitive functions. (If a different cognitive model had been used, the details of the couplings would, of course, be different; the overall principle would nevertheless remain the same.)

The simplest way of accounting for the effects of the CPCs is to use the probable control mode. As described above, the control mode can be determined on the basis of the CPC score. It is therefore only necessary to determine how the control mode may affect the CFP. This can be done by a simple table, as shown below. The weights shown in this table are a simple way of summarising the weighting factors listed in Table 12.

Table 10: Weighing factors for control modes.

Control Mode	Average weighing factor
Scrambled	2.3E+01
Opportunistic	7.5E+00
Tactical	1.9E+00
Strategic	9.4E-01

Table 10 defines the adjustment of the nominal value for a CFP for each of the four control modes. For instance, if the control mode is determined to be opportunistic, the CFP value must be multiplied by 7.5.

Conversely, if the strategic control mode is seen as more likely, the CFP value must be multiplied by 0.94, corresponding to a reduced CFP. (The values shown in Table 10 represent the averages across COCOM functions, calculated from Table 12, and aggregated according to the mapping shown in Figure 3. They should therefore be considered as rough approximations to the results from the more detailed analysis.)

In cases where this simple adjustment does not seem to provide the necessary validity, a somewhat more complex approach may be used. The basis for describing the relationship between CPCs and the cognitive functions of COCOM is the current knowledge about cognitive ergonomics and cognitive functions. As such it requires a considerable degree of expertise in these fields. Table 11 indicates the main functional couplings and the assignments are based on a general understanding of the nature of cognitive the functions.

Table 11: Couplings between CPCs and cognitive processes

Common Performance Conditions	COCOM functions			
	Observation	Interpretation	Planning	Execution
Adequacy of organisation	Weak	Weak	Medium	Medium
Working conditions	Medium	Medium	Weak	Medium
Adequacy of MMI and operational support	Strong	Weak	Weak	Strong
Availability of procedures / plans	Medium	Weak	Strong	Medium
Number of simultaneous goals	Medium	Medium	Strong	Medium
Available time	Strong	Strong	Strong	Strong
Time of day	Weak	Weak	Weak	Weak
Adequacy of training and experience	Medium	Strong	Strong	Medium
Crew collaboration quality	Strong	Strong	Strong	Strong

The meaning of the assignments can be illustrated by going through one of the cases. Take, for instance, the case of "Availability of procedures / plans".

♦ This CPC is likely to have a **medium** influence on **observation**, since procedures to some extent may define or determine which observations the operator makes. On the other hand, the influence is limited because observations mainly are data or event driven.

♦ There is only a **weak** influence on **interpretation**, because the interpretation is usually not covered by procedures. Operators may be trained in the principles of interpretation, but it is very much a function that cannot be regulated by procedures. As the table shows, the cognitive function **interpretation** is instead strongly influenced by the available time and by the amount of training.

♦ With respect to **planning** the influence is assumed to be **strong**, the reason being that the planning of what to do depends on the information that is available about action alternatives, more than information about plant states. The action alternatives are typically described in the procedures, which in that sense define the set of alternatives that an operator may consider. The availability, and format, of the procedure may also help the operator in structuring his way of planning.

♦ Finally, this CPC is likely to have a **medium** influence on **execution**. This is due to the fact that while execution is guided by the plans that have been made and the instructions that are given, the procedures are rarely described on a level of detail that corresponds to the actual actions. Other CPCs, such as available time and adequacy of MMI and operational support, are likely to be a much stronger influence that the availability of procedures and plans.

Having defined the basic relations, the next step is to turn the assignments shown in Table 11 into an operational relationship that can be used to adjust the values for the generic cognitive function failures. As a first approximation, we can assume that the weak influence is so small that it is negligible. There is certainly an influence, but since there in no case are weak influences only, it is reasonable to assume that the effect of a weak influence always will be dominated by the medium / strong influences. There is consequently no need to consider the weak influences further. (They are, however, retained in the table because it may later be possible to increase the differentiation of influences, hence refine the categorisation. It is also possible that the interactions between the CPCs require consideration, for instance using the dependencies described in Chapter 4.)

For the medium / strong influences there is clearly both a positive and a negative aspect. In each case, the influence may either be beneficial, leading to a lower CFP value, or detrimental, leading to a higher CFP value. Considering again the case of "availability of procedures / plans", it is clear that inappropriate procedures will degrade the situation while appropriate procedures will improve it. The influence need, however, not be completely symmetrical; it is reasonable to assume that the negative influence on the whole is stronger than the positive. This reflects the fact that there are usually more ways of failure than of success.

If the values of the CPCs were continuous, conforming to a ratio scale, it might be feasible to define an asymmetrical but continuous function to account for the influence of the CPC on a specific CFP. Since, however, the CPCs are described by qualitative terms, we need to use these terms as the starting point for the transformation of the CFPs. The simplest way to do that is to define *a weighting factor* for each CPC and its categories, which then is applied to the corresponding CFPs. Based upon a previous review of HRA techniques, especially HEART (Williams, 1988) supplemented by general experience from this field, a database was developed which specifies appropriate weighting factors for all cognitive function failures (see Table 12).

The bases for the assignments in Table 12 are the expected effects of the CPCs on performance reliability (Table 3) and the couplings between the CPCs and cognitive processes (Table 11). The former table determines whether there will be an effect, and whether it will lead to reduced or improved performance reliability. In the case where the expected effect is "not significant", the weight is set to be 1, which means that the basic CFP will not change. In cases where there is an expected effect, Table 11 is used to determine whether the effect will be strong, medium, or weak. Weak effects correspond to a weight of 1, while medium and strong effects are given different numerical weights, in accordance with the principles described above.

The use of Table 12 can be illustrated by the following example. Assume that the CPC for availability of procedures / plans has been rated as "inappropriate". Assume further that the task analysis has shown that the tasks include a step of planning, and that a possible cognitive function failure is priority error. Table 9 provides a basic CFP of 0.01 for this specific cognitive function failure. It follows from Table 12 that the adjustment to the CFP is by a factor 5.0, i.e., that the resulting CFP becomes 0.05 (with the upper and lower limits adjusted accordingly). In the same example, the adjustment of CFPs for cognitive function failures in the interpretation function would have been 1, i.e., that no changes would have been made. This follows from the indication in the table that only a weak interaction is assumed here - as generally shown by greying the cells.

In extreme cases it is possible that the value of the adjusted CFP becomes larger than 1. This is a consequence of the way in which the adjustments are made, and the problem can be found in other HRA approaches that use the same principle, e.g. HEART. A value greater than 1 can result if the CPCs are assessed as being generally unfavourable, which will lead to large value for the calculated total influence of the CPCs. The simple solution to this problem is simply to treat all values greater than 1 as equal to 1, since a probability by definition cannot be greater than 1. A more complex solution is to calculate the range of possible values and select an appropriate mathematical function to map the adjusted CFP values onto the interval [0, 1].

CPC name	Level	COCOM function			
		OBS	INT	PLAN	EXE
Adequacy of organisation	Very efficient	1.0	1.0	0.8	0.8
	Efficient	1.0	1.0	1.0	1.0
	Inefficient	1.0	1.0	1.2	1.2
	Deficient	1.0	1.0	2.0	2.0
Working conditions	Advantageous	0.8	0.8	1.0	0.8
	Compatible	1.0	1.0	1.0	1.0
	Incompatible	2.0	2.0	1.0	2.0
Adequacy of MMI and operational support	Supportive	0.5	1.0	1.0	0.5
	Adequate	1.0	1.0	1.0	1.0
	Tolerable	1.0	1.0	1.0	1.0
	Inappropriate	5.0	1.0	1.0	5.0
Availability of procedures / plans	Appropriate	0.8	1.0	0.5	0.8
	Acceptable	1.0	1.0	1.0	1.0
	Inappropriate	2.0	1.0	5.0	2.0
Number of simultaneous goals	Fewer than capacity	1.0	1.0	1.0	1.0
	Matching current capacity	1.0	1.0	1.0	1.0
	More than capacity	2.0	2.0	5.0	2.0
Available time	Adequate	0.5	0.5	0.5	0.5
	Temporarily inadequate	1.0	1.0	1.0	1.0
	Continuously inadequate	5.0	5.0	5.0	5.0
Time of day	Day-time (adjusted)	1.0	1.0	1.0	1.0
	Night-time (unadjusted)	1.2	1.2	1.2	1.2
Adequacy of training and preparation	Adequate, high experience	0.8	0.5	0.5	0.8
	Adequate, low experience.	1.0	1.0	1.0	1.0
	Inadequate.	2.0	5.0	5.0	2.0
Crew collaboration quality	Very efficient	0.5	0.5	0.5	0.5
	Efficient	1.0	1.0	1.0	1.0
	Inefficient	1.0	1.0	1.0	1.0
	Deficient	2.0	2.0	2.0	5.0

4. EXTENDED CREAM METHOD: AN EXAMPLE

The extended method will be illustrated by using the same example that was used for the basic method. The basic method ended by determining the probable control mode and the generic action failure probability. In the example it was found that the operator was expected to be in an opportunistic control mode and that the general action failure probability was in the range [1.0 E-2, 0.5 E-0]. This result was taken as a clear indication that the analysis should be continued with the extended method, since the general action failure probability was unacceptably high.

4.1 Build A Cognitive Demands Profile

The cognitive demands profile begins by characterising each step of the event sequence in terms of a cognitive activity. This involves the consideration of each step of the event sequence (Table 4) vis-à-vis the list of cognitive activities provided in Table 6. A decision must be taken as to the most suitable descriptive term for the activity on the basis of the definitions shown in column 2 of this table. In some circumstances it may not be possible to identify a single activity as corresponding to a particular step of the event sequence. In such cases the analyst must make a decision as to which is the predominant activity implicated in the event. If a decision is not possible, then the construction of the event sequence must be resumed for that particular step or segment until an appropriate level of detail has been reached. This process continues until such time as an unambiguous attribution of the predominant cognitive activity can be made. In the example this is illustrated by the additional decomposition of step 0.4.4 into two component actions, corresponding to the cognitive activities of *verify* and *execute* respectively. The results of the analysis in relation to the hypothetical furnace warm-up task are shown in Table 13.

Table 13: Cognitive activities in the furnace warm-up task.

Step #	Goal	Task step or activity	Cognitive activity
0.1.1	Prepare plant	Ensure plant is in a ready state	Evaluate
0.1.2	and services	Ensure that gas-oil supply is available	Verify
0.1.3		Ensure that O_2 analysis system is working	Verify
0.2	Start air blower	Start air blower	Execute
0.3	Start oil pump	Start oil pump	Execute
0.4.1	Heat to 800 C	Increase temperature controller as per chart	Regulate
0.4.2		Monitor O2	Monitor
0.4.3		Monitor temperature	Monitor
0.4.4		When temperature = 800 C,	Verify
		then switch furnace to automatic control	Execute

When the cognitive activities have been described, the next step is to determine the corresponding cognitive functions. This can be done by matching the cognitive activities of Table 13 with the cognitive demand matrix in Table 7. Since each cognitive activity can be found in Table 7, the matching is simple to make and could therefore easily be done automatically (i.e., by a small program). The result of the matching process is shown in Table 14.

In this case the number of task steps is so small that it would be meaningless to divide the event sequence into segments. The cognitive demands can therefore be found by simply adding the entries for the four last columns of Table 14. It may, however, be more useful to convert the entries to a bar-chart representation of the cognitive demands, using relative rather than absolute values. When this is done a graph such as that shown in Figure 7 is produced. (The preparation of a bar chart that summarises cognitive demand is an optional part of the method.) Figure 7 indicates that the cognitive demands are almost equally distributed between observation, interpretation and execution and that there are few demands to planning.

Table 14: Cognitive demands table for furnace warm-up task.

Step #	Task step or activity	Cognitive activity	Obs	Int	Plan	Exe
0.1.1	Ensure plant is in a ready state	Evaluate		♦	♦	
0.1.2	Ensure that gas-oil supply is available	Verify	♦	♦		
0.1.3	Ensure that O_2 analysis system is working	Verify	♦	♦		
0.2	Start air blower	Execute				♦
0.3	Start oil pump	Execute				♦
0.4.1	Increase temperature controller as per chart	Regulate	♦			♦
0.4.2	Monitor O2	Monitor	♦	♦		
0.4.3	Monitor temperature	Monitor	♦	♦		
0.4.4	When temperature = 800 C,	Verify	♦	♦		
	then switch furnace to automatic control	Execute				♦

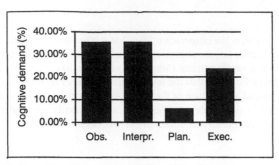

Figure 7: Cognitive Demands profile of the furnace warm-up task.

4.2 Identify Likely Cognitive Function Failures

When the cognitive demands profile has been constructed, the next step of the extended CREAM method is to identify the likely cognitive function failures. To commence this step of the extended method the analyst can enter each cognitive activity of the scenario in the left hand columns of an appropriate table, for instance as shown in Table 15. The sequence for the furnace warm-up task involved one *evaluation* and two *verification* activities required to prepare the system for operation. Two *execution* steps were needed to start the ancillary systems, while the furnace warm-up sub-task required the operator to first *regulate* the system and then to *monitor* temperature and oxygen levels, before *verifying* the target condition and *executing* the action required to switch the system to automatic control.

Beginning with the first task element Table 15, it can be seen that *evaluation* activity involves the cognitive functions of interpretation and planning. Thus, the predominant cognitive function failure must be one of the five function failures associated with these activities (see Table 8). Using the information provided by the task analysis and the Common Performance Conditions the investigator must choose which is the most likely cognitive function failure given the work context. In this case, it is assumed that the most credible failure mode is to make the wrong decision about the readiness of the plant; this is recorded in the appropriate cell using a ✦.

Table 15: Credible failure modes for furnace warm-up task.

Step #	Cognitive activity	Observation		Interpretation				Planning		Execution				
		O1	O2	O3	I1	I2	I3	P1	P2	E1	E2	E3	E4	E5
0.1.1	Evaluate					✦								
0.1.2	Verify			✦										
0.1.3	Verify			✦										
0.2	Execute											✦		
0.3	Execute												✦	
0.4.1	Regulate							✦						
0.4.2	Monitor			✦										
0.4.3	Monitor			✦										
0.4.4	Verify					✦								
	Execute											✦		
Totals				4		2				1		2	1	

Going on to the next task elements, the two *verify* activities involve the cognitive functions of observation and interpretation. Again, considering the context of the task it is assumed that the most credible cognitive function failure is associated with the observation, and that furthermore the specific type of failure is not making the required observation, i.e., an omission. This is recorded as shown in Table 15. In contrast to that, the third occurrence of a *verify* activity (Step 0.4.4) is more likely to fail in the sense of making a decision error, i.e., incorrectly deciding that the temperature has reached 800 C°. This assignment process is repeated for each activity and an estimate of the accumulated error tendency is attained by counting the number of entries recorded in each column. The data can optionally be used to construct a bar-chart representation of the predominant error tendencies in the task (as in Figure 8).

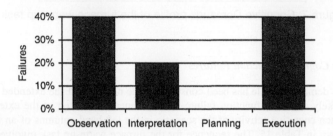

Figure 8: Cognitive function failure profile of the furnace warm-up task.

It is worth noting that the cognitive demands profile and the cognitive function failure profile normally will be different. This is because the cognitive function failure profile represents a choice of the most likely cognitive function failure for each activity. The difference can clearly be seen by comparing Figure 7 and Figure 8. It follows that even if two tasks have the same cognitive demands profile, their cognitive function failure profiles may be different due to different performance conditions.

4.3 Determine Failure Probability

The third step in the extended CREAM method is to determine the Cognitive Failure Probability. When the HRA is performed as part of a PSA, the event tree defines the activities for which a failure probability must be calculated. These activities will typically only represent a subset of the steps in the event sequence. In the context of the example used here, it is assumed that the interesting part of the event is task 0.4: heat to 800 C° (cf. Figure 4). As shown above, this was first decomposed into four steps (0.4.1, 0.4.2, 0.4.3, 0.4.4.) of which the latter again was broken down in more detail to two sub-steps. In compliance with the description of the extended method, the first thing to do is to assign the nominal Cognitive Failure Probability for each of the likely cognitive function failures. This can be done by combining part of Table 15 with the nominal values provided by Table 9, leading to the assignments shown in Table 16.

Table 16: Nominal CFPs for part of furnace warm-up task.

Task element	Error mode	Nominal CFP
0.4.1. Increase temperature controller as per chart	E1. Action of wrong type	3.0E-3
0.4.2. Monitor oxygen	03. Missed observation	3.0E-3
0.4.3. Monitor temperature	03. Missed observation	3.0E-3
0.4.4 Switch furnace to automatic	I2. Decision error	1.0E-2
	E2. Action at wrong time	3.0E-3

Following that, the effects of the CPCs on the nominal CFP values must be assessed. In the illustration of the basic CREAM method, the CPCs for the furnace warm-up task were provided in Table 5. This information can be used to select the weighting factors from Table 12, and apply these to the nominal CFP values in Table 16 to provide the situation specific weights as shown in Table 17.

Table 17: Assessment of the effects of CPCs on cognitive function failures

CPC name	Level	T 0.4.1	T 0.4.2	T 0.4.3	T 0.4.4	
		E1	O3	O3	I2	E2
Adequacy of org.	Inefficient	1.2	1.0	1.0	1.0	1.2
Working conditions	Compatible	1.0	1.0	1.0	1.0	1.0
Adequacy of MMI	Tolerable	1.0	1.0	1.0	1.0	1.0
Procedures/ plans	Inappropriate	2.0	2.0	2.0	1.0	2.0
Number of goals	Matching current capacity	1.0	1.0	1.0	1.0	1.0
Available time	Adequate	0.5	0.5	0.5	0.5	0.5
Time of day	Day-time	1.0	1.0	1.0	1.0	1.0
Training & preparation	Inadequate	2.0	2.0	2.0	5.0	2.0
Crew collaboration	Efficient	1.0	1.0	1.0	1.0	1.0
Total influence of CPC		**2.4**	**2.0**	**2.0**	**2.5**	**2.4**

For cognitive function failures associated with more than one CPCs, only the strong and medium influences are taken into account; this is achieved by having a weight of 1.0 for the weak influences. The total effect of the influence from the CPCs for each cognitive function failure is found by multiplying them. The result is shown in the bottom row of Table 17. When the combined weighting factors and the nominal CFPs are taken into account, the analysts can specify the adjusted CFPs that correspond to the work conditions of the specific scenario used throughout this report (see Table 18).

Table 18: Adjusted CFPs for cognitive function failures

Task element	Error mode	Nominal CFP	Weighting factor	Adjusted CFP
0.4.1. Increase temperature controller as per chart	E1. Action of wrong type	3.0E-3	2.4	7.2E-3
0.4.2. Monitor oxygen	O3. Missed observation	3.0E-3	2.0	6.0E-3
0.4.3. Monitor temperature	O3. Missed observation	3.0E-3	2.0	6.0E-3
0.4.4 Switch furnace to automatic	I2. Decision error	1.0E-2	2.5	2.5E-2
	E2. Action at wrong time	3.0E-3	2.4	7.2E-3

Instead of using this detailed procedure to calculate the adjustment to the CFPs we could have used the table showing the weights for the four control modes (Table 10). In the furnace-warm-up example the analysis showed that the opportunistic control mode was the expected one. According to Table 10 this corresponds to a weight of 7,5. Although this is somewhat higher than the average weights shown at the bottom line of Table 17, the general uncertainty of the CFP values means that the difference is not serious. It may therefore be concluded that if only an approximate indication of the magnitude of the effect is sought, the values from Table 10 may be sufficient.

4.4 Incorporating Adjusted CFPs Into Event Trees

The final step is to incorporate the adjusted CFP value into the PSA event tree. This requires that a single value is produced. In the case where the value is found by means of a fault tree, the calculation follows the structure of the fault tree. In the present case, however, the calculations have led to a set of adjusted CFPs, rather than a single value. The problem is therefore how these five values (Table 18) can be combined into one.

The answer to that is provided by considering the structure of the event sequence as it was given by the task analysis. As seen from Figure 4, the five steps that constitute task 0.4 all need to be performed correctly in order to achieve the goal of completing the warm-up sequence. This means that task 0.4 fails if just one of the steps fails. The overall probability of a failure of the task is therefore:

$$Max(CFP_i), i = 1, n$$

where CFP_i is an adjusted CFP value and n is the number of values calculated.

If the task had been structure in a different manner, the calculation of the overall probability would have changed accordingly. In general, the calculation corresponds to deriving the truth value of the corresponding sub-tree of the task hierarchy. In most cases this will correspond to a disjunctive condition, hence to finding the largest of the constituent CFPs. If instead a conjunctive condition existed, the overall probability would be defined as the product of the constituent CFPs. The principle of the calculation of the overall probability is thus to follow the structure identified by the task analysis, rather than the structure imposed by more or less arbitrary models of cognition or human information processing.

In the case of the example, the maximum value is 2.5E-2, which is the value for the decision error (I2). This should be compared with the outcome of the basic CREAM method, which for this example yielded a probability interval of [1.0E-2, 0.5E-0]. The value provided by the extended CREAM method therefore effectively reduces the uncertainty that was attached to the outcome of the basic method. This shows the advantage of applying the extended CREAM method. However, this need only be done in cases where the basic method has indicated: (1) that the general action failure probability is unacceptably high, and/or (2) where the uncertainty is unacceptably large.

The advantage of having a detailed task analysis as the basis for the extended method is that it may obviate the need to construct a separate fault tree. The event sequence described by the task analysis is *de facto* a fault tree. This means that the principles for combining a set of adjusted CFP values are provided by the structure of the task. There is accordingly no need to develop a separate cognitive model or cognitive fault tree, which may make assumptions about the structure of cognition that are arbitrary in the sense that they are not otherwise related to the task at hand. The cognitive model is brought into the analysis by means of the cognitive-activity-by-cognitive-demand matrix (Table 7). The calculation of the adjusted CFP value takes into account the effect of the Common Performance Conditions for the task. These were also the basis for determining the likely control mode in the basic CREAM method. The adjusted CFP value therefore indirectly reflects the control mode, and the final value will fall within the general failure probability interval produced by the basic method.

5. CHAPTER SUMMARY

The last chapter provides a detailed description of how CREAM can be used for performance prediction, hence presents CREAM as a full-blown second-generation HRA approach.

The method has two versions. One is a basic type of performance prediction that effectively can serve as a screening tool to identify the events or situations that may require further analysis. The other is an extended type of performance prediction that goes into more detail with specific events. Both can be performed as a qualitative analysis or be extended to provide a quantified result.

The basic performance prediction uses a task analysis and an assessment of the common performance conditions to identify the likely control mode (cf. the description of COCOM in Chapter 6). It describes how the characterisation of the performance conditions can be used to "map" the determining features onto the control modes. This indicates whether there is a reason to be concerned about the expected performance reliability. If necessary, this can be supplemented by a quantitative estimate of the probability of making an incorrect action that corresponds to each control mode.

The extended performance prediction uses a cognitive task analysis to identify the cognitive activities that are required of the operator. By using the simplified set of cognitive functions that are part of COCOM, this can provide a profile of the cognitive demands of the task, which gives a first indication of where potential problem areas may be. In the next step the cognitive demands profile is combined with a simplified description of the possible error modes to identify the likely cognitive function failures. These provide the qualitative results of the analysis, and may be shown in a graphical form as a cognitive function failure profile. A further step is the quantification of the cognitive failure probabilities (CFP). For practical reasons this is accomplished by using a table of nominal probabilities, based on the commonly used reference works. An important step is how the effects of the common performance conditions on the CFPs can be described. This can be done either by using the likely control mode found by the basic method, or by a more detailed - but also less certain - calculation of specific adjustment values or weights. Finally, the chapter describes how the adjusted CFPs may be incorporated into the standard PSA event tree. The various steps of the performance prediction are illustrated by an example.

References

Allen, J. (1983). Maintaining knowledge about temporal intervals. *Communication of the ACM, 26,* 832-843.

Altman, J. W. (1964). Improvements needed in a central store of human performance data. *Human Factors, 6,* 681-686.

Alty, J. L., Elzer, P., Holst, O., Johannsen, G. & Savory, S. (1985). *Literature and user survey of issues related to man-machine interfaces for supervision and control systems.* Kassel, Germany: Gesamthochschule Kassel, Fachbereich maschinenbau.

Amalberti, R. & Deblon, F. (1992). Cognitive modeling of fighter aircraft process control: A step towards an intelligent onboard assistance system. *International Journal of Man-Machine Studies, 36,* 639-671.

Amendola, A. (1988). Accident sequence dynamic simulation versus event trees. In G. E. Apostolakis, P. Kafka & G. Mancini (Eds.), *Accident sequence modeling.* London: Elsevier Applied Science.

Anderson, J. R., (1980). *Cognitive Psychology And Its Implications.* San Francisco. W.H. Freeman.

Annett, J. (1996). Recent developments in hierarchical task analysis. In: S. A. Roberston (Ed.), *Contemporary ergonomics 1996.* London: Taylor & Francis.

Annett, J. & Duncan, K. D., (1967) Task Analysis and Training Design. *Occupational Psychology,* 41, pp. 211-221

Apostolakis, G. (1984). *Methodology for time-dependent accident sequence analysis including human actions* (UCLA-ENG-P-5547-N-84). Los Angeles, CA: UCLA.

Ashby, W. R. (1956). *An introduction to cybernetics.* London: Methuen & Co.

Attneave, F. (1959). *Applications of information theory to psychology: A summary of basic concepts, methods, and results.* New York: Holt, Rinehart & Winston.

Baddeley, A. (1990). *Human memory: Theory and practice.* London: Lawrence Erlbaum Associates.

Bagnara, S., Di Martino, C., Lisanti, B., Mancini, G. & Rizzo, A. (1989). *A human taxonomy based on cognitive engineering and on social and occupational psychology.* Ispra, Italy: Commission of the European Communities Joint Research Centre.

Bainbridge, L. (1991). Mental models in cognitive skill: The example of industrial process operation. In A. Rutherford & Y. Rogers (Eds.), *Models in the mind.* London: Academic Press.

Bainbridge, L. (1993). *Building up behavioural complexity from a cognitive processing element.* London: University College, Department of Psychology

Barriere, M., Luckas, W., Whitehead, D. & Ramey-Smith, A. (1994). *An analysis of operational experience during low power and shutdown and a plan for addressing human reliability issues* (NUREG/CR-6093). Washington, DC: Nuclear Regulatory Commission.

Barriere, M. T., Wreathall, J., Cooper, S. E., Bley, D. C., Luckas, W. J. & Ramey-Smith, A. (1995). *Multidisciplinary framework for human reliability analysis with an application to errors of commission and dependencies* (NUREG/CR-6265). Washington, DC: Nuclear Regulatory Commission.

Beare, A. N., Dorris, R. E., Bovell, C. R., Crowe, D. S. & Kozinsky, E. J. (1983). *A simulator-based study of human errors in nuclear power plant control room tasks* (NUREG/CR-3309). Washington, DC: U. S. Nuclear Regulatory Commission.

Billings, C. E. & Cheany, E. S. (1981). *Information transfer problems in the aviation system* (NASA Technical Paper 1875). Moffett Field, CA: NASA.

Boy, G. A., (1987). Operator assistant systems. *International Journal of Man-Machine Studies, 27,* 541-554.

Bruce, D., (1985). The how and why of ecological memory. *Journal of Experimental Psychology. General. 114,* 78-90.

Brunswik, E. (1956). *Perception and the representative design of psychological experiments.* Berkeley: University of California Press.

Bryant, A. (1996). From an airliner's black box, next-to-last words. *The New York Times,* Sunday, April 21.

Buratti, D. L. & Godoy, S. G. (1982). *Sneak analysis application guidelines* (RADC-TR-82-179). New York: Rome Air Development Center.

Cacciabue, P. C. (1991). *Understanding and modelling man-machine interaction.* Principal Division Lecture, Proceedings of 11th SMiRT Conference, Tokyo, August 1991.

Cacciabue, P. C., Cojazzi, G., Hollnagel, E. & Mancini, S. (1992). Analysis and modelling of pilot-airplane interaction by an integrated simulation approach. In H. G. Stassen (Ed.), *Analysis, design and evaluation of man-machine systems 1992.* Oxford: Pergamon Press.

Cacciabue, P. C., Cojazzi, G. & Parisi, P. (1996). A dynamic HRA method based on a taxonomy and a cognitive simulation model. In P. C. Cacciabue & I. Papazoglou (Eds.), *Probabilistic safety assessment and management '96.* Berlin: Springer Verlag.

Cacciabue, P. C., Decortis, F., Drozdowicz, B., Masson, M. & Nordvik, J.-P. (1992). COSIMO: A cognitive simulation model of human decision making and behavior in accident management of complex plants. *IEEE Transactions on Systems, Man, Cybernetics, 22(5),* 1058-1074.

Cacciabue, P. C. & Hollnagel, E. (1992). Simulation of cognition: Applications. In J.-M. Hoc, P. C. Cacciabue & E. Hollnagel (Eds.), *Expertise and technology: Cognition and human-computer cooperation.* New York: Lawrence Erlbaum Associates.

Cacciabue, P. C., Pedrali, M. & Hollnagel, E. (1993). *Taxonomy and models for human factors analysis of interactive systems: An application to flight safety* (ISEI/IE 2437 / 93). Paper presented at the 2nd ICAO Flight Safety and Human Factors Symposium, Washington D. C., April 12-15.

Caeser, C. (1987). *Safety statistics and their operational consequences.* Proceedings of the 40th International Air Safety Seminar, Tokyo, Japan.

Canning, J. (1976). *Great disasters*. London: Octopus Books.

Caro, P. W. (1988). Flight training and simulation. In E. L. Wiener & D. C. Nagel (Eds.), *Human factors in aviation*. San Diego: Academic Press.

Casey, S. (1993). *Set phasers on stun*. Santa Barbara, CA: Aegean Publishing Company.

Clancey, W. J. (1992). Representations of knowing: In defense of cognitive apprenticeship. *Journal of Artificial Intelligence in Education, 3*(2), 139-168.

Cojazzi, G. (1993). *Root cause analysis methodologies. Selection criteria and preliminary evaluation* (ISEI/IE/2442/93). JRC Ispra, Italy: Institute for Systems Engineering and Informatics.

Cojazzi, G., Pedrali, M. & Cacciabue, P. C. (1993*). Human performance study: Paradigms of human behaviour and error taxonomies* (ISEI/IE/2443/93). JRC Ispra, Italy: Institute for Systems Engineering and Informatics.

Cojazzi, G. & Pinola, L. (1994). *Root cause analysis methodologies: Trends and needs*. In G. E. Apostolakis & J. S. Wu (Eds.), Proceedings of PSAM-II, San Diego, CA, March 20-25, 1994.

Comer, M. K., Seaver, D. A., Stillwell, W. G. & Gaddy, C. D. (1984*). General human reliability estimates using expert judgment* (NUREG/CR-3688). Washington, DC: U. S. Nuclear Regulatory Commission.

Cooper, S. E., Ramey-Smith, A. M., Wreathall, J., Parry, G. W., Bley, D. C., Luckas, W. J., Taylor, J. H. & Barriere, M. T. (1996). *A technique for human error analysis (ATHEANA)* (NUREG/CR-6350). Washington, DC: US Nuclear Regulatory Commission.

Corker, K., Davis, L., Papazian, B. & Pew, R. (1986). *Development of an advanced task analysis methodology and demonstration for army aircrew / aircraft integration (BBN R-6124)*. Boston, MA.: Bolt, Beranek & Newman.

De Keyser, V., Masson, M., Van Daele A. & Woods, D. D. (1988). *Fixation errors in dynamic anc omplex systems: Descriptive forms, psychological mechanisms, potential countermeasures*. Liege, Belgium: Universite de Liege, Working Paper Series No. 1988-014.

Dorieau, P., Wioland, L. & Amalberti, R. (1997). La détection des erreurs humaines par des opérateurs extérieurs a l'action: Le cas du pilotage d'avion. *Le Travail humain, 60*(2), 131-153.

Dougherty, E. M. Jr. (1981). *Modeling and quantifying operator reliability in nuclear power plants* (TEC Internal Report). Knoxville, TN: Technology for Energy Corporation.

Dougherty, E. M. Jr. (1990). Human reliability analysis - Where shouldst thou turn? *Reliability Engineering and System Safety, 29*(3), 283-299.

Dougherty, E. M. Jr., & Fragola, J. R. (1988). *Human reliability analysis. A systems engineering approach with nuclear power plant applications*. New York: John Wiley & Sons.

Embrey, D. E. (1980). *Human error. Theory and practice*. Conference on Human Error and its Industrial Consequences. Birmingham, UK: Aston University.

Embrey, D. E. (1992). *Quantitative and qualitative prediction of human error in safety assessments.* Major Hazards Onshore and Offshore, Rugby IChemE.

Embrey, D. E. (1994). Evaluating and reducing human error potential in off-shore process installations. *Journal of Loss Prevention in the Process Industries* (special issue on Loss Prevention on Offshore Process Installations), in press.

Embrey, D. E., Humphreys, P. Rosa, E. A., Kirwan, B. & Rea, K. (1984). *SLIM-MAUD. An approach to assessing human error probabilities using structured expert judgment* (NUREG/CR-3518). Washington, D.C.: USNRC.

ESA PSS-01. (1988). *System safety requirements for ESA space systems and associated equipment* (ESA PSS-01-40 Issue 2). Paris: European Space Agency.

Feggetter, A. J. (1982). A method for investigating human factors aspects of aircraft accidents and incidents. *Ergonomics, 25,* 1065-1075.

Feigenbaum, E. A. (1961). *The simulation of verbal learning behavior.* Proceedings of the Western Joint Computer Conference, 19, 121-132.

Fitts, P. M. (Ed). (1951). *Human engineering for an effective air navigation and traffic-control system.* Ohio State University Research Foundation, Columbus, Ohio.

Fleming, K. N., Raabe, P. H., Hannaman, G. W., Houghton, W. J., Pfremmer, R. D. & Dombek, F. S. (1975). *HTGR accident investigation and progression analysis status report (Vol. II): AIPA risk assessment methodology* (GA/A13617 Vol. II UG-77). San Diego: General Atomic Co.

Fullwood, R. R. & Gilbert, K. J. (1976). *An assessment of the impact of human factors on the operations of the CRBR SCRS.* SAI-010-76-PA.

Furuta, T. (1995). *Management of maintenance outages and shutdowns.* Proceedings of Joint OECD/NEA-IAEA Symposium on Human Factors and Organisation in NPP Maintenance Outages: Impact on Safety. Stockholm, Sweden, June 19-22, 1995.

Gagne, R. M. (1965). *The conditions of learning.* New York, Rinehart and Winston.

Gertman, D. I. (1992). *Representing cognitive activities and errors in HRA trees* (EGG-HFRU-10026). Idaho Falls, ID: Idaho National Engineering Laboratory.

Gertman, D. I. (1993). Representing cognitive activities and errors in HRA trees. *Reliability Engineering and Systems Safety, 39,* 25-34.

Gertman, D. I. & Blackman, H. S. (1994). *Human reliability & safety analysis data handbook.* New York: John Wiley & Sons, Inc.

Gertman, D. I., Blackman, H. S. & Hahn, H. A. (1992). Intent: A method for estimating human error probabilities for decision based errors. *Reliability Engineering and System Safety, 35,* 127-137.

Gertman, D. I., Blackman, H. S., Haney, L. N., Seidler, K. S., Sullivan, C & Hahn, H. A. (1990). *INTENT: A method for estimating human error probabilities for errors of intention* (EGG-SRE-9178). Idaho Falls, ID: Idaho National Engineering Laboratory.

Gore, B. R., Dukelow, T. M. Jr. & Nicholson, W. L. (1995). *A limited assessment of the ASEP human reliability analysis procedure using simulator examination results* (NUREG/CR-6355). Washington, DC: Nuclear Regulatory Commission.

Hahn, A. H. et al. (1991). Applying sneak analysis to the identification of human errors of commission. *Reliability Engineering and System Safety, 33*, 289-300.

Hall, R. E., Fragola, J. & Wreathall, J. (1982). *Post event human decision errors: Operator action tree / time reliability correlation* (NUREG/CR-3010). Washington, DC: U. S. Nuclear Regulatory Commission.

Haney, L. N., Blackman, H. S., Bell, B. J., Rose, S. E., Hesse, D. J., Minton, L. A. & Jenkins, J. P. (1989). *Comparison and application of quantitative human reliability analysis methods for the risk method integration and evaluation program* (RMIEP) (NUREG/CR-4835). Washington, DC: US Nuclear Regulatory Commission.

Hannaman, G. W. & Spurgin, A. J. (1984). *Systematic human action reliability procedure (SHARP)* (EPRI NP-3583). Palo Alto, CA: Electric Power Research Institute.

Hannaman, G. W., Spurgin, A. J. & Lukic, Y. D. (1984*). Human cognitive reliability model for PRA analysis* (NUS-4531). Palo Alto, CA: Electric Power Research Institute.

Helander, M. (Ed.), (1988). *Handbook of human-computer interaction.* Amsterdam: North-Holland.

Helmreich, R. L. (1997). Managing human error in aviation. *Scientific American, May 1997*, 40-45.

Heslinga, G. & Arnold, H. (1993). *Human reliability: To what extent can we consider humans as system components.* International ENS Topical Meeting Towards the next Generation of Light Water Reactors, April 25-28, The Hague, Netherlands.

Hoc, J.-M., Amalberti, R. & Boreham, N. (1995). Human operator expertise in diagnosis, decision-making, and time management. In J.-M. Hoc, P. C. Cacciabue & E. Hollnagel (Eds.), *Expertise and technology.* Hillsdale, NJ: Lawrence Erlbaum.

Hoc, J.-M., Cacciabue, P. C. & Hollnagel, E. (Eds.), (1995). *Expertise and technology: Cognition and human-computer cooperation.* Hillsdale, N. J.: Lawrence Erlbaum Associates.

Hollnagel, E. (1984). Inductive and deductive approaches to modelling of human decision making. *Psyke & Logos, 5(2)*, 288-301.

Hollnagel, E. (1988). Plan recognition in modelling of users. In G. E. Apostolakis, P. Kafka, & G. Mancini (Eds.), *Accident sequence modelling: Human actions, system response, intelligent decision support.* London: Elsevier Applied Science.

Hollnagel, E. (1990). The phenotype of erroneous actions: Implications for HCI design. In G. Weir & J. Alty (Eds.), *Human-computer Interaction and complex systems.* London: Academic Press, 1991.

Hollnagel, E. (1993a). *Human reliability analysis: Context and control.* London: Academic Press.

Hollnagel, E. (1993b). The phenotype of erroneous actions. *International Journal of Man-Machine Studies, 39*, 1-32.

Hollnagel, E. (1993c). Requirements for dynamic modelling of man-machine interaction. *Nuclear Engineering and Design, 144*, 375-384.

Hollnagel, E. (1993d). Modelling of cognition: Procedural prototypes and contextual control. *Le Travail Humain, 56(1)*, 27-51.

Hollnagel, E. (1995a). The art of efficient man-machine interaction: Improving the coupling between man and machine. In J.-M. Hoc, P. C. Cacciabue & E. Hollnagel (Eds.), *Expertise and technology: Cognition & human-computer cooperation.* . Hillsdale, N. J.: Lawrence Erlbaum Associates.

Hollnagel, E. (1995b). *Latent failure conditions and safety barrier integrity.* Joint OECD/NEA-IAEA Symposium on Human Factors and Organisation in NPP Maintenance Outages: Impact on Safety. Stockholm, Sweden, 19-22 June.

Hollnagel, E. (1995c). *Summary of approaches to dynamic analysis of man-machine interaction* (NKS/RAK-1(95)R1. Roskilde, Denmark: Risø National Laboratory.

Hollnagel, E. (1997). Context, cognition, and control. In Y. Waern, (Ed.). *Co-operation in process management - Cognition and information technology.* London: Taylor & Francis

Hollnagel, E. & Cacciabue P. C. (1991). *Cognitive modelling in system simulation.* Proceedings of Third European Conference on Cognitive Science Approaches to Process Control, Cardiff, September 2-6, 1991

Hollnagel, E. & Cacciabue, P. C. (1992). *Reliability Assessment Of Interactive Systems With The System Response Analyser.* European Safety and Reliability Conference '92, Copenhagen, 10-12 June, 1992.

Hollnagel, E., Cacciabue, P. C. & Rouhet, J.-C. (1992). *The use of integrated system simulation for risk and reliability assessment.* Paper presented at the 7th International Symposium on Loss Prevention and Safety Promotion in the Process Industry, Taormina, Italy, 4th-8th May, 1992.

Hollnagel, E., Edland, A. & Svenson, O. (1996). *A cognitive task analysis of the SGTR scenario* (NKS/RAK-1(96)R3). Roskilde, Denmark: Risø National Laboratory.

Hollnagel, E. & Embrey, D. E. (1993). *Human reliabilty analysis in safety assessment: Adressing cognitive and action errors.* IBC Conference on PSA/PRA for the Nuclear Industry, November 1-2, 1993, London.

Hollnagel, E. & Marsden, P. (1995). *Further development of the phenotype-genotype classification scheme for the analysis of human erroneous actions.* Dalton, UK: HRA Ltd.

Hollnagel, E., Pedersen, O. M., & Rasmussen, J. (1981). *Notes on human performance analysis* (Risø-M-2285). Roskilde, Denmark: Risø National Laboratory, Electronics Department.

Hollnagel, E., Rosness, R. & Taylor, J. R. (1990). *Human Reliability And The Reliability of Cognition.* Proceedings of 3rd International Conference on 'Human Machine Interaction And Artificial Intelligence In Aeronautics And Space' Toulouse-Blagnac, 26-28 September.

Hollnagel, E. & Woods, D. D. (1983). Cognitive systems engineering: New wine in new bottles. *International Journal of Man-Machine Studies, 18*, 583-600.

Hollnagel, E. & Wreathall, J. (1996). HRA at the turning point? In P. C. Cacciabue & I. Papazoglou (Eds.), *Probabilistic safety assessment and management '96.* Berlin: Springer Verlag.

Holmes, B. (1994). Fantastic voyage into the virtual brain. *New Scientist*, *143*, (1932), 26-29.

Hudson, P. T. W., Reason, J., Wagenaar, W. A., Bentley, P., Primrose, M. & Visser, J. (1994). Tripod delta: Proactive approach to enhanced safety. *Journal of Petroleum Technology*, *46*, 58-62.

Human Factors Reliability Group. (1985) *Guide to Reducing Human Error in Process Control*. Safety and Reliability Directorate. Culcheth, Warrington, UK.

Hutchins, E. (1995). *Cognition in the wild*. Cambridge, MA: MIT Press.

INPO 83-030 (1982). *Analysis of steam generator tube rupture events at Oconee and Ginna*. Atlanta, GA: Institute of Nuclear Power Operations.

INPO (1984). *An analysis of root cause failures in 1983 significant event reports*. Atlanta, GA: Institute of Nuclear Power Operations.

INPO (1985). *A maintenance analysis of safety significant events*. Atlanta, GA: Institute of Nuclear Power Operations.

Johannsen, G. (1990). Fahrzeugführung. In C. G. Hoyos & B. Zimolong (Eds.), *Ingenieurpsychologie*. Göttingen, FRG: Verlag für Psychologie.

Johnson, W. G. (1980). *MORT safety assurance systems*. New York: National Safety Council, Marcel Dekker.

Johnson, W. E. & Rouse, W. B. (1982). Analysis and classification of human errors in troubleshooting live aircraft power plants. *IEEE Transactions on Systems, Man, and Cybernetics*, *SMC-12*, 389-393.

Kantowitz, B. H. & Fujita, Y. (1990). Cognitive theory, identifiability and human reliability analysis (HRA). *Reliability Engineering and System Safety*, *29*, 317-328.

Kirwan, B. (1994). *A guide to practical human reliability assessment*. London: Taylor & Francis.

Kirwan, B. (1996). *The requirements of cognitive simulations for human reliabiltiy and probabilistic safety assessment*. In H. Yoshikawa & E. Hollnagel (Eds.), Proceedings of Cognitive Systems Engineering in Process Control (CSEPC 96), November 12-15, 1996, Kyoto, Japan.

Kirwan, B. & Ainsworth, L. A. (Eds). (1992). *A guide to task analysis*. London: Taylor & Francis.

Klein, G. A., Oramasu, J., Calderwood, R. & Zsambok, C. E. (1993). *Decision making in action: Models and methods*. Norwood, NJ: Ablex.

Kowalsky, N. B., Masters, R. L., Stone, R. B., Babcock, G. L. & Rypka, E. W. (1974). *An analysis of pilot error related to aircraft accidents* (NASA CR-2444). Washington, DC: NASA.

Kruglanski, A. W. & Ajzen, I. (1983). Bias and error in human judgement. *European Journal of Social Psychology*, *13*, 1-44.

Legge, D. (1975). *An introduction to psychological science*. London. Methuen and Co.

Leplat, J. & Rasmussen, J. (1987). Analysis of human errors in industrial incidents and accidents for improvement of work safety. In J. Rasmussen, K. Duncan & J. Leplat (Eds.), *New technology and human error*. London: Wiley.

Leveson, N. G. & Turner, C. S. (1992). *An investigation of the Therac-25 accidents* (UCI TR #92.108). Irvine, CA: University of California.

Lind, M. & Larsen, M. N. (1995). Planning and the intentionality of dynamic environments. In J.-M. Hoc, P. C. Cacciabue & E. Hollnagel (Eds.*), Expertise and technology: Cognition and human-computer interaction.* Hillsdale, N. J. Lawrence Erlbaum Associates.

Lindblom, C. E. (1959). The science of "muddling through." *Public Administration Quaterly, 19,* 79-88.

Lindsay, P .H. & Norman, D. A. (1976). *Human information processing.* New York, Academic Press.

Lucas, D. (1988). *State of the art review of qualitative modelling techniques.* Dalton, Lancs., UK: Human Reliability Associates.

Lucas, D. (1992). Understanding the human factor in disasters. *Interdisciplinary Science Reviews, 17*(2), 185-190.

Macwan, A. & Mosleh, A. (1994). A methodology for modelling operator errors of commission in probabilistic risk assessment. *Reliability Engineering and Systems Safety,* 45, 139-157.

Macwan, A., Wieringa, P. A. & Mosleh, A. (1994). *Quantification of multiple error expressions in following emergency operating procedures in nuclear power plant control room.* In G. E. Apostolakis & J. S. Wu (Eds.), Proceedings of PSAM-II, March 20-25, San Diego, CA.

Macwan, A., Bos, J. F. T. & Hooijer, J. S. (1996). Implementation of cause-based pilot model for dynamic analysis of approach-to-landing procedure: Application of human reliability to civil aviation. In P. C. Cacciabue & I. A. Papazoglou (Eds.), *Probabilistic safety assessment and management '96.* Berlin: Springer Verlag.

Maier, N. R. F. (1930). Reasoning in humans. I. On direction. *Journal of Comparative Psychology, 10,* 115-143.

Malin, J. T., Schreckenghost, D. L. & Woods, D. D. (1990). *Making intelligent systems team players.* LBJ Research Center, Houston, TX, NASA TM 104738.

Mandler, G. (1975). *Mind and emotion.* New York: Wiley.

McCormick, E. J. & Tiffin, J. (1974). *Industrial psychology.* London. George Allen and Unwin Ltd.

Meric, M., Monteau, M. & Skekely, J. (1976). *Technique de gestion de la securite* (234/RE). Vandoeuvre, France: INRS.

Miller, D. P. & Swain, A. D. (1987). *Human Error and Human Reliability.* In G. Salvendy (Ed.) Handbook of Human factors. New York: Wiley.

Miller, G. A., Galanter, E. & Pribram, K. H. (1960). *Plans and the structure of behavior.* New York: Holt, Rinehart & Winston.

Miller, T. E. & Woods, D. D. (1996). Key issues for naturalistic decision making researchers in systems design. In C. Zambok & G. Klein (Eds.) *Naturalistic decision making.* Hillsdale, NJ: Lawrence Erlbaum Associates.

Modarres, M. (1993). *Reliability and risk analysis: What every engineer should know about reliability and risk analysis.* New York: Marcel Dekker, Inc.

Neisser, U. (1967). *Cognitive psychology.* New York, Appleton Century Crofts.

Neisser, U. (1976). *Cognition and reality.* San Francisco: W. H. Freeman.

Neisser, U. (1982). *Memory observed. Remembering in natural contexts.* San Francisco, Freeman.

Newell, A. (1990). *Unified theory of cognition.* Cambridge, MA.: Harvard University Press.

Newell, A. F. (1993). HCI for everyone. In S. Ashlund, K. Mullet, A. Henderson, E. Hollnagel & T. White (Eds). *Proceedings of ACM INTERCHI'93 Conference on Human Factors in Computing Systems.* New York: ACM SIGCHI.

Newell, A. F. (1993). Interfaces for the ordinary and beyond. *IEEE Software, 10*(5) Sept, 76-78.

Newell, A. & Simon, H. A. (1972). *Human problem solving.* Englewood Cliffs, NJ.: Prentice-Hall.

Newell, A. & Simon, H. A. (1976). Computer science as empirical inquiry: Symbols and search. *Communications of the ACM, 19* (March).

Norman, D. A. (1981). Categorization of action slips. *Psychological Review, 88,* 1-15.

Norman, D. A. (1986). New views of information processing: Implications for intelligent decision support systems. In E. Hollnagel, G. Mancini & D. D. Woods (Eds.), *Intelligent decision support in process environments.* Berlin: Springer.

Norman, D. A. (1988). *The psychology of everyday things.* New York: Basic Books.

NUREG-0909. (1982). *NRC Report on the January 25, 1982 SGTR Event at R. E. Ginna NPP* Washington, DC: US Nuclear Regulatory Commission. (April).

NUREG-1154. (1985). *Loss of main and auxilliary feedwater event at the Davis-Besse plant on June 9, 1985.* Washington, DC: US Nuclear Regulatory Commission.

Orvis, D. D., Moieni, P., Murray, R. & Prassinos, P. (1991). Seismic margins systems analysis for Hatch unit 1 nuclear power plant. In G. Apostolakis (Ed.*), Probabilistic safety assessment and management* (pp. 103-108). New York: Elsevier Science Publishing Co.

Otway, H. J. & Misenta, R. (1980). Some human performance paradoxes of nuclear operations. *Futures.* October. pp. 340-357.

Parasuraman, R. (1979). Memory loads and event rate control sensitivity decrements in sustained attention. *Science, 205,* 924-927.

Park, K. S. (1987). *Human reliability. Analysis, prediction, and prevention of human errors.* Amsterdam: Elsevier.

Parry, G. W. & Mosleh, A. (1995). *Control room crew operations research project* (EPRI TR-105280). Palo Alto, CA: Electrical Power research Institute.

Payne, D. & Altman, J. W. (1962). *An index of electronic equipment operability*. Report of Development, AIR-C43-1/62/FR, Pittsburgh, PA.

Pedersen, O. M. (1985). *Human risk contributions in process industry. Guides for their pre identification in well-structured activities and for post-incident analysis* (Risø-M-2513). Roskilde, Denmark: Risø National Laboratories.

Pedrali, M. (1996). *Vers un environnement multimédia pour l'analyse vidéo des causes d'erreurs humaines. Application dans les simulateurs d'avions*. Thèse de Doctorat en Informatique, Université Toulouse 1, France

Pedrali, M., Bastide, R. & Cacciabue, P. C. (1996). A *multimedia environment for retrospective analysis of accidents and incidents involving human factors* (ISI/SMA 3120/96). Ispra, Italy: Joint Research Centre.

Pew, R. W., Miller, D. C. & Feeher, C. E. (1981). *Evaluation of proposed control room improvements through analysis of critical operator decisions* (NP-1982). Palo Alto, CA: Electric Power Research Institute.

Phillips, L. D., Humphreys, P. C., & Embrey D. E. (1983). *A socio-technical approach to assessing human reliability* (83-4). Oak Ridge, TN: Oak Ridge National Laboratory.

Potash, L. M., Stewart, M., Dietz, P. E., Lewis, C. M. & Dougherty, E. M. Jr. (1981). *Experience in integrating the operator contributions in the PRA of actual operating plants*. (ANS/ENS Topical Meeting on Probabilistic Risk Assessment, Port Chester, NY.) LaGrange, IL: American Nuclear Society.

Rasmussen, J. (1986). *Information processing and human-machine interaction: An approach to cognitive engineering*. New York: North-Holland.

Rasmussen, J., Duncan, K. & Leplat, J. (Eds.) (1978). *New technology and human error*. London: John Wiley & Sons.

Rasmussen, J. & Jensen, A. (1974). Mental procedures in real-life tasks. A case study of electronic troubleshooting. *Ergonomics, 17,*. 193-207.

Rasmussen, J., Pedersen, O. M., Mancini, G., Carnino, A., Griffon, M. & Gagnolet, P. (1981). *Classification system for reporting events involving human malfunctions* (Risø-M-2240, SINDOC(81)14). Risø National Laboratory, Roskilde, Denmark.

Reason, J. T. (1976). Absent minds. *New Society*, 4. pp. 244-245.

Reason, J. T. (1979). Actions not as planned. The price of automatization. In G. Underwood & R. Stevens (Eds). *Aspects of consciousness* (Vol. 1). *Psychological issues*. London: Wiley.

Reason, J. T. (1984). Absent-mindedness. In J. Nicholson & H. Belloff, (Eds) *Psychology Survey No. 5.* Leicester. British Psychological Society.

Reason, J. T. (1985). Recurrent errors in process control environments: Some implications for the design of Intelligent Decision Support Systems. In E. Hollnagel, G. Mancini & D. D. Woods (Eds.), *Intelligent decision support in process environments*. Heidelberg, F. R. Germany: Springer Verlag.

Reason, J. T. (1986). *The classification of human error*, unpublished manuscript. University of Manchester.

Reason, J. T. (1987). Generic error-modelling system (GEMS): A cognitive framework for locating human error forms. In J. Rasmussen, K. Duncan & J. Leplat (Eds.), *New technology and human error*. London: John Wiley.

Reason, J. T. (1990). *Human error*. Cambridge, U.K.: Cambridge University Press.

Reason, J. T. (1991). Reducing the risks of organisational accidents in complex systems. *Paper presented to the Colloquium on Human Reliability in Complex Systems*, Nancy 17th-18th April, 1991.

Reason, J. T. (1992). The identification of latent organisational failures in complex systems. In J. A. Wise, V. D. Hopkin & P. Stager (Eds.), *Verification and validation of complex systems: Human factors issues*. Berlin: Springer Verlag.

Reason, J. T. (1997). *Managing the risks of organizational accidents*. Aldershot, UK: Ashgate.

Reason, J. T. & Embrey, D. (1985). *Human factors principles relevant to the modelling of human errors in abnormal conditions of nuclear power plants and major hazard installations*. Parbold, Lancs, Human Reliability Associates.

Reason, J. T. & Mycielska, K. (1982). *Absent-minded? The psychology of mental lapses and everyday errors*. Englewood Cliffs, NJ: Prentice-Hall.

Reitman, W. R. (1965). *Cognition and thought*. New York: Wiley.

Rigby, L. V. (1970) *The nature of human error*. Annual Technical Conference Transactions of the American Society for Quality Control, Milwaukee, WI, ASQC.

Rockwell, T. H. & Giffin, W. C. (1987). *General aviation pilot error modelling - again?* Proceedings of the Fourth International Symposium on Aviation Psychology. Columbus: Ohio.

Rogers, W. P. et al. (1986). *Report of the presidential commission on the space shuttle Challenger accident*. Washington, D. C.

Rook, L. W. (1962). *Reduction of human error in industrial production*. SCTM 93-62(14), Alberquerque, NM.

Roth, E. M., Woods, D. D. & Pople, H. E. Jr. (1992). Cognitive simulation as a tool for cognitive task analysis. *Ergonomics, 35*, 1163-1198.

Rouse, W. B. (1981). Human-computer interaction in the control of dynamic systems. *ACM Computing Survey, 13*(1), 71-99.

Rouse, W. B. (1983). Models of human problem solving: Detection, diagnosis, and compensation for system failures. *Automatica, 19*, 613-625.

Rouse, W. B. & Rouse, S. H. (1983). Analysis and classification of human error. *IEEE Transactions on Systems, Man and Cybernetics, SMC-13*, 539-549.

SAE - Society of Automotive Engineers. (1987). *Human error avoidance techniques*. Conference Proceedings P-204. Washington, DC.

Samanta, P. K., O'Brien, J. N. & Morrison, H. W. (1985). *Multiple-sequential failure model: Evaluation of and procedures for human error dependency* (NUREG/CR-3837). Washington, DC: U. S. Nuclear Regulatory Commission.

Seaver, D. A. & Stillwell, W. G. (1983). *Procedures for using expert judgment to estimate HEPs in nuclear power plant operations* (NUREG/CR-2743). Washington, DC: U. S. Nuclear Regulatory Commission.

Senders, J. W. & Moray, N. P. (1991). *Human error. Cause, prediction, and reduction*. Hillsdale, NJ.: Lawrence Erlbaum.

Senders, J. W., Moray, N. & Smiley, A. (1985). *Modelling Operator Cognitive Interactions in Nuclear Power Plant Safety Evaluations*. Atomic Energy Control Board of Canada. Ottawa, Canada.

Sheridan, T. B. (1982). *Supervisory control: Problems, theory and experiment for application to human-computer interaction in undersea remote systems*. Dept. of Mechanical Engineering, MIT.

Sheridan, T. B. (1986). *Forty-five years in man-machine systems: history and trends*, Proceedings of 2nd IFAC Conf. on Analysis, Design and Evaluation of Man-Machine Systems, Varese, Italy, 10-12 September 1985, Pergamon Press, Oxford.

Sheridan, T. & Hennessy, R. (Eds) (1984). *Research and modeling of supervisory control behaviour*. Hillsdale, NJ. Erlbaum.

Siegel, A. I., Bartter, W. D., Wolff, J. J., Knee, H. E. & Haas, P. M. (1984). *Maintenance personnel performance simulation (MAPPS) model* (NUREG/CR-3626). Washington, DC: U. S. Nuclear Regulatory Commission.

Silverman, B. G. (1992). *Critiquing human error: A knowledge based human-computer collaboration approach*. London: Academic Press.

Simon, H. A. (1957). *Administrative behaviour* (2nd ed.). New York: Free Press.

Simon, H. A. (1972). *The sciences of the artificial*. Cambridge, MA.: The M. I. T. Press.

Simon, H. A. (1979). *Models of Thought. Vol.2*. New Haven. Yale University Press.

Singleton, W. T. (1973). Theoretical approaches to human error. *Ergonomics*, 16. pp. 727-737.

Speaker, D. M., Voska, K. J. & Luckas, W. J. Jr. (1982). *Identification and analysis of human errors underlying electrical / electronic component related events reported by nuclear power plant licensees* (NUREG/CR-2987). Washington, DC: U. S. Nuclear Regulatory Commission.

Spurgin, A. J. & Moieni, P. (1991). An evaluation of current human reliability assessment methods. In G. Apostolakis (Ed.), *Probabilistic safety assessment and management*. New York: Elsevier.

Stassen, H. G. (1988). Human supervisor modelling: Some new developments. In E. Hollnagel, G. Mancini & D. D. Woods (Eds.), *Cognitive engineering in complex dynamic worlds*. London: Academic Press.

Stoklosa, J. H. (1983). *Accident investigation of human performance factors*. Proceedings of the Second Symposium on Aviation Psychology. Columbus, Ohio.

Swain, A. D. (1963). *A method for performing a human factors reliability analysis*. Monograph SCR-685, Alberquerque, NM.

Swain, A. D. (1967). Some limitations in using the simple multiplicative model in behavioural quantification. in W. B. Askren (Ed). *Symposium on Reliability of Human Performance in Work*. AMRL-TR-67-88. Wright-Patterson Air Force Base. OH.

Swain, A. D. (1982*). Modelling of response to nuclear power plant transients for probabilistic risk assessment*. Proceedings of the 8th Congress of the International Ergonomics Association. Tokyo, August, 1982.

Swain, A. D. (1987). *Accident sequence evaluation program: Human reliability analysis procedure* (NUREG/CR-4772). Washington, DC: U. S. Nuclear Regulatory Commission.

Swain, A. D. (1989). *Comparative evaluation of methods for human reliability analysis* (GRS-71). Garching, FRG: Gesellschaft für Reaktorsicherheit.

Swain, A. D. (1990). Human reliability analysis: Need, status, trends and limitations. *Reliability Engineering and System Safety, 29*, 301-313.

Swain, A. D. & Guttman, H. E. (1983). *Handbook of human reliability analysis with emphasis on nuclear power plant applications* (NUREG CR-1278). Washington, DC: NRC.

Thompson, R. (1968). *The Pelican history of psychology*. Harmondsworth. Penguin Books Ltd.

Turing, A. M. (1950). Computing machinery and intelligence. *Mind*, October, *59*, 433-460.

Tversky, A. & Kahneman, D. (1974). Judgment under uncertainty: Heuristics and biases. *Science, 185*, 1124-1131.

Van Eckhout, J. M. & Rouse, W. B. (1981). Human errors in detection, diagnosis, and compensation for failures in the engine control room of a supertanker. *IEEE Transactions on Systems, Man and Cybernetics. SMC-11*, 813-816.

Vaughan, W. S. & Maver, A. S. (1972) Behavioural characteristics of men in the performance of some decision-making task component. *Ergonomics, 15*, 267-277.

Wickens, C. D. (1984). *Engineering psychology and human performance*. Columbus, Ohio, Merrill.

Wickens, C. D. (1987). Information processing, decision-making and cognition. in G. Salvendy (Ed) *Handbook of human factors*. New York, John Wiley and Sons Ltd.

Williams, J. C. (1988). *A data-based method for assessing and reducing human error to improve operational performance*. Proceedings of IEEE 4th Conference on Human factors in Power Plants, Monterey, CA, 6-9 June.

Williams, J. C. (1989). Human reliability data - The state of the art and the possibilities. In *Proceedings Reliability' 89*, United Kingdom, June 14-16, Vol. 1.

Woods, D. D. (1986). Paradigms for intelligent decision support, in E. Hollnagel, G. Mancini & D. D. Woods (Eds) *Intelligent decision support in process environments*. New York. Springer Verlag.

Woods, D. D., Johannesen, L. J., Cook, R. I. & Sarter, N. B. (1994). *Behind human error: Cognitive systems, computers and hindsight*. Columbus, Ohio: CSERIAC.

Woods, D. D. & Roth, E. M. (1988). Cognitive Engineering. Human Problem-solving with Tools. *Human Factors, 30*, 415-430.

Woods, D. D., Roth, E. & Pople, H. (1987). *Cognitive environment simulation: System for human performance assessment* (NUREG-CR-4862). Washington, DC: US Nuclear Regulatory Commission.

Woods, D. D., Roth, E. M. & Pople, H. Jr. (1988). Modeling human intention formation for human reliability assessment. In G. E. Apostolakis, P. Kafka, & G. Mancini (Eds.), *Accident sequence modelling: Human actions, system response, intelligent decision support*. London: Elsevier Applied Science.

Woods, D. D., Rumancik, J. A. & Hitchler, M. J. (1984). Issues in cognitive reliability. In: *Anticipated and abnormal plant transients in light water reactors*, Vol. 1. New York: Plenum Press.

Wortman, D. W., Duket, S. D., Seifert, D. J., Hann, R. L. & Chubb, G. P. (1978). *The SAINT user's manual* (AMRL-TR-77-62). Wright-Patterson Air Force Base, OH.

Wreathall, J. (1982). *Operator action trees. An approach to quantifying operator error probability during accident sequences*, NUS-4159. San Diego, CA: NUS Corporation.

Zapf, D., Brodbeck, F. C., Frese, M., Peters, H. & Prumper, J. (1990). Error working with office computers. In J. Ziegler (Ed) *Ergonomie und Informatik*. Mitteilungen des Fachausschusses 2.3 heft, 9, 3-25.

A

Accident analysis 4; 7; 36; 77; 86; 152; 190; 207; 218; 250;
Action Error Mode Analysis 8;
adaptation 107; 141;
Ainsworth, L. A. 222; 236; 268;
AIPA .... 122 - 125; 131; 136; 137;
Ajzen, I. 268;
Allen, J. 164; 262; 269;
Altman, J. W. 54; 139; 262; 271;
Alty, J. L. 6; 262; 266;
Amalberti, R. 96; 262; 264; 266;
Amendola, A. 8; 262;
analysed event data 90; 91;
anatomy of an accident 34; 35; 36;
Anderson, J. R. 59; 262;
Annett, J. 236; 262;
antecedent .... 7; 161; 176 - 186; 189; 191; 193 - 196; 198; 203 - 207; 213 -
                                                      215; 225; 229; 231; 232;
Apostolakis, G. E. 123; 262; 264; 266; 269; 270; 273; 275;
Arnold, H. 220; 266;
Ashby, W. R. 217; 262;
Ashlund, S. 270;
Askren, W. B. 274;
ATHEANA 153; 146; 147; 148; 149; 219;
Attneave, F. 59; 262;
automation 5; 92; 121;
auxiliary feedwater 208; 209;
available time .... 12; 20; 54; 112; 115 - 117; 156; 199; 200; 237; 238; 239; 253;

B

Babcock, G. L. 268;
Baddeley, A. 103; 262;
Bagnara, S. 67; 262;
Bainbridge, L. 72; 138; 139; 154; 155; 262;
Barriere, M. T. 45; 112; 247; 262; 263; 264;
Bartter, W. D. 273;
basic method 21; 142; 234; 239; 242; 245; 246; 255; 260;
Bastide, R. 271;
Beare, A. N. 123; 252; 263;
Bell, B. J. 266;
Belloff, H. 271;
Bentley, P. 268;
Billings, C. E. 57; 263;
Blackman, H. S. 84; 123; 124; 135; 138; 252; 265; 266;
Bley, D. C. 263; 264;

blunt end 30; 227;
Boreham, N. 266;
Bos, J. F. T. 269;
Bovell, C. R. 263;
Boy, G. A. 71; 263;
Brodbeck, F. C. 275;
Bruce, D. 60; 263;
Brunswik, E. 263;
Bryant, A. 217; 263;
Buratti, D. L. 7; 263;

C

Cacciabue, P. C. 7; 8; 19; 72; 86; 93; 96; 148; 153; 154; 217; 247; 263; 264; 266;
267; 269; 271;
Caeser, C. 57; 263;
Calderwood, R. 268;
Canning, J. 101; 264;
Carnino, A. 271;
Caro, P. W. 112; 264;
Casey, S. 1; 264;
causal chain 77; 187; 214; 217;
CES 71; 87; 140; 141; 142; 148; 149;
Cheany, E. S. 263;
Chubb, G. P. 275;
circadian rhythm .... 109; 112 - 114; 116; 211; 230; 241; 244;
Clancey, W. J. 107; 264;
COCOM 72; 154; 155; 156; 162; 187; 189; 225; 235; 247; 248; 252; 253;
255; 261;
COGENT 138; 143; 144; 149;
cognitive error 17; 18; 20; 37; 46; 47; 49; 50; 77; 87; 106; 138; 140; 143; 144;
148; 150;
cognitive modelling 65; 92; 118; 139;
cognitive systems engineering .... 16; 20; 52; 54; 71 - 73; 75; 79; 80; 81; 83; 92; 106; 154; 217;
222;
Cojazzi, G. 3; 7; 19; 88; 192; 218; 247; 263; 264;
Comer, M. K. 123; 124; 264;
commission .... 14; 17; 42 - 47; 49; 50; 56; 58; 87; 104; 105; 118; 126; 136 -
138; 142 - 146; 149; 151; 157; 162; 166; 209;
common mode 112; 199;
common performance condition 21; 111; 119; 188; 215; 233; 261;
competence 20; 91; 100; 152; 154; 156; 163; 174; 187; 189;
completeness 8; 41; 217;
confusion matrix 125; 126; 132; 133;
consistency 9; 87; 198; 207; 214;
context dependence 111; 139; 159; 187;
contextual control mode 72;
control mode .... 21; 71; 72; 154 - 157; 235; 238 - 242; 245; 252; 255; 259 - 261;
Cooper, S. E. 146; 263; 264; 275;
core cooling 210;
Corker, K. 72; 264;

correct action 46; 47;
Crowe, D. S. 263;
curve fitting 115;
cybernetics 217;

D
data analysis... 84; 91; 160;
data collection.. 2; 89; 131; 139;
Davis, L. 264; 270;
De Keyser, V. 156; 264;
Deblon, F. 96; 262;
Decortis, F. 263;
determinism... 77; 94;
Di Martino, C. 262;
Dietz, P. E. 271;
Dombek, F. S.. 265;
Dorieau, P.. 264;
Dorris, R. E. 263;
Dougherty, E. M. Jr. 8; 14; 19; 32; 40; 52; 84; 106; 122; 123; 140; 150; 264; 271;
Drozdowicz, B.. 263;
Dukelow, T. M. Jr. 266;
Duket, S: D. 275;
Duncan, K. D... 23; 236; 262; 268; 271; 272;

E
Edland, A... 267;
Elzer, P. 262;
Embrey, D. E. 23; 33;54; 55; 64; 70; 102; 123; 124; 264; 265; 267; 271; 272;
empirical data 69; 80; 99; 100;
engineering approach................................ 39; 106; 122; 222;
EPRI 144; 148; 149;
ergonomics 16; 18; 39; 54; 83; 201; 244; 250; 253;
erroneous action .... 1; 2; 3; 5; 7; 8; 11; 16 - 20; 24 - 31; 40; 45 - 50; 52 - 56; 59; 60;
                                                      62; 63; 65; 67; 68; 70 - 75; 77; 80; 81; 83; 84; 87; 93; 99; 100 -
                                                      103; 105; 106; 120; 122; 126; 127; 129; 130; 132 - 136; 140;
                                                      141; 151; 153; 157; 158; 162; 164; 166; 167; 170; 172; 174;
                                                      176; 178; 184; 185 - 187; 189; 199; 202; 204; 214; 222; 228;
                                                      244; 249;
error mechanism...................................... 28; 31; 94; 102; 103;
error mode .... 14; 21; 23; 29; 43; 45; 49; 54 - 56; 58; 74; 75; 79 - 81; 99; 110;
                                                      111; 114; 118; 130; 132 - 134; 140; 141; 143; 144 - 146; 148 -
                                                      151; 159; 161 - 167; 169; 176; 177 - 179; 184 - 186; 188; 189;
                                                      191; 193; 194; 196; 198; 200 - 205; 210 - 213; 215; 224; 225;
                                                      227; 229; 231; 233; 261;
error psychology 8; 52; 75; 82;
event tree .... 10; 12; 13; 15; 33; 34; 35 - 43; 45 - 47; 50; 76; 77; 79; 86; 103;
                                                      104; 126; 127; 129; 130; 132; 135; 137; 138; 143 - 148; 150;
                                                      151; 217 - 221; 225; 227; 229; 233; 236; 245; 251; 258; 260;
expert estimation .... 129; 131 - 133; 137;
expert judgment....................................... 139;

extended method... 21; 206; 234; 235; 245; 246; 255; 257; 258; 260;
external validity 93;

F

failure mode.. 7; 13; 14; 46; 66; 83; 100; 164; 196; 217; 228; 249; 257;
fallible machine 28; 121;
fault tree.. 34; 37; 40; 41; 127; 136; 217; 220; 221; 224; 251; 260;
Feeher, C. E. 271;
Feggetter, A. J... 57; 265;
Feigenbaum, E. A. 98; 265;
Fitts, P. M. 1; 27; 71; 265;
Fleming, K. N. 123; 124; 265;
FMEA.. 7; 217;
FMECA 32; 217;
focus gambling 141;
Fragola, J. R... 32; 40; 52; 84; 122; 123; 264; 266;
Frese, M... 275;
Fujita, Y... 115; 268;
Fullwood, R. R. 123; 265;
Furuta, T. 68; 69; 265;

G

Gaddy, C. D.. 264;
Gagne, R. M. 265;
Gagnolet, P. 271;
Galanter, E... 269;
genotype 19; 48; 106; 203; 212; 222; 249;
Gertman, D. I.. 84; 123; 124; 135; 138; 142; 143; 252; 265;
Giffin, W. C. 272;
Gilbert, K. J. 123; 265;
GMTA 236;
Godoy, S. G. 7; 85; 263;
Gore, B. R... 123; 129; 266;
Griffon, M. 271;
Guttman, H. E.. 10; 16; 32; 70; 103; 105; 107; 123; 124; 129; 252; 274;

H

Haas, P. M. 273;
Hahn, H. A.. 85; 265; 266;
Hall, R. E. 10; 32; 42; 45; 123; 126; 266; 270; 272;
Haney, L. N. 123; 124; 265; 266;
Hann, R. L. 275;
Hannaman, G. W. 12; 64; 80; 123; 124; 265; 266;
HAZOP.. 8;
HCR... 12; 13; 80; 123; 124; 133; 134; 136; 137; 138;
HEART... 123; 254;
HEAT 67;
Helander, M.. 93; 266;
Helmreich, R. L. 3; 266;

Henderson, A. 270;
Hennessy, R. 71; 273;
HEP 10; 11; 13; 14; 20; 31; 37; 107; 119; 130; 142; 251;
Heslinga, G. 220; 266;
Hesse, D. J. 266;
Hitchler, M. J. 275;
HITLINE 145; 146; 148; 149;
Hoc, J.-M. 93; 263; 266; 267; 269;
holistic 128; 132;
Hollnagel, E. 1; 2; 3; 8; 15; 16; 18; 19; 23; 24; 31; 32; 48; 68; 71; 72; 84; 87;
88; 90; 93; 94; 96; 97; 99; 137 - 139; 148; 152 - 154; 162; 166;
197; 236; 247; 248; 263; 266 - 271; 273; 275;
Holmes, B. 97; 268;
Holst, O. 262;
homo economicus 28;
Hooijer, J. S. 269;
Hopkin, D. 272;
Houghton, W. J. 265;
Hoyos, C. G. 268;
HRA .... 2; 8; 9; 10; 12; 14 - 22; 26; 28; 30; 31; 32; 34; 36; 37; 39; 40 -
43; 45 - 47; 50; 59; 64; 75 - 84; 85; 87; 100; 103; 104; 106 -
108; 110; 112; 114; 115; 117 - 120; 122 - 153; 155; 183; 184;
187; 188; 199; 216; 217; 219; 220 - 222; 224; 225; 227; 228;
232; 234 - 236; 241; 243; 251; 254; 258; 260;
Hudson, P. T. W. 183; 268;
human action .... 1; 2; 3; 4 - 8; 10; 14; 15; 17; 19; 22; 24; 31; 32; 34 - 43; 46 - 48;
52; 54; 57; 58; 70; 73; 75; 77; 83 - 86; 88; 96; 101; 107; 114;
120; 121; 130; 147; 152; 154; 157; 158; 160; 161; 162; 169;
188; 204; 216; 220; 222; 227;
human cognition .... 15; 16; 20; 26 - 28; 31; 41; 46; 47; 59; 66; 69; 72; 73; 75; 81;
83; 88; 93; 96; 99; 100; 104; 106; 107; 114; 120; 138; 139; 140;
153; 154; 166; 252;
human error probability 11; 137;
human factor 1; 3; 20; 22; 32; 39; 48; 52; 54; 56; 74; 75; 83; 104; 106; 114;
121; 152; 172; 176; 201; 226; 244; 250;
human factors engineering 75; 176;
human performance .... 1; 2; 7; 9; 10; 13; 15 - 18; 22; 24; 26 - 28; 30; 31; 34; 36; 37;
39; 41; 46; 48; 50; 57; 60; 66; 69; 70; 71; 72; 75; 81; 84; 86;
88; 90; 94; 96; 101; 102; 105; 107; 110; 111; 115; 121; 135;
138; 139; 146; 152; 154; 171; 200; 218; 219; 224; 236; 239;
human reliability analysis 18; 22; 26; 28; 31; 48; 56; 68; 103; 122; 157; 216; 232; 250;
human reliability assessment 19; 84; 191;
Humphreys, P. C. 265; 271;
Hutchins, E. 98; 268;

I

identifiable model 73; 136;
individual action 20; 64; 234; 235;
information processing mechanism 30; 65; 71; 134;
INPO 2; 207;

INTENT.. 138; 142; 143; 144; 149;
interface design................................. 172;
intermediate data format................. 90; 91;
introspection 47; 94; 95; 153;

J

Jenkins, J. P. 266;
Jensen, A. 34; 61; 62; 271;
Johannesen, L. J............................... 275;
Johannsen, G. 139; 262; 268;
Johnson, W. E................................... 59; 268;
Johnson, W. G. 183; 268;
joint cognitive system...................... 26; 71; 72; 117;
joint system....................................... 118; 121;

K

Kafka, P. 262; 266; 275;
Kahneman, D..................................... 169; 274;
Kantowitz, B. H................................. 115; 268;
Kirwan, B. 8; 23; 84; 123; 124; 222; 236; 265; 268;
Klein, G. A. 94; 268; 269;
Knee, H. E. 273;
Kowalsky, N. B. 58; 268;
Kozinsky, E. J.................................... 263;
Kruglanski, A. W............................... 69; 268;

L

lapse... 64; 87; 143;
Larsen, M. N...................................... 236; 269;
latent failure..................................... 5; 6; 100; 197;
Legge, D. 59; 268;
Leplat, J. 23; 123; 268; 271; 272;
Leveson, N. G.................................... 165; 269;
Lewis, C. M. 271;
Lind, M... 236; 269;
Lindblom, C. E. 28; 269;
Lindsay, P. H. 59; 139; 269;
Lisanti, B. 262;
Lucas, D... 197; 123; 124; 269;
Luckas, W. J. Jr. 262; 263; 264; 273;
Lukic, Y. D. 266;

M

Macwan, A. 145; 146; 269;
Maier, N. R. F.................................... 98; 269;
Malin, J. T. 71; 269;
Mancini, G.. 262; 266; 270; 271; 273; 275;
Mancini, S. 263;
Mandler, G... 94; 269;

manifestation .... 5; 46; 48; 53 - 55; 58; 65; 78; 79; 106; 153;
MAPPS 123; 124; 135; 136; 137;
Marsden, P. 154; 267;
Masson, M. 263; 264;
Masters, R. L. 268;
Maver, A: S. 27; 274;
McCormick, E. J. 54; 139; 269;
MCM framework .... 20; 88 - 90; 118; 120; 122; 125; 132; 150; 222;
Meric, M. 123; 269;
Miller, D. C. 271;
Miller, D. P. 25; 39; 222; 269;
Miller, G. A. 59; 269;
Miller, T. E. 98; 269;
minimal modelling manifesto 152;
Minton, L. A. 266;
Misenta, R. 101; 270;
MMI .... 69; 71; 112; 113 - 117; 174; 185; 189; 201 - 203; 211; 212; 230;
                                                                        237; 238; 241; 242; 244; 253; 255; 259;
Modarres, M. 7; 270;
model of cognition .... 49; 81; 89; 91; 100; 143; 152 - 154; 162; 166; 167; 169; 176;
                                                                        189; 235;
Moieni, P. 123; 124; 139; 270; 273;
Monteau, M. 269;
Moray, N. P. 18; 23; 65; 70; 78; 273;
Morrison, H. W. 273;
Mosleh, A. 144; 145; 269; 270;
MTO 68; 85; 100; 114; 158; 173; 188; 215; 223;
Mullet, K. 270;
Murray, R. 270;
Mycielska, K. 26; 78; 186; 272;

N

Nagel, D. C. 264;
Neisser, U. 59; 60; 98; 153; 270;
Newell, A. 59; 96; 98; 270;
Newell, A. F. 169; 270;
Nicholson, W. L. 266; 271;
Nordvik, J.-P. 263;
Norman, D. A. 2; 26; 46; 56; 58; 59; 68; 96; 139; 143; 144; 269; 270;
nuclear power plant 2; 5; 6; 14; 32; 34; 68; 92; 135; 207; 210; 211; 217; 248;
NUPEC 68; 69; 102;
NYC subway .... 195 - 198; 204; 205;

O

OAT 122; 123; 124; 126; 127; 136; 137;
O'Brien, J. N. 273;
omission .... 14; 16; 42; 43 - 47; 49; 50; 56; 58; 67; 87; 103; 104; 105; 111;
                                                                        118; 126; 131; 137; 138; 143; 144; 151; 157; 162; 166; 178;
                                                                        179; 204; 250; 258;

operational support 112; 113; 115; 116; 170; 202; 203; 211; 212; 230; 237; 238; 244; 245; 253;

opportunistic control.............................. 154; 156; 240; 242; 245; 255; 259;

Oramasu, J. 268;

organisation related genotype 159; 163; 173; 176; 178; 182; 203;

Orvis, D. D. 33; 270;

Otway, H. J. 101; 270;

P

Papazian, B. 264;

Papazoglou, I. A. 263; 267; 269;

Parasuraman, R. 61; 270;

Parisi, P. 263;

Park, K. S. 84; 122; 270;

Parry, G. W. 144; 264; 270;

Payne, D. 54; 139; 271;

Pedersen, O. M. 62; 63; 66; 105; 106; 222; 267; 271;

Pedrali, M. 19; 263; 264; 271;

performance analysis 52; 83; 88; 96; 152; 218;

performance characteristic 24; 129; 138; 139; 160;

performance prediction .... 19; 20; 21; 45; 48; 52; 57; 75; 79; 80; 83; 86; 87; 90; 100; 103; 139; 152; 155; 158 - 161; 164; 176; 184; 185; 188 - 190; 198; 200; 216 - 220; 222 - 229; 231 - 233; 260; 261;

performance shaping factor 9; 10; 11; 12; 102; 106; 109; 115; 119; 130; 147; 199; 215; 226;

person related genotype 158; 162; 166; 169; 176; 178; 179; 186; 189; 203;

Peters, H. 275;

Pew, R. W. 123; 264; 271;

Pfremmer, R. D. 265;

phenotype 19; 48; 73; 106; 161; 204; 222; 249;

Phillips, L. D. 123; 124; 226; 271;

Pinola, L. 3; 88; 192; 218; 264;

planning 14;

planning .... 27; 35; 49; 56; 69; 94; 96; 138; 148; 153; 154; 156; 167; 169; 174; 176; 179; 180; 185; 187; 213; 216; 224; 225; 229; 231; 247; 249 - 251; 253; 254; 256; 257;

POET 68; 71; 272; 275;

possible manifestation 49; 54; 68; 157;

Pople, H. E. Jr. 71; 272; 275;

Potash, L. M. 123; 124; 271;

PRA 2; 17; 45; 135; 146; 147; 148;

Prassinos, P. 270;

prediction .... 8; 19; 20; 21; 45; 48; 50; 52; 53; 57; 69; 75; 77; 79 - 83; 86; 87; 90; 91; 100; 101; 103; 121; 139; 151; 152; 154 - 156; 158 - 161; 164; 168; 176 - 180; 183; 184 - 191; 198; 200; 207; 216 - 220; 222 - 233; 260; 261;

Pribram, K. H. 269;

Primrose, M. 268;

privileged knowledge 94;

probable cause 65;

probable cause 189; 86; 87; 99; 191; 198; 200; 202; 205; 218; 222; 233;

procedural prototype 99; 156;
Prumper, J. 275;
PSA .... 2; 10; 12 - 15; 17; 20; 31; 32 - 43; 45 - 47; 50; 76; 77; 79; 85;
103; 110; 120; 122; 127; 129; 139; 140; 142 - 145; 146; 148;
150; 151; 216; 217; 219; 220 - 222; 225 - 227; 233; 234; 236;
244; 245; 251; 258; 260;
PSA-*cum*-HRA 14; 15; 31; 32; 33; 34; 37; 50;
PSF 12; 22; 107; 112; 115; 136; 142; 149; 199;
psychological mechanism 59; 77;

Q
qualitative analysis 31; 82; 144; 148; 185; 223; 228; 233; 234; 246; 261;
quantification 78; 31; 36; 39; 45; 139; 146; 147; 148; 223; 228; 234; 246; 251;
261
quantitative analysis 31; 61;

R
Raabe, P. H. 265;
Rasmussen, J. 12; 23; 34; 52; 58; 61; 62; 67; 68; 93; 102; 123; 130; 133; 134;
137 - 139; 143; 144; 153; 267; 268; 271; 272;
raw data 9;
raw data 89; 90; 91;
Rea, K. 265;
Reason, J. T. .... 2; 4; 6; 23; 26; 28; 30; 35; 56; 58; 59; 63 - 66; 68; 70; 73; 78;
84; 101; 105; 134; 137; 139; 143; 144; 147; 148; 156; 166; 183;
186; 197; 226; 268; 271; 272;
Reitman, W. R. 59; 272;
reliability analysis 18; 22; 26; 28; 31; 39; 41; 48; 52; 56; 68; 93; 103; 121; 122;
139; 157; 172; 184; 216; 222; 232; 235; 250;
reliable performance 111; 189;
replacement 247;
residual erroneous action 7;
reversal 42;
Rigby, L. V. 25; 272;
Rizzo, A. 262;
Roberston, S: A. 262;
Rockwell, T. H. 54; 272;
Rogers, W. P. 192; 217; 262; 272;
Rogers, Y. 262;
Rook, L. W. 54; 139; 272;
root cause 7; 77; 121; 192; 218;
Rosa, E. A. 265;
Rose, S. E. 266;
Rosness, R. 267;
Roth, E. M. 60; 71; 72; 272; 275;
Rouhet, J.-C. 267;
Rouse, S: H. 59; 65; 105; 224; 272;
Rouse, W: B. 59; 65; 66; 103; 105; 123; 224; 247; 268; 272; 274;
Rumancik, J. A. 275;
Rutherford, A. 262;

Rypka, E. W. 268;

S

Salvendy, G. 269; 274;
Samanta, P. K. 123; 273;
Sarter, N. B. 275;
Savory, S. 262;
schema 65;
Schreckenghost .... 269; , D. L.
scrambled control 240; 154; 155;
Seaver, D. A. 131; 264; 273;
Seidler, K. S. 265;
Seifert, D. J. 275;
Senders, J. W. 18; 23; 65; 70; 78; 273;
sensitivity 129;
sequence model 33; 34; 37;
sequentiality 96; 97;
SHARP 123;
sharp end 24; 29; 30; 31; 73; 88; 101; 227;
Sheridan, T. B. 71; 92; 139; 273;
Siegel, A. I. 123; 124; 273;
Silverman, B. G. 169; 273;
similarity 125; 156;
Simon, H. A. 28; 59; 96; 98; 166; 270; 273;
simulation 8; 121; 135; 136; 137; 140; 141; 145; 146; 148; 149; 154;
simulator 9; 36; 134; 141; 211;
simultaneous goals .... 112; 113; 115 - 117; 203; 211; 212; 230; 238; 239; 244; 253; 255;
Singleton, W. T. 23; 105; 273;
Skekely, J. 269;
SLIM 32; 123; 124; 132; 133; 136; 137;
slip 26; 46; 87; 64; 68; 78; 80; 138; 143; 171;
SLM 61;
Smiley, A. 70; 273;
software fault 186; 187;
Speaker, D. M. 123; 273;
specificity 57; 176;
Spurgin, A. J. 64; 123; 124; 139; 266; 273;
Stager, P. 272;
STAHR .... 123; 124; 127 - 129; 133; 136; 137; 226;
Stassen, H. G. 139; 263; 273;
Stevens, R. 271;
Stewart, M. 271;
Stillwell, W. G. 131; 264; 273;
Stoklosa, J. H. 57; 274;
Stone, R. B. 268;
stop rule .... 21; 86; 177; 191 - 194; 198; 205 - 207; 215;
strategic control 154; 156; 240; 253;
stress 57; 69; 70; 103; 109; 112; 115; 129; 134; 135; 169; 170; 181; 225;

structural model 118;
subjectively available time 12;
subsumed 236;
Sullivan, C. 265;
Svenson, O. 267;
Swain, A. D. 9; 10; 16; 23; 25; 32; 39; 42; 54; 56; 70; 102; 103; 105; 107; 108; 123; 124; 129; 133; 137; 139; 222; 252; 269; 274;
system design 17; 31; 40; 68; 209; 216; 236;
system induced erroneous action 7;

T

tactical control 154; 156; 240;
task allocation 121; 174; 183; 184;
task analysis 21; 40; 41; 50; 129; 130; 133; 217; 221; 222; 226; 227; 233; 235; 236; 243; 244; 249; 254; 257; 260; 261;
task description 131; 235; 236; 251;
task element 242; 251; 257;
task load 69;
task representation 222; 236;
Taylor, J. H. 264;
Taylor, J. R. 19; 267;
technical system 4; 7; 14; 22; 23; 29; 31; 32; 34; 41; 42; 48; 85; 127; 173; 222; 226;
Technique for Human Error Rate Prediction   10; 107; 123; 124; 129;
technology related genotype 163; 170; 176; 178; 181; 202;
terminal cause 193; 196;
THERP .... 10; 11; 13; 32; 39; 107 - 109; 110; 122; 123; 124; 129; 130; 136; 137; 142; 143; 144; 149; 219; 222;
Thompson, R. 59; 274;
Tiffin, J. 54; 139; 269;
time horizon 155; 156;
trust 113; 244;
tube rupture 207; 208; 209;
Turing, A. M. 95; 98; 274;
Turner, C. S. 165; 269;
Tversky, A. 169; 274;

U

Underwood, G. 271;
unexpected event 34; 35;
unwanted consequences 84; 85; 100; 184; 223;

V

validation 192;
validity 56; 75; 83; 93; 119; 120; 131; 222; 223; 253;
Van Daele, A. 264;
Van Eckhout, J. M. 59; 274;
Vaughan, W. S. 27; 274;
Visser, J. 268;

Voska, K. J. 273;

W
Waern, Y. 267;
Wagenaar,, W. A. 268;
Weir, G. 266;
White, T. 117; 270;
Whitehead, D. 262;
Wickens, C. D. 60; 61; 274;
Wiener, E. L. 264;
Wieringa, P. A. 269;
Williams, J. C. 123; 252; 254; 274;
Wioland, L. 264;
Wise, J. A. 272;
Wolff, J. J. 273;
Woods, D. D. .... 1; 2; 16; 23 - 25; 30; 31; 60; 70; 71; 72; 77; 84; 87; 93; 98; 123; 139; 140; 141; 160; 264; 267; 269; 270 - 273; 275;
work condition 243; 244; 259;
working memory 61;
workload 112; 116;
Wortman, D. W. 123; 275;
Wreathall, J. 15; 32; 123; 124; 126; 263; 264; 266; 267; 275;
Wu, J. S. 264; 269;

Y
Yoshikawa, H. 268;

Z
Zambok, C. 269;
Zapf, D. 29;275;
Ziegler, J. 275;
Zimolong, B. 268;
Zsambok, C. E. 268;

Printed and bound by CPI Group (UK) Ltd, Croydon, CR0 4YY

03/10/2024

01040320-0008